CHEESE-MAKING TECHNOLOGY

CHEESE-MAKING TECHNOLOGY

M.E. Schwartz

NOYES DATA CORPORATION
Park Ridge, New Jersey London, England
1973

Library of Congress Catalog Card Number: 73-75545
ISBN: 0-8155-0502-7
Printed in the United States

Published in the United States of America by
Noyes Data Corporation
Noyes Building, Park Ridge, New Jersey 07656

FOREWORD

The detailed, descriptive information in this book is based on U.S. patents since 1960 relating to the manufacture of cheeses.

This book serves a double purpose in that it supplies detailed technical information and can be used as a guide to the U.S. patent literature in this field. By indicating all the information that is significant, and eliminating legal jargon and juristic phraseology, this book presents an advanced, commercially oriented review of cheese-making technology.

The U.S. patent literature is the largest and most comprehensive collection of technical information in the world. There is more practical, commercial, timely process information assembled here than is available from any other source. The technical information obtained from a patent is extremely reliable and comprehensive; sufficient information must be included to avoid rejection for "insufficient disclosure."

The patent literature covers a substantial amount of information not available in the journal literature. The patent literature is a prime source of basic commercially useful information. This information is overlooked by those who rely primarily on the periodical journal literature. It is realized that there is a lag between a patent application on a new process development and the granting of a patent, but it is felt that this may roughly parallel or even anticipate the lag in putting that development into commercial practice.

Many of these patents are being utilized commercially. Whether used or not, they offer opportunities for technological transfer. Also, a major purpose of this book is to describe the number of technical possibilities available, which may open up profitable areas of research and development. One should have to go no further than this condensed information to establish a sound background before launching into research in this field.

Advanced composition and production methods developed by Noyes Data are employed to bring our new durably bound books to you in a minimum of time. Special techniques are used to close the gap between "manuscript" and "completed book." Industrial technology is progressing so rapidly that time-honored, conventional typesetting, binding and shipping methods are no longer suitable. We have bypassed the delays in the conventional book publishing cycle and provide the user with an effective and convenient means of reviewing up-to-date information in depth.

The Table of Contents is organized in such a way as to serve as a subject index. Other indexes by company, inventor and patent number help in providing easy access to the information contained in this book.

15 Reasons Why the U.S. Patent Office Literature Is Important to You —

1. The U.S. patent literature is the largest and most comprehensive collection of technical information in the world. There is more practical commercial process information assembled here than is available from any other source.

2. The technical information obtained from the patent literature is extremely comprehensive; sufficient information must be included to avoid rejection for "insufficient disclosure."

3. The patent literature is a prime source of basic commercially utilizable information. This information is overlooked by those who rely primarily on the periodical journal literature.

4. An important feature of the patent literature is that it can serve to avoid duplication of research and development.

5. Patents, unlike periodical literature, are bound by definition to contain new information, data and ideas.

6. It can serve as a source of new ideas in a different but related field, and may be outside the patent protection offered the original invention.

7. Since claims are narrowly defined, much valuable information is included that may be outside the legal protection afforded by the claims.

8. Patents discuss the difficulties associated with previous research, development or production techniques, and offer a specific method of overcoming problems. This gives clues to current process information that has not been published in periodicals or books.

9. Can aid in process design by providing a selection of alternate techniques. A powerful research and engineering tool.

10. Obtain licenses — many U.S. chemical patents have not been developed commercially.

11. Patents provide an excellent starting point for the next investigator.

12. Frequently, innovations derived from research are first disclosed in the patent literature, prior to coverage in the periodical literature.

13. Patents offer a most valuable method of keeping abreast of latest technologies, serving an individual's own "current awareness" program.

14. Copies of U.S. patents are easily obtained from the U.S. Patent Office at 50¢ a copy.

15. It is a creative source of ideas for those with imagination.

CONTENTS AND SUBJECT INDEX

INTRODUCTION

Cheese-making is an art that has changed little over the course of its history. The basic process in which curd is formed from milk either by bacterial fermentation or addition of rennet or a combination of these to develop lactic acid, the subsequent removal of whey from the curd and the further treatment of the curd to produce the many varieties of cheeses remains much the same.

Cheese contains nearly all the fat, casein, calcium and vitamin A and a part of the milk sugar and salts of the milk from which it was made. As such, since it preserves much of the food value of milk for long periods of time, it is a highly significant food supply. With increasing demands for nutritional supplies to satisfy a growing consumer market, it has become important to try to transform the art to a science.

This has produced a significant cheese technology. This technology takes into account all aspects of the cheese-making art with attacks focused on attempts to standardize the resultant product to satisfy heightened consumer awareness, to aid storageability and preservation and to increase yields to enhance the productivity of the industry.

Current needs also dictate that the industry should tackle a variety of other problems. The emphasis on ecology focuses on the great waste problem of whey and how best to utilize it rather than dispose of it since its disposal creates serious water pollution problems. Concern with good nutrition and high cholesterol levels dictate that a consumer acceptable low fat cheese be found and made available since cheese remains one of the most attractive sources of high protein in the diet.

In addition to the search for better and more reproducible cheese processes, a further search is being conducted by food scientists for a more economical and efficient means of producing similar materials. This work is finding some measure of success as witnessed by the advent of "nondairy" dairy products.

These many aspects of the science of cheese-making have been taken into account and are covered in this review of the cheese-making technology. The field is active and current; the wealth and scope of the material is testimony to this fact.

COTTAGE CHEESE

APPARATUS

Washing Curd

The process of R.B. Wolf; U.S. Patent 2,980,538; April 18, 1961; assigned to The Borden Company concerns handling the curd in making cottage cheese. It provides a means for rapidly washing and cooling the curd. This is accomplished by moving the curd and whey in layer form alternately up and down and continuously forward in a mass of cold water, subjecting the layer at intervals to sprays of refrigerated water applied in such manner as to drain through the layers of the curd between successive immersions in the cold water. The process also includes means for maintaining the layers of curd during the washing and cooling in loose condition so that liquid passes freely over and between the individual lumps of the curd. The process comprises means for pressing the washed, firmed curd against a draining surface to decrease its moisture content.

Figure 1.1 is a side sectional view of a machine for washing, chilling and pressing the curd. The mix (10) is shown as delivered from the usual curd making vats to the hopper (12) with an outlet at the bottom above a woven wire belt (14) or any type of filtering support with shoulders (16) extending upward from each side to a distance of 2 inches (preventing the mix from falling over the sides). The cross bar or blade (18) restricts the backward flow of the wet curd delivered to the belt. The adjustable blade (20) spreads the curd on the belt as a layer of desired thickness. The eccentric roller (21) vibrates the belt and shakes out excessive whey or water before the curd passes to the next step.

The belt with the curd then passes over a series of eccentrically mounted, rotated, power driven rollers (22). These subject the belt and the curd to a combined vibrating and alternate raising and lowering, to alternately immerse the curd layer in the refrigerated water tank (32) for cooling and raise the curd for draining. Disposed both above and below the belt and the curd layer are a series of water sprays. Part of the recovered water (30) is used in sprays (24) and (25) as a prerinse. The rest of this cold water (30) goes to the drain. Line (26) delivers refrigerated water from coolers to sprays (27), and line (31) delivers refrigerated water to sprays (29).

FIGURE 1.1: RAPID PROCESS FOR WASHING COTTAGE CHEESE CURD

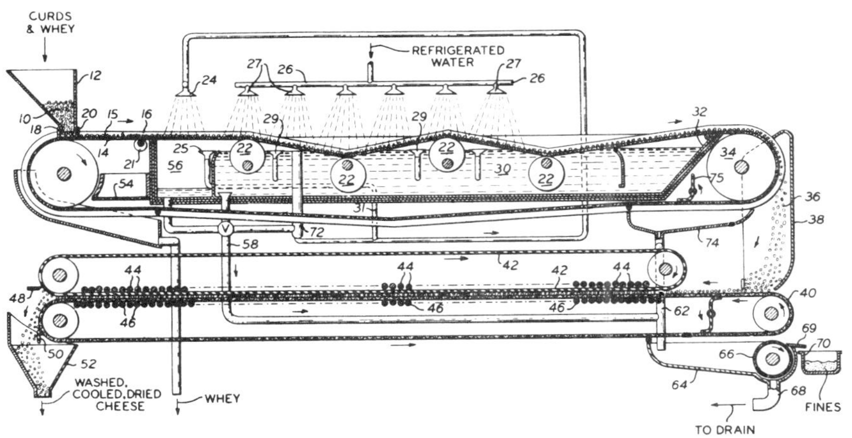

Source: R.B. Wolf; U.S. Patent 2,980,538; April 18, 1961

The eccentrically mounted rollers are at such a level that, at their upper parts, they alternately dip below and rise above the level of the cold water. At the positions of sinking of the belt on the rollers, it submerges the curd in the refrigerated wash water which contains some whey, and alternately raises the curd out of the water for direct contact with the refrigerated water from the sprays (27) and (29). After repeated up and down movement by the rollers with accompanying submerging in the refrigerated water and subsequent draining and washing with the sprayed water, the curd is delivered over the change of direction drum (34), so that the curd falls, as shown at (36), in large part through chute (38) to a perforated or second woven wire belt (40) for passage between this belt serving as a draining conveyor, and an over-riding belt converging with it, as the two belts at their adjacent reaches are moved towards the left end of the mechanism shown in the Figure 1.1.

The belts are forced together against the wet curd, sandwiched between by the series of coacting rollers (44) and (46). The resulting compression produces the required dewatering of the curd. The thus-pressed and largely dewatered curd (48) is delivered finally by the conveyor (40) against the scraper knives (50) and (48) and into the collecting hopper (52) where the curd is ready for mixing with fresh cream or other usual components, to make the finished cottage cheese.

A screen (54) mounted below the part of the belt (14) immediately to the right of the hopper (12), receives the whey or water and fine curd suspended in it. The liquor drains from the curd caught on this screen, the screen being of mesh size to allow the whey or water to pass through but to retain the fine particles of curd. The accumulated curd on the screen and fine settlings on the bottom of tank (32) are removed at intervals. A part of the fines which pass through the belt at positions above tanks (32) and (56) do not settle, particularly since the rollers cause some agitation. The unsettled fines go through outlet line (58) and finally into the line (62) which delivers the wet fines to collecting vessel (64). Here a rotary filter with filtering surface (66) filters water away to outlet (68) and collects the curd particles on (66) for delivery against the scraper knife (69) to tank (70) for receiving fines. The recovered fines are placed in hopper (52).

The excess of refrigerated wash water that accumulates in the tank from the sprays (27) and (29) overflows through overflow line (72). Part of this overflow enters line (58) and returns as a prerinse through sprays (24) and (25). The balance of this water is discarded through line (62). Flapper (75) dislodges adhering curd from the belt (14) when the major part of the curd has been removed. Conventional means (not shown) drive this flapper, so that it strikes the belt (14) from above, on the return or or lower reach thereof. The curd then falls into the collector (74) and passes through line (62) from this collector to the final curd recovery tank (64).

The temperature of the water supplied to sprays (27) and (29) should be very low. The nearer the temperature is to the freezing point of water so long as the water is fluid, the more quickly the water chills and firms the curd and, as a result, the more it speeds up the washing and increases the shelf life of the finished cottage cheese. For best results, a temperature of 2°C. is preferred.

The pressure forcing the squeeze rollers (44) and (46) in the direction of each other and against the layer of curd between them supported on belt (40) is varied with the thickness of the layer of curd being dewatered by the squeezing action. The thicker the layer of curd at this stage, the higher is the pressure forcing rollers (44) down and rollers (46) up. The pressure is adjusted in any case so that the finished dewatered curd going into the hopper contains the proper proportion of moisture for subsequent blending with cream. This is the proportion that causes the cream to film completely over each particle of the curd in making the finished cottage cheese. The weight of the rollers (44), or even of the belt (42) alone, is often adequate to dewater to the desired percentage of moisture in the curd to be mixed with the cream.

The up and down and forward movement to which the layer of curd on the belt is subjected to in the alternating immersions and spraying with the refrigerated water give a quick and uniform cooling of the cheese curd over all of the surfaces. The vibrating action which the rollers create continuously exposes fresh surfaces of the lumps of curd to the action of the refrigerated water. The belt with the curd are moved up and down at a mean rate of 20 to 30 rpm, and with a horizontal or forward component of velocity, that is, to the right for the upper reach of the belt of 10 to 20 ft./min.

Washing and Cooling Curd

The process of R.E. Osborne and M.E. Powell; U.S. Patent 3,044,863; July 17, 1962; assigned to Knudsen Creamery Co. of California is designed to provide a method and apparatus for washing and cooling cottage cheese curd which has a more efficient utilization of the refrigerant properties of the cooling water. The process also provides for draining water from cottage cheese in which the step of manually removing the "dry" curd from the vat is eliminated.

Figure 1.2 illustrates an arrangement of apparatus (10) for manufacturing cottage cheese curd. The equipment includes a series of conventional vats (11), three of which are shown here and identified as (A), (B) and (C); a final washing and cooling tower (12) of the process; a portable final curd draining hopper (13); and a tank (14) for storing water from the cooling tower prior to its reuse in the cheese vats. Production of cheese curd in the vats is regulated so that the curd in one vat reaches the washing and cooling stage before that in the next vat, whereby a successive production of curd takes place in the vats which creates a continuous operation of washing and cooling the curd in the tower. The water used for washing and cooling the curd in the vats originates from three sources, namely, the usual domestic water supply for the plant, pipe (15), the refrigerated water supply, pipe (16), and the water from tank (14) (accumulated from the outflow of water from the cooling tower) which is delivered through pipe (17) to the vats.

FIGURE 1.2: APPARATUS FOR WASHING AND COOLING COTTAGE CHEESE CURD

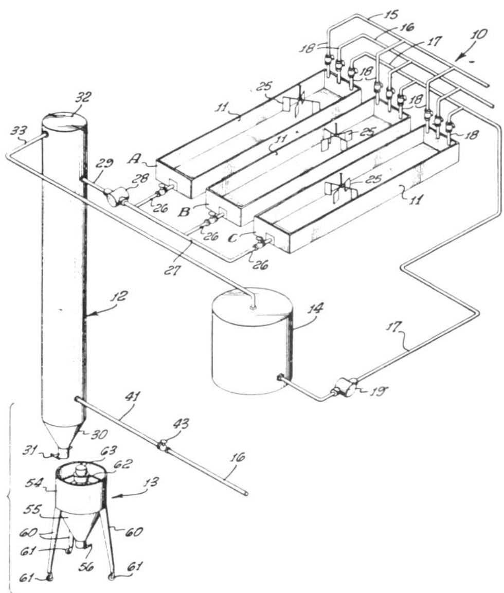

Source: R.E. Osborne and M.E. Powell; U.S. Patent 3,044,863; July 17, 1962

The delivery of water from the pipes to the vats is controlled by valved delivery pipes (18) which lead from each of the main pipes to each of the vats. A pump (19) is provided to withdraw water from the tank and discharge this under pressure into pipe (17). Each vat is equipped with a conventional power operated rotary curd mixing paddle

(25) which is supported by an overhead conveyor for stirring the curd during its production. Leading from the opposite end of each vat (11), from the end receiving water from delivery pipes (18), is a valved curd discharge pipe (26) through which curd may be withdrawn from the vat, these pipes connecting to a pipe (27) leading to the suction end of a pump (28) which has a discharge pipe (29).

The curd washing and cooling tower (12) is cylindrical in form and may be approximately 2 feet in diameter and 30 feet high. The dimensions of the tower may vary widely according to the capacity it is desired to have. The tower is appropriately supported in an erect position and has a conical bottom (30) which terminates axially at its lower end in a curd discharge valve (31). A removable lid (32) is provided on the tower to facilitate cleaning. An outflow pipe (33) connects with the upper end of the tower just below the lid. The other end of this pipe discharges into the storage tank (14) as shown in Figure 1.2. Spaced downwardly from the outflow pipe (33) and opening into the tower is a sanitary pipe coupling (34) with which pipe (29) connects, and which in turn connects with a down-curved inlet pipe (35) through which curd is delivered into the tower.

Spaced downwardly a substantial distance from the pipe (35) and mounted within the tower a relatively short distance above the tower bottom (30) is a perforated cold water manifold (40) in the shape of a ring connected to a chilled water pipe (41) which enters the wall of the tower to connect with the manifold. Perforations (42) are provided in the manifold in its upper surface to direct chilled water discharged from this manifold in an upward direction. The pipe (41) has a valve (43) and connects with the chilled water line (16).

The apparatus also includes a series of like curd draining hoppers (13). Each of these has a cylindrical body (54) and conical hopper bottom (55) which terminates at its lower end in a discharge valve (56) having an opening (57) of about 6 inches in diameter. Each of the hoppers is supported on legs (60) provided at their lower ends with casters (61). Slidable vertically in an axial collar (62), provided in the upper end of the hopper body (54), is a perforated tube (63). When the tube is allowed to gravitate downward, the lower end rests on the hopper bottom so as to close the opening and trap any curd contained in the hopper from flowing downwardly out through the opening. Perforations (64) provided in the tube permit drainage of liquid from the contents of the hopper when this tube is in closed position, this liquid flowing downward through the tube and out through the opening.

A preferred method of the process uses the cooling tower to perform the final washing and cooling step in the production of cottage cheese curd formerly done in the vats. The method involves performance in these vats of the previous steps in the usual method of producing cottage cheese curd and then transferring the curd from these vats to the tower for the final cooling and washing step. This final washing and cooling step goes on continuously in the tower by providing a sufficient number of vats so that there will always be a supply of cheese curd in one of these vats which has just been completed up to the beginning of the final cooling and washing step, and thus be in readiness to be transferred to the tower. In Figure 1.2 it has been assumed that three of the vats (11) will satisfactorily supply curd for maintaining the tower (12) in continuous operation.

Fresh chilled water is constantly supplied to the lower end of the tower and an approximately equal amount of water flows out from the upper end of the tower into the storage tank throughout the operation of the apparatus. The valve at the lower end of the tower is kept closed except when opened to remove curd from the lower end of the tower. This constant delivery of chilled water through the pipe (41) and perforated manifold (40) into the lower portion of the tower causes a continuous upward current in the water occupying this tower. By manipulation of valve (43), this current is maintained at a rate of flow which is slower than the rate at which particles of curd delivered into the water occupying the tower will sink in the water owing to the relatively higher density of the curd.

The temperature of successive baths of water to which the curd is subjected in the vats is progressively lower and the method makes maximum use of the refrigerated water used in the tower by collecting this in the storage tank and then withdrawing water from this tank and delivering it successively to the vats for use in partially chilling the curd prior to the final cooling.

Let us assume that the batch production cycles have started in vats (A), (B) and (C) in the order mentioned, and curd has reached the stage in vat (A) where it is ready to be delivered to the tower (12) while the batch cycles in (B) and (C) are still uncompleted. When this occurs, the valve in discharge pipe (26) of vat (A) is opened and the pump (28) started so that the curd just produced in vat (A), while still mixed with a portion of the final washing water applied in this vat to the curd, is sucked out through the outlet pipe (26) of this vat and delivered through pipe (29) into the final cooling and washing tower (12). Any curd remaining in the vat will be flushed into the suction line with water from tank (14).

By the time the curd produced in vat (A) has been completely withdrawn and delivered to the tower the cheese curd produced in the cycle being performed in vat (B) is ready to be withdrawn from the latter and delivered to the tower. This is accomplished merely by shutting the valve in pipe (26) of vat (A) and opening the valve in pipe (26) of vat (B). Similarly, after the curd has been withdrawn from vat (B), the curd previously being produced in vat (C) is ready for withdrawal, and this is accomplished by closing the valve in pipe (26) of vat (B) and opening the valve in pipe (26) of vat (C).

As the discharge pipe of each of the vats is thus closed following a discharge from that vat of curd produced in a batch operation cycle, the vat may be cleaned and sanitized at this time in order to receive a new batch of skim milk. All of the wash water employed in the series of washing baths to which the curd is subjected in batch cycles performed in vats (11) is selectively drawn either from the tap water line (15) the fresh chilled water line (16), or the line (17) which connects with the tank (14).

Production in Sealed Sterilized Vat

L.J. Collins; U.S. Patent 3,433,648; March 18, 1969; assigned to Grace Machinery Corporation describes both the method and apparatus for processing milk culture contained in a sealed and sterilized vat. Purified air is applied under a pressure slightly above atmospheric and maintained throughout the process to produce a high-quality cheese curd less susceptible to spoilage. The process is completed in 12 hours or less, depending on the capacity of the vat, the amount of culture added; the less culture, the longer the process. The milk processing method to produce the curd entails:

(a) automatically timing and controlling sterilization of the apparatus;

(b) sealing the sterilized apparatus;
(c) applying purified air to the apparatus under a pressure slightly above atmospheric and maintaining such pressure throughout the process to isolate the sterilized apparatus from atmospheric air and prevent contamination from harmful bacteria;
(d) sequentially delivering milk and culture to the sealed apparatus;
(e) heating and agitating the mixture so formed;
(f) cooking the mixture for a predetermined interval to set the mixture to form a curd;
(g) draining of the whey from the curd so-formed;
(h) washing the curd, and then
(i) cooling the washed curd while it is retained sealed within the apparatus.

The apparatus designed to carry out the process is depicted and completely described in the patent. The apparatus includes a vat having a water jacket; a cover and means for sealing the cover on the vat; an agitator within the vat; a drain, means connecting the vat to sources of steam, wash water, sterilizing water and chlorinated water for sterilizing and heating the vat; means connecting the vat to sources of milk, culture, ice water and purified air under a pressure greater than atmospheric. The apparatus as described provides a system in which a multiple timer unit functions as a master unit for controlling operation of a plurality of circuit control timer units for sequential and parallel operations of the various control elements used in the process.

Solids Content of Milk Used

A.H. Wakeman; U.S. Patent 3,645,751; February 29, 1972; assigned to Wisconsin Alumni Research Foundation describes a method for making cheese curd on a continuous basis in which a water-soluble food-processing grade acid is utilized to adjust the pH of the milk to within the isoelectric range of the casein present in the milk. The process produces cheese curd from milk on a continuous basis over an extended period of time.

If these ends are to be realized, i.e., if continuous curd production over an extended period of time is to be achieved with concomitant consistency in curd size, texture, and creaming characteristics, careful attention must be given to four of the steps in the process. In all cases the practice of these four steps is essential to the achievement of the desired results. The four essential steps in the order in which they occur in the process as set forth are:

(1) deaeration of the milk to be processed;
(2) substantially instantaneous acidification of the milk with a food grade acid without localized overacidification;
(3) a periodical momentary reversal of the flow of the coagulating milk during passage through the tubes of the heat exchanger;
(4) cooking the curd at a temperature in the range of from about 170° to 184°F. The other steps of the process as outlined can be carried out under conventional conditions.

The cooking temperature range specified in step (4) above is essential to the making of cottage cheese curd having the desired characteristics. This cooking temperature range will be different if the process is to be used for preparing cheese curd other than cottage cheese curd. In the preparation of cheddar cheese curd the cooking temperature must be controlled so that the desirable organisms present in the curd and which are essential to the later curing process are not destroyed. It is also to be appreciated that the equilibrium pH of the acidified milk will vary depending upon the particular cheese curd being prepared.

A more detailed description of the process of making a satisfactory cheese with this method is given in the listing below:

(1) substantially completely deaerating a stream of milk containing from about 8 to about 40% nonfat milk solids;
(2) cooling the stream of milk to a temperature in the range from about 35° to 50°F.;
(3) continuously adding to the stream of cooled milk and mixing intimately therewith, a water-soluble, food-processing grade acid, e.g., hydrochloric acid, in an amount sufficient to acidify the milk to an equilibrium pH in the range from about 4.5 to about 4.7 when at a temperature of about 70°F., under conditions which substantially avoid any local overacidification and which provide substantially instantaneous acidification of the milk stream;
(4) substantially immediately following the acidification, heating the stream of acidified milk to a temperature in the range from about 90° to about 125°F. by passing the stream in nonturbulent flow and in the substantial absence of interparticle motion through heat exchange means whereby the acidified milk is coagulated and the resulting curd is set;
(5) periodically and momentarily reversing the flow of the coagulating and coagulated stream of acidified milk during passage through the tubes of the heat exchange means whereby a reverse shear action between the coagulum and the walls of the tubes is created and sticking of the coagulum to the walls is prevented;
(6) cutting the curd into substantially uniform segments as it emerges from the heat exchange means;
(7) cooking the cut curd by passage of the cut curd and whey through heat exchange means wherein the temperature of the curd and whey is raised to from about 170° to about 184°F.;
(8) cooling the curd and whey to a temperature below about 145°F.;
(9) draining the whey from the curd; and
(10) washing the curd.

C.A. Ernstrom; U.S. Patent 3,089,776; May 14, 1963; assigned to Wisconsin Alumni Research Foundation devised a process in which a cheese-making curd can be achieved by the direct addition of acid to milk, i.e., duplicate the pH achieved by bacterial fermentation. The method in which a firm curd suitable for making cheese consists of adding a water-soluble, nontoxic acid to cool liquid milk, agitating the cool mixture to mix the acid with the milk, and then allowing the mixture to warm without agitation, providing that the temperature of the milk in the first two steps was below the point at which coagulation takes place and the temperature of the milk in the third step was at or above the point at which coagulation takes place.

The liquid milk used can be skim milk which will produce a cheese-making curd low in milk fat or whole milk which produces a cheese-making curd high in milk fat. The milk can also be a mixture of skim and whole milk to produce cheese-making curds with intermediate or varying contents of milk fat.

Example: Fresh raw skim milk is pasteurized by heating to 143°F. for 30 minutes. The pasteurized milk is cooled to 43°F. or lower. Four liters of the cold skim milk are treated with 18 ml. of concentrated hydrochloric acid (36.5 to 38.0% HCl) and stirred until the acid is uniformly mixed with the milk. The amount of acid added is such that the pH of the acidified milk is about 4.6 when measured at 70°F. The acidified milk is warmed to a temperature of 75°F. while the milk remains in a quiescent state. Agitation of the milk during warming must be avoided. Warming may be accomplished by standing at room temperature, by steam or hot water, or electrically by dielectric or resistance heating. The curd which forms during warming is cut with curd knives into 1/4" cubes or into any size cubes desired. The process then follows a conventional cottage cheese-making operation.

In a continuation of work done on curd formation by direct acidification C.A. Ernstrom; U.S. Patent 3,298,836; January 17, 1967; assigned to Wisconsin Alumni Research Foundation describes a curd-making process using high nonfat milk solids milk. The process previously described poses a problem due to the time required to heat up large commercial vats of the cool acidified milk to the coagulation temperature in the absence of agitation. Allowing a large vat of cool acidified milk at about 40°F. to warm up to room temperature (70° to 75°F.), was found to take a relatively long period of time, since the curd initially formed around the sides of the vat (much like a big doughnut) acted as an insulator for the cooler milk in the center of the vat.

Heating the cool milk by means of hot water coils in the vats speeded up coagulation but was also found to take a considerable period of time, the curd first forming around the hot coils again acting as an insulator and preventing the rapid transfer of heat to the cool milk some distance from the coils.

In continued investigations it was found that measurable time is needed to form curd at 70° to 75°F. With skim milk containing about 9% nonfat milk solids and acidified to pH 4.64, the time was found to be about 3 to 6 seconds. It was anticipated that this short time period might increase with skim milk of lower nonfat milk solids content, however, the reverse took place. For example, skim milk containing about 13% solids and acidified to pH 4.5, took about 50 to 60 seconds to coagulate at 70°F. and skim milk containing about 16% solids and acidified to pH 4.5 took 400 seconds to coagulate at 70°F.

Since the addition of concentrated acid to cold (35° to 50°F.) skim milk, containing 13% or more solids tends to produce local coagulation, it was found when using concentrated acid that the acid should be added slowly with agitation, as for example, to a pump pumping the cold milk. Local coagulation can also be avoided by diluting the acid with water and then adding the water-acid mixture to a skim milk concentrate, i.e., one containing about 20 to 40% nonfat milk solids, to obtain the acidified (pH 4.4 to 4.8) skim milk containing 13% or more nonfat milk solids.

Once the coagulation temperature of the skim milk is reached, the curd strength of the curd produced from the high nonfat milk solids milk (containing 13+% nonfat milk solids) increases with temperature much more rapidly than that produced from low nonfat milk solids milk which forms a curd at a lower temperature. Once coagulated, the curd from high nonfat milk solids milk quickly becomes firm enough to cut. Curd shrinkage was also found to be materially less than curd shrinkage with normal skim milk containing 9% nonfat milk solids. These features were found to make the acidified cool high nonfat milk solids skim milk product adaptable for rapid heating, with excellent curd formation, by passage through a heat exchanger.

In addition, it was found that when using high nonfat milk solids liquid milk, the formed curd can be cooked to the desired moisture levels specified in the standards for cottage cheese as established by the Federal Food and Drug Administration, and in accordance with standards of identity for cottage cheese as established by the various states in much less time, i.e., 10 to 15 minutes, than is possible with other processes.

Thus, the use of high nonfat milk solids liquid milk permits: acidified milk to be handled at temperatures from 40° to 70°F. without instant coagulation; a firmer curd to be formed when the acidified milk is heat coagulated; curd to be cooked to desired moisture levels in much less time than is possible in other processes; and material reduction in curd shrinkage over that which occurs when normal skim milk containing about 9% nonfat milk solids is used.

The general procedure for this process can be described in three steps. (1) Skim milk containing 14 to 16% nonfat milk solids made either by adding nonfat milk solids to normal skim milk or water to make reconstituted skim milk or by concentrating normal skim milk is first cooled to 35° to 50°F.

(2) A water-soluble, nontoxic acid, e.g., hydrochloric, phosphoric or lactic acid, in concentrated or dilute form is added to the cool high nonfat milk solids milk product with sufficient agitation to prevent localized coagulation. The amount of acid added will normally provide the milk, when at about 70°F., with a pH of 4.5 to 4.7. (It has been observed that after the acid is added to the cool milk, reaction with the milk constituents is not instantaneous. This is evidenced by the fact that immediately after addition of the acid, the pH of the acid-milk mixture is frequently as low as about 4.3. After the acid-milk mixture has come to equilibrium, however, the pH rises to 4.5 to 4.7. The time in which equilibrium is reached is temperature-dependent and decreases as temperature increases. With the milk-acid mixture at 40°F., equilibrium is achieved within 20 minutes). The addition of a concentrated food processing grade acid with the vigorous agitation provided by metering the acid through a pump pumping the cool milk, is a preferred procedure.

(3) The cool acidified high nonfat milk solids milk product is then warmed to the coagulation temperature, without agitation during coagulation and the resultant formation of the curd. The rapid heating of the cool acidified high nonfat milk solids milk product to 90° to 125°F. by passage through tubes of a heat exchanger with flow controlled to avoid turbulence and interparticle motion is a preferred procedure for curd formation.

Since penetration of heat for coagulation purposes through large quantities of acidified milk involves many problems because of the insulating effects of the milk which coagulates initially next to the heated surfaces, the preferred heat exchanger procedure above involves the use of relatively small diameter tubes through which the acidified milk passes. Usually in making cottage cheese the curd is cut into roughly cubical particles from 1/4" to 3/4", and as a practical matter, the diameter of the heat exchanger tubes would be roughly equivalent to these or the desired particle sizes. The use of relatively small diameter heat exchanger tubes permits formation of the curd without internal turbulence (interparticle motion) in the flowing acidified milk stream.

The walls of the heat exchanger tubes to which the acidified milk is exposed are heated by means of a hot water jacket and as the acidified milk contacts the heated wall it coagulates immediately adjacent the wall. Upon coagulation, some whey is expelled between the coagulum and the wall to form a thin layer of liquid which then functions as a lubricant between the wall and the formed coagulum allowing the formed coagulum (curd) to slip easily through the tube. As the acidified milk progresses from the inlet end to the outlet end of the heat exchanger tube, coagulation of the acidified milk progresses from the portion nearest the tube wall inwardly toward the center of the tube as heat is conducted toward the center of the flowing stream.

Thus, the coagulum contrives to form a sleeve around the still liquid center portion of the coagulating milk as it progresses through the heat exchanger tube. The speed with which the acidified milk is forced through what may be termed the coagulation tube and the temperature of the water jacket are controlled so that the milk is completely coagulated by the time it reaches the outlet end of the coagulation tube, and the coagulum emerges from the tube as a continuous pencil of firm curd which can be cut to the length desired to approximate the desired curd size.

Following formation of the curd by this procedure, the cut curd along with the whey is passed into a continuous cooker (recycling of the whey in the cooker is carried out to insure the presence of sufficient liquid in the cooker to support movement of the curd through the cooker), where the temperature is raised to 110° to 160°F. Heating is carried out so that the increase in temperature of the curd in whey is slow at the beginning of the cooking process but rapid toward the end. The total cooking time normally requires from 10 to 25 minutes.

As the hot curd and whey emerge from the cooker, the whey is drained from the curd and the curd is held for 10 to 90 seconds without further treatment to control the moisture content, since under such condition the curd expels moisture rapidly. The curd is then sprayed with or immersed in wash water at a temperature of 80° to 120°F. for 1 to 10 minutes. Optionally, it is then further washed with water at a temperature of 55° to 80°F. for 1 to 10 minutes, and subsequently with cold water (about 40°F.) for a time just sufficient to cool the curd. After the cold water wash the curd is allowed to drain for a time sufficient to bring its moisture content to below 80%.

It was found in the washing procedure that the higher the temperature of the initial wash water the more effective was the washing and the more quickly the pH of the curd increased to the desired pH of about 4.7 to 4.8. In addition, it was found that the pH of the wash water should be held below about 6.5 if partial dissolution of the curd is to be avoided, but should not be too acid or it will not have the desired effect of raising the pH of the curd. Although unnecessary, rennet can be used in the preparation of the curd with some attendant advantages. When rennet extract is used, either of the following procedures are preferred.

The high nonfat solids liquid milk, after pasteurization, is brought to a temperature from 80° to 105°F. and rennet extract is added in an amount from 1 to 20 ml. per 1,000 pounds of milk. As soon as the rennet has been added, cooling of the milk is begun and continued until it has reached a temperature of about 40°F., when a nontoxic, food processing grade acid is added and the milk is then further treated in the general method outlined.

The described rennet treatment is readily adaptable to a continuous process, for example, suitable rennet action on the pasteurized milk could be achieved by injecting the rennet into a holding tube of adequate length installed between the regenerating and cooling section of a high temperature-short time pasteurizer, the time required for the renneted milk to transverse the holding tube being adequate for the desired rennet activity before the milk is cooled. Alternatively, the rennet extract can be added to the cooled skim milk obtained in step (1) of the general method outlined before or to the cooled milk just after the addition of the acid according to step (2) of the general method.

Continuous Process

A.H. Wakeman, D.C. Roahen and R.L. Knox; U.S. Patent 3,585,721; June 22, 1971; assigned to St. Regis Paper Company conceived an apparatus which lends itself to the continuous production of curd and which can treat the curd thus formed to produce cottage cheese. By means of this apparatus uniform curd and cottage cheese in a closed system is produced thus preventing contamination. The process is also accomplished in a time cycle of the order of one hour or less.

In essence this apparatus for the continuous production of curd and cheese, comprises a curd former for the formation of curd and whey, fluid delivery means for delivery of milk to the curd former, a cooker or processor arranged for timed temperature treatment of curd, means for delivering curd and whey from the curd former to the cooker, a conditioner for separating curd and whey and adjusting the solid content and if desired, other characteristics of the curd, and means for delivering the curd and whey from the cooker to the conditioner.

The curd former may take the form of a heat exchanger including a number of tubes into which the milk is delivered and in which coagulation takes place due to heating of the milk as it passes through the tubes. Thus, curd and whey are discharged continuously from the outlet end of the tubes. In the vicinity of the outlet end of the tubes, means are provided for reducing the curd to discrete particles. For this purpose, rotating blades positioned at the tube ends to cut the rodlike curds into small slugs are used.

To treat the curd to produce cottage cheese, arranged in the cooker or processor are a plurality of temperature treating zones by which the curd and whey delivered to the processor from the cooker are successively treated. Actually, it is preferred to change the temperature of the curd and whey incrementally and for controlled time periods to produce cottage cheese of the desired quality and consistency.

To maintain the curd suspended in the whey and to prevent matting of the curd as it advances from the cooker through the processor, it is desirable to induce agitation in the curd and whey and for this purpose, means, including a surge tank upstream of the processor and surge-inducing means such as a piston and cylinder downstream of the processor for moving whey into and out of the conduit of flowing curd and whey are provided while maintaining the discharge of the curd and whey from the conduit constant. As an alternative, means for injecting a quantity of whey periodically into the conduit by means of a pump can be provided. A further feather of the apparatus adds a conditioner for separation of the curd and whey and for adjusting the solid content and acidity of the curd which comprises a conveyor, perforated baskets carried by the conveyor and drive means for advancing the baskets in a closed circuit between a plurality of stations and for maintaining the basket at each station for a predetermined dwell time. The stream of curd and whey from the processor is delivered to the baskets at one of the stations, and as the baskets advance, the whey is drained off, and the curd is washed at least once and preferably, a number of times at subsequent stations, and means are provided for pressing the curd in the baskets after which it may be discharged for subsequent treatment such as creaming and packaging.

Means are also provided for adjusting the pH of the curd at one of the stations, if necessary; and for agitating the curd during washing at at least one of the wash stations. Actually, the agitation may be accomplished by injecting into the curd intermittent blasts of fluid such as air or water.

There is a broad description of the working method of apparatus designed for continuous production of cottage cheese, which also pertains to similar equipment design by A.H. Wakeman, D.C. Roahen and R.L. Knox; U.S. Patent 3,465,439; September 9, 1969; assigned to St. Regis Paper Co. Both patents offer very detailed diagrams and operational descriptions.

CURD FORMATION

Noncoagulated Starter Culture

Inoculation with coagulated, fully ripened plant cultures, having titratable acidities of 0.75 and 0.95% (calculated as lactic acid) is essential to the production of unripened cheese (e.g., cottage cheese). A.T. Dahlstrom; U.S. Patent 3,017,274; January 16, 1962; assigned to Safeway Stores, Incorporated has found that equivalent or better results can be obtained through the use of noncoagulated cultures developed to a relatively low acidity of 0.45% or less. Specifically, in the preparation of bulk starters, it has been found that two to three times the volume of the low acid noncoagulated cultures will more than equal the activity of the higher acid cultures, and will reduce the time required for starter preparation by almost half.

Carrying this concept into cheese manufacture, a plant starter ranging in acidity from 0.25 to 0.45% will reduce the time to coagulate or "set-up" the curd in substantial proportion to the amount of low acid starter used. Best results require the use of at least 10% of such low acid starter (based on the weight of the milk) with amounts ranging to 50% or more preferred. Furthermore, the use of low noncoagulating acidities makes possible a continuous step in fermentation processing in which a removed portion of the culture is replaced with fresh milk that is not possible with coagulated, high acid cultures.

In the production of cottage cheese from skim milk, the method can be carried out substantially as illustrated in Figure 1.3. Step (10) represents the inoculation of a quantity of skim milk with the mother culture of lactic acid bacteria. Inoculation can be with either a high or a low acid culture. When a low acid culture is used, it should be added in amounts approximately two to three times that used with a high acid culture (e.g., 2 to 3% on the weight of the milk).

Step (12) illustrates incubation of the inoculated milk at constant temperature (70°F.) to obtain the low acid starter. Incubation of 6 to 9 hours is generally sufficient to obtain a desired noncoagulating acidity between 0.25 and 0.45%. This compares with 15 or 16 hours required to produce a high acid starter (e.g., 0.75 to 0.95%). The resulting low acid starter can be used as is, or preserved for future use by cooling as indicated in step (14). For this latter purpose, chilling to a temperature well below 50°F. is essential, although cooling to below 60°F. will substantially arrest the activity of the starter for a relatively short period of storage.

FIGURE 1.3: REDUCTION IN FERMENTATION TIME USING LOW ACID STARTER

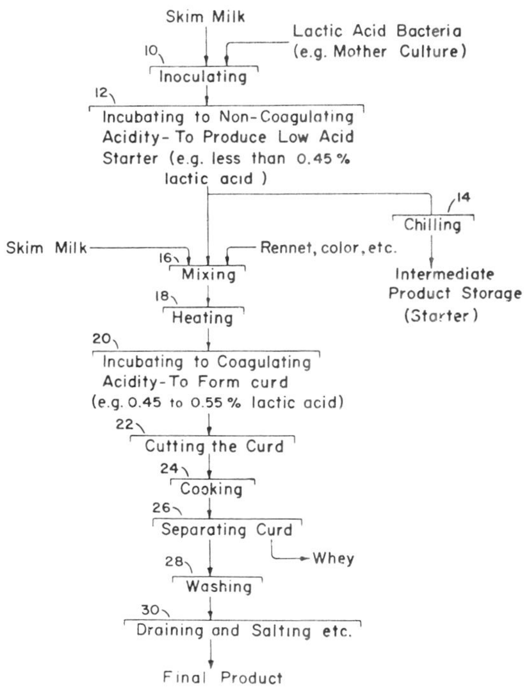

Source: A.T. Dahlstrom; U.S. Patent 3,017,274; January 16, 1962

In step (16) the low acid starter is added to a supply of fresh pasteruized skim milk in a cheese vat or other suitable container. The proportion of starter added is between 10 and 50% of the weight of the milk being treated. Agents to induce settling (e.g., rennet), color, and other desired ingredients may also be added at this time. The dispersed ingredients can now be heated in step (18) to induce a more rapid curd formation. Heating to a temperature of 85° to 90°F. is usually used in the quick set procedure. The ingredients may be separately heated before mixing in the step (16). In step (20) time is allowed for incubation to achieve a coagulating acidity of 0.48% (acidity of clear whey) to obtain setting of the curd. The precise acidity of the coagulated product will depend on whether a rennet curd or an acid curd (without rennet) is obtained, with acidities ranging from 0.45 to 0.55% (acidity of whey).

As indicated, the time required for setting or coagulation in step (20) depends to a large extent on the temperature of incubation (85° to 90°F. being preferred). It has also been found that use of the low acid starter in amounts greater than 10% has the supplementary effect of reducing the set-up time almost in direct relation to the amount of low acid starter used. This relation is illustrated in the following table which compares the set-up times obtained with a low acid starter (0.45%) with the five-hour period experienced with high acid starters.

% Addition Low Acid Starter	Set-up Time, hours at 85° to 90°F.	Reduction in Set-up Time percent	Reduction in Set-up Time hours
10	5	--	--
25	4	20	1
50	2 1/2	50	2 1/2
100	1	80	4

The curd produced in step (20) can now be processed in conventional fashion to produce cottage cheese. This includes cutting the curd in step (22) to obtain different uniform sizes as required for various markets. The curd is heated or cooked in step (24), typically by a steam jacket surrounding the cheese vat. After the whey has been drawn or separated from the curd in step (26), the curd is washed, drained and salted to produce the final product.

The processing as described relates to the production of separate batches of the bulk starter. However, the starter in noncoagulated low acid form can be continuously used and replaced by fresh skim milk in a continuous fermentation process. Skim milk can be continuously added to a freshly prepared quantity of low acid starter, with portions of the starter being simultaneously removed for use in the cheese vats. Such processing assumes an addition of fresh skim and withdrawal of starter at a predetermined proportional rate to maintain the acidity of the mixture of added milk and starter at a preferred noncoagulating acidity between 0.25 and 0.45%. Subsequent processing can be substantially as before or, if desired, the skim milk can be preheated to 85° to 90°F. and mixed with a quantity of the removed starter. Coagulation and setting of the curd can proceed in conventional manner.

Example: A low acid bulk starter was prepared by inoculating a quantity of fresh pasteurized skim milk with a laboratory culture (0.90% lactic acid) and allowing it to incubate at 74°F. for about 9 hours. This produced a starter having an acidity of 0.42%. To 19,500 pounds of fresh pasteurized skim milk (16 seconds at 163°F.) at 93°F. with an acidity of 0.15%, were added 1,400 pounds of the above starter at 45°F. Also added were 20 ml. of rennet. The resulting mixture, at a temperature of 82°F. and an initial acidity of 0.17%, was allowed to incubate in a cheese vat until the titratable acidity of the whey reached 0.50%. The elapsed time to setting of the curd was 5 hours and 35 minutes.

High Heat Treatment of Milk and Curd

In the process described by R.V. Hussong, E.H. Marth and D.G. Vakaleris; U.S. Patent 3,117,870; January 14, 1964; assigned to National Dairy Products Corporation which is designed to extend the shelf life of cottage cheese, skim milk to be used in the manufacture of the cottage cheese is treated under special conditions of temperature and time to effect a desired conditioning of the milk. The treated milk is then inoculated with the usual starter organisms and rennet is added. A relatively low cutting acid is developed and an acidity between 0.40 and 0.44%, expressed as lactic acid, is developed.

This cutting acidity is considerably less than the usual cutting acid for cottage cheese make procedures, and the pH of the curd is substantially higher. The curd is then subjected to heat treatment at 130°F. for 45 minutes up to 145°F. for equivalently diminished periods of time at the higher temperatures. Thus, the curd is subjected to conditions of temperature and time which effect destruction of most microorganisms. This ultra-high heat treatment is such that more than about 40% of the serum protein is denatured. For best results, more than 60% of the serum protein is denatured. This can be accomplished by holding the milk at 200°F. for 40 seconds; 260°F. for 12 seconds; or at 300°F. for zero time.

The heat-treated milk is then converted into cottage cheese by setting the milk and cutting the resultant curd when the acid has developed into the previously indicated range, which is substantially lower than the conventional cutting acidity. In this connection, harmless acid-producing bacteria and rennet are added to the heat-treated milk to cause desired acid development and setting. The acid-producing bacteria are normally added as a lactic acid starter. Acid development is allowed to proceed to an acidity of 0.40 and 0.44%. This particular acidity is chosen to render the curd amenable to subsequent heat treatment.

The cutting operation is carried out in the usual manner so that the curd is not broken into fine particles but remains in cube form. After cutting, the curd is cooked to a higher temperature than is usually used. The curd is cooked to a temperature between 130° and 145°F. The cooking step is normally carried out over 1 1/2 hours. If the curd is cooked to temperatures higher than 130°F., the time during which it is subjected to such higher temperatures must be taken into consideration in connection with the heat treatment of the method. Following heat treatment, the curd may be given additional water washings, may be salted, and is then in substantially finished form. The resultant cottage cheese may be creamed and packaged for distribution.

As a consequence of the heat treatment of the curd, the number of bacteria in the curd is substantially reduced so that extended shelf life in excess of 8 weeks may be provided. Of course, care should be taken that the cottage cheese is not recontaminated following the heat treatment. In this connection, following the heat treatment recognized sanitary practices should be observed to insure that the curd is not recontaminated. Desirably, the curd is maintained at temperatures above about 130°F. throughout processing subsequent to heat treatment and until the cheese is packaged and sealed. Alternatively, it is possible to maintain aseptic conditions throughout this processing. All of the wash water should be pasteruized, and, if the cheese is to be creamed, the cream should be pasteurized. Preferably, the cream is at elevated temperature, i.e., above 140°F.

By adding hot cream, the curd may be better maintained at temperatures which inhibit microbial survival and growth until the cheese is packaged and the packages are sealed. The containers in which the cheese is packaged should be virtually free from microbial contamination. Despite the temperature conditions to which the curd is exposed during the heat treatment, the product has desirable organoleptic properties.

Example: 475 pounds of raw skim milk were subjected to ultra-high heat treatment to denature the serum protein. The ultra-high heat treatment was effected in a tubular heat exchanger comprising 25 feet of 1 inch diameter stainless steel pipe, the outer surface of which was steam heated so that the milk was heated to 220°F. The residence time of the skim milk in the heat exchanger was 30 seconds, following which the milk was immediately cooled to ambient temperatures. Analysis of the serum protein nitrogen content by the Harland and Ashworth method indicated that 70% of the serum protein was denatured by the foregoing heat treatment.

The milk was introduced into a conventional cottage cheese vat, and its temperature was adjusted to 90°F. 25 pounds of lactic acid starter were added to the milk. When the acidity had reached 0.28%, about 0.5 ml. of rennet extract was added to the milk. About 4 hours after addition of the starter, the acidity had reached 0.44%. At this acid level, the curd was cut, and the curd was then cooked up to a temperature of 140°F. within a period of about 1 1/2 hours. The curd was at temperatures greater than 130°F. for the last 15 minutes of the cooking period.

After completion of the cooking step, the curd was allowed to remain in the whey at a temperature of 140°F. for an additional 15 minutes, whereupon a first portion of the curd was removed from the whey and was given 3 washes with 140°F. water to provide a first portion of cottage cheese. Each wash required 5 minutes, for a total washing time of 15 minutes. All of the wash water had previously been pasteurized by holding it at 150°F. for 30 minutes. A second portion of the curd was allowed to remain in the whey at 140°F. for an additional 60 minutes after removal of the first portion.

The second portion of the curd was also given 3 washes with 140°F. pasteurized water in the same manner as the first portion. A third portion of the curd was allowed to remain in the whey at 140°F. for an additional 45 minutes after removal of the second portion of the curd, and the third portion was also washed with 140°F. water in the same manner as the first portion, to provide a third portion of cottage cheese.

A creaming mixture was prepared by blending cream and skim milk to provide a mixture comprising 12% fat. Salt was added to the mixture at a level of 3%, and the salted mixture was then pasteurized by holding at 180°F. for 30 minutes. The creaming mixture was then homogenized at a pressure of 3,500 psi, and was refrigerated until ready for use at which time it was again pasteurized immediately before use.

Each of the first, second, and third portions of cottage cheese was creamed immediately after washing by adding to the cheese a quantity of the foregoing pasteurized creaming mixture in a proportion of 1 part of the creaming mixture to 2 parts of cheese. The creaming mixture was at a temperature of 160°F. at the time of addition to the cheese, so that the temperature of the creamed cottage cheese was 138° to 140°F. Immediately after creaming, each of the first, second and third portions of cottage cheese was packaged in sterilized containers. The containers were sealed and cooled by submerging them in ice water, and were thereafter stored at refrigerator temperatures. Samples of each of the first, second, and third portions of cottage cheese were evaluated for organoleptic properties 1 day after making. Table 1 summarized the results of this evaluation in relation to the time period during which the curd comprising each cottage cheese sample was held at temperatures above 130°F. The first and second portions of cheese had good body, texture, and flavor 1 day after preparation. The third portion of cheese was more firm than desirable, indicating that the curd from which the cheese was prepared had been subject to heat-treatment for an excessive period of time.

TABLE 1: EFFECT OF TREATMENT TIME ON BODY AND TEXTURE

Cottage Cheese Sample	Time (min.) During Which Curd was Held at or Above 130°F.					Organoleptic Properties		
	During Cook	In Whey After Cook	During Wash	During Creaming and Packaging	Total	Body	Texture	Flavor
First	15	15	15	10	55	Good	Good	Good
Second	15	75	15	10	115	Good	Good	Good
Third	15	120	15	10	160	Too Firm	Acceptable	Good

In addition, creamed cottage cheese samples prepared in accordance with this process were stored at 45°F. for 60 days. Table 2 summarizes the bacteriological status of the cottage cheese during the storage period, in comparison with a conventional cottage cheese product. The conventional cottage cheese product was completely unsalable at the end of 2 weeks due to spoilage, while the cottage cheese portions of this process were still good after 8 weeks of refrigerated storage.

TABLE 2: EFFECT OF HEAT TREATMENT ON STORAGE LIFE

Cottage Cheese Sample	Time (min.) During Which Curd was Held at or Above 130°F.				Numbers of Bacteria per gram After Storage at 45°F.								
	During Cook	After Cook and During Wash	During Packaging and Creaming	Total	1 Day	1 Week	2 Weeks	3 Weeks	4 Weeks	5 Weeks	6 Weeks	7 Weeks	8 Weeks
First	15	55	10	80	350	---	--	20	500	---	---	---	340
Second	15	70	10	95	360	---	--	360	370	360	340	250	280
Conventional Cottage Cheese	--	--	--	--	6,000	1.8 million	36 million	---	---	---	---	---	---

D.G. Vakaleris; U.S. Patent 3,039,879; June 19, 1962; assigned to National Dairy Products Corporation developed a process which yields increased amounts of cottage cheese solids of high quality. The increase in yield is accomplished by recovering additional protein from the milk increasing the yield up to 14%. This increase in yield reduces the amount of protein in the whey so that it is more readily disposed of and is in better condition for further use facilitating its use in the manufacture of various other products. The process also provides for control of bacteriophage in cottage cheese manufacture.

The process includes the treatment of milk under special conditions of temperature and time to effect a conditioning of the protein in the milk. The treated milk is then subjected to a low acid cottage cheese make utilizing a cutting acid in the range of

0.36 to 0.4%, which provides a curd having a pH in the range of 5.0 to 4.8. The cutting acid is considerably less than the usual cutting acid for cottage cheese make procedures, and the curd pH is substantially higher than the usual procedures.

More particularly, the process comprises subjecting the milk to high temperature treatment which will be referred to as "ultra high" treatment. Such treatment comprises effecting more than 40% denaturation of the milk serum protein. To obtain best results, the serum protein should be denatured to more than 60%. Such denaturation can be accomplished by holding the milk at 260°F. for 12 seconds, or at 300°F. for zero time, i.e., bringing the milk up to 300°F. and immediately cooling it. This treatment can be practiced with a Roswell type heater. This treatment of the milk is to be distinguished from the usual pasteurization treatment.

The heat-treated milk is then converted into cottage cheese by setting of the milk and cutting of the resulting curd when the acid has developed into the indicated range. Thereafter, the curd is subjected to washing, draining, salting, etc. To aid in establishing the desired moisture content and to assure production of cottage cheese having an analysis within federal standards, calcium chloride may be added during the make procedure. A few hundredths of a percent of calcium chloride may be added to the milk. Because of the low acid make procudure and because of the desire for some increased acidity in the end product, it has been found that a desired end product can be provided by culturing the cream which is added to the curd.

The milk is first given the ultra high temperature treatment. This treatment has the effect of conditioning the milk serum protein to maximize the recovery in the make procedure. Moreover, the treatment has the effect of sterilizing the milk so that bacteria are not available in viable form to deleteriously influence the cottage cheese during the make. In addition, such sterilizing tends to extend shelf life of the end product. The treatment also has the effect of controlling bacteriophage in the cottage cheese manufacture and in the whey. To the milk is added calcium chloride in an amount of not more than 0.02% (calculated as anhydrous calcium chloride).

To the milk are added harmless acid-producing bacteria and rennet to cause acid development and setting of the milk. The acid-producing bacteria are added in a starter, the starter being a usual lactic acid starter. The amount of starter which may be added is conventional. The amount of starter added and the rate of growth of the organisms determine, to a considerable extent, the period of setting. Whatever the setting technique during the setting period, the starter organisms increase in the milk at the setting temperature with the production of lactic acid so that the pH of the milk gradually becomes more acid. The acid develops in accordance with this process to a level between 0.36 and 0.4%.

When the acidity exceeds the specified range, the resulting clot retains whey very strongly. An insufficient amount of whey is released upon cutting and during cooking. Thus, draining of whey and subsequent washing of the curd are extremely poor. As a consequence, the final curd is excessively wet and pasty. At acid levels below 0.36%, the clot of the milk does not firm to the desired degree and a soft curd results. Such curd breaks early during cooking and there is excessive loss of fine curd particles. In addition, at lower acid levels there is a loss of flavor, the curd tends to exhibit rubbery and mealy characteristics, and its keeping quality is adversely affected. Accordingly, the specified range of acidity is an important feature of this process.

The rennet is added at a level of about 1 ml./1,000 lbs. of milk. At higher levels the whey expelling characteristics of the curd are impaired and a wet and fragile product results. Therefore, the amount of rennet should be less than 1 ml./1,000 lbs. of milk.

After setting, the cutting operation is carried out in the usual manner so that the curd is not broken up into fine particles but remains in discrete cubed form. The cut curd is then cooked at higher than conventional temperature (130° to 140°F.) for a suitable period of time. The whey, after the cooking operation, is drained from the curd. This draining is often conducted through strainers to limit the loss of solids. Accordingly, it is important that the curd not clog the straining material. The curd is then washed first with cool water. Several washing steps may be used and usually are used. After washing, the curd is lightly salted. Because the cottage cheese resulting from the make procedure of this process is somewhat low in acid, the end product acidity can be adjusted by using cultured cream in the creaming operation. The cream is cultured and added to provide a pH for the product, which can be that obtained during conventional make procedures.

Example: 451 lbs. of skimmed cow's milk were subjected to ultra high temperature treatment by heating to 260°F. and holding this temperature for 12 seconds. This effected about 75% denaturation of the serum protein. This was accomplished in a Roswell heater. The milk was then cooled to 89°F. and passed to a vat. To the milk was added calcium chloride at a level of 0.02% and a Streptococcus lactis starter at a level of 5% of the skim milk. After 90 minutes, rennet was added at a level of 1 ml./1,000 lbs. of milk. The milk was maintained in the vat at a temperature of 90°F., and after 4 hours and 15 minutes from the time of addition of the starter, the acid was 0.39%, curd pH 4.83.

At this acid level, the curd was cut and the vat heated until the temperature of the whey was 130°F. The curd was heated to this temperature in 1 hour and 30 minutes, and it was held at this temperature for 15 minutes, and the whey was then drained. The curd was then given 3 successive wash treatments, 2 with tap water and 1 with chilled water at about 45°F.

Simultaneously with the start of the above vat, a vat was started with the same amount of pasteurized milk to make cottage cheese by a normal make procedure. Calcium chloride was added, and the same amount and type of starter was added. Likewise, rennet was added at the same time and in like amount. The acid was developed to 0.50%, but this required 45 minutes additional time. After this acid development, the curd was cut and heated to a temperature of 120°F. within 1 hour and 35 minutes. Thereupon, the whey was drained and the curd washed as in the case of the curd of this process. The cottage cheese of this process and the conventionally made cottage cheese had the following analysis.

	Cheese of Process	Conventional Cheese
Pounds of cheese	86.9	72.8
Pounds of solids (dry basis)	16.34	14.58
Percent increase of yield (cheese solids)	12.8	-----
Nitrogen in milk serum, mg./ml. (Harland & Ashworth method)	0.210	0.710

(continued)

	Cheese of Process	Conventional Cheese
Total nitrogen in whey, mg./ml.	0.67	1.29
Nonprotein nitrogen in whey, mg./ml.	0.41	0.40
Total protein nitrogen in whey, mg./ml.	0.26	0.89
Whey protein nitrogen recovered in curd, percent	70.0	----

It can be seen that this process produced a substantial increase in cheese solids, with a corresponding increase in yield, and a substantial decrease in protein in the whey, as reflected by the nitrogen values. The cheese was salted and to it was added cultured cream, the cream being added at a level of 1 part of cream per 2.5 parts of curd. The cream comprised 14% butterfat and had a pH of 4.64. The resultant cream and curd mixture had a pH of 4.88. The creamed curd was indistinguishable in flavor from cottage cheese made by the usual make procedures.

Addition of Proteolytic Enzymes

L.L. Little; U.S. Patent 3,406,076; October 15, 1968; assigned to The Battelle Development Corporation describes a direct acid process in which a curd of excellent texture is obtained at refrigerated temperatures without the necessity of warming the milk. A high quality cheese curd is obtained from milk by introducing acid and proteolytic enzymes into refrigerated milk in quantities large enough to effect casein coagulation. Particular success is achieved using the proteolytic enzyme rennin in excessive quantities.

In this process, milk is first refrigerated. The maximum operable temperature includes the highest temperature at which the milk could be acidified without coagulation. This varies from batch to batch depending on the composition of the milk and particularly on milk solids content. For example, for cottage cheese made from skim milk which contains 8 to 9% nonfat milk solids, the maximum temperature is about 55°F. Fortifying the milk with additional nonfat milk solids raises the maximum temperature at which the milk can be acidified without coagulation. At nonfat milk solids concentrations of 16 to 20% by weight, the maximum temperature may be as high as 60°F. Generally, the preferred maximum temperature is about 55°F. It is, of course, undesirable to freeze milk so that the minimum useful temperature is just above the freezing point of milk. The freezing point of milk varies to some extent within the range of 30° to 32°F., but may be regarded as being 31°F.

Once the milk has reached the necessary refrigeration temperature, it is "set" with the acid and enzyme additions. Such setting constitutes putting the milk in a condition where casein coagulation occurs. In this process, the time of setting may be accurately controlled since the time required for complete casein coagulation is dependent to a large degree on such controllable parameters as acidity, the quantity of enzyme addition, the temperature of the milk, and the nonfat milk solids content of the milk. Ordinarily, the acid addition will be made first; however, coagulation will not occur at the refrigerated temperatures to any appreciable extent until both additions have been made so that both additions may be made simultaneously or part or all of the enzyme addition may be made first.

Preferably, the milk is vigorously stirred or agitated during and after the acid and enzyme additions to obtain a uniform mixture prior to casein agglomeration. The additions are made in such quantities to effect substantially complete coagulation within about 4 hours; however, they may be such as to effect a delayed coagulation over a period of 24 hours or longer, since at the refrigerated temperatures, undesirable bacterial growth is not a problem.

Any material capable of acidifying milk to the desired pH which does not render the curd deleterious for human consumption may be employed as the acid addition. In addition to lactic acid, readily usable acids include hydrochloric, phosphoric, citric, tartaric, fumaric, etc. Such acids should be used in a relatively concentrated form, and should be of such concentration as to acidify the milk to prevent a thin, watery product. Preferably the concentration should not exceed about 5% of the weight of the milk.

In this method, the milk is brought to about 40°F., and the desired amount of lactic acid added. Sufficient enzyme or rennet is added to bring about coagulation within the desired time. When the coagulation is firm enough to cut, it can be determined easily by visual observation. In the manufacture of cottage cheese by the culture process, small additions of proteolytic enzymes, particularly rennet extract, are frequently made to alter or enhance casein coagulation. However, these additions are not made in amounts which would cause casein coagulation at refrigerated temperatures. In this process, the milk is acidified and one or more proteolytic enzyme is added in sufficient quantity to effect casein coagulation at the refrigerated temperature.

When a slight excess of rennet is used in the culturing method, the finished cottage cheese is soft and lacks the characteristic firm texture desired. When a significant excess of rennet is used in the range of 1 1/2 to 2 cc (Hansen's standard) per 1,000 pounds of milk, the milk coagulates at an abnormally high pH and the curd is so sticky and rubbery and mats so badly it cannot be cooked. In this process, adding excessive amounts of rennet also produces a soft curd; however, considerably greater quantities per 1,000 pounds of milk can be utilized before this characteristic becomes objectionable.

The practical and preferred range of rennet addition for obtaining a commercially attractive cottage cheese product is found to be from 10 to 66 cc/1,000 lbs. of milk (Hansen's standard). Adding greater than 60 cc of rennet produces a curd that is softer than is desirable for cottage cheese, though perfectly acceptable and sometimes desirable for other cheese curd products. The use of less than 10 cc of rennet (per 1,000 pounds of milk) generally entails an excessive delay in obtaining a coagulum. In the manufacture of cheese curd where a soft curd is desired or may be tolerated, up to 100 cc of such rennet can be used in conjunction with this process. Though soft, the cheese curd does not mat during cooking providing the pH is about 4.95 or lower. This feature of the process makes it readily adaptable to making other varieties of soft, unripened cheese, such as baker's cheese, cream cheese, and Neufchatel cheese. Since these cheeses are essentially a soft, fine-textured curd, excessive amounts of rennet assure such characteristic.

The preferred quantity of enzyme varies widely and is dependent on the time requirements as well as the degree of acidity, refrigeration temperature and cheese curd product being made. When the pH of the milk is adjusted to 4.50 to 4.80 and the milk is set at refrigeration temperatures, it is difficult to add too much rennet providing the desired product is not large curd cottage cheese. Under these conditions, increasing the amount of rennet hastens the time of coagulation but does not alter the type of coagulation or its cooking characteristics. This is significant since in the culturing process the customary 1 ml. of rennet (Hansen's standard) per 1,000 pounds of milk is sufficient to bring about coagulation almost instantly when the milk is in the temperature range of 70° to 90°F. and a pH range of 4.6 to 4.8. However, 4 to 18 hours of setting time are required by the culturing process for acid development to reach a pH of 4.95 or below.

With this process, the pH may be brought to the desired level in little or no time and the setting time can be controlled by the amount of rennet used to obtain setting times varying from 30 minutes to 24 hours. Where the desired acidity is within the 4.50 to 4.95 pH range, it is preferable to use at least 10 ml. of rennet (or other proteolytic enzyme) per 1,000 pounds of milk to avoid excessive setting times, although it may be possible to use as little as 1.5 ml. of rennet for such purpose, and as stated above, as much as 100 cc or rennet per 1,000 pounds of the refrigerated milk-acid mixture can be used to obtain rapid coagulation without adverse effects, depending on the product being made. The preferred pH range of acidity for making large curd cottage cheese with this process is from 4.5 to 4.8.

An advantage of this process is that the balance between acid coagulation and rennet coagulation characteristics of the curd can be precisely controlled. If small curd-type cottage cheese is desired by utilizing the culturing process, acid coagulation is paramount and little or no coagulator or enzyme is used. This same characteristic can be obtained in this process by acidifying to a lower pH, i.e., 4.20 to 4.50. Then, even when a great excess of rennet is used to coagulate the milk, the finished cottage cheese has the characteristic of that made by acid coagulation with little or no rennet. By this process the pH to which the milk is acidified determines the characteristic of the curd and this can be controlled precisely. Then the amount of rennet used can range from 0.5 to 1 cc per 1,000 pounds of milk, substantially the amount normally used in the culturing process.

Example: Three batches of pasteurized skim milk were cooled to a temperature of about 40°F. and acidified to a pH of about 4.7 with concentrated hydrochloric acid. While maintaining the 40°F. temperature, the curd was coagulated by adding 30 ml. of Hansen's single strength liquid rennet per 100 gallons of milk. The milk was coagulated and firm enough to cut in 1 hour. One lot was cut at the end of the hour and cooked. The second lot was held at the 40°F. temperature for 12 hours after it had coagulated firm enough to cut before it was actually cut. The third lot was held 48 hours at the refrigerated temperature before it was cut and cooked. All three batches cooked out very nicely and made excellent quality cottage cheese. There was no indication of bacterial growth or deterioration of quality due to the prolonged delay in cutting and cooking.

Acidification of Whey

The process of J.L. Holder and D.R. Morgan; U.S. Patent 3,411,920; Nov. 19, 1968; assigned to Safeway Stores, Incorporated is predicated upon the fact that if the pH of the whey immediately after cutting is lowered by the addition of acid, a number of benefits are obtained. After such adjustment of the pH value of the whey, the curd is at one pH value and the whey at a lower pH. Under such conditions, the desirable effects of both low and high acid curd can be obtained.

The desirable features of the process are that it affords high yields, it produces a form of cottage cheese having superior texture and more uniform curd size, and it enables more rapid cooking of the curd resulting in time-saving.

As illustrated in the flowsheet, Figure 1.4, the initial steps for coagulating milk may follow conventional practice. Thus, milk is shown being supplied to mixing step (10) where a suitable starter is introduced together with one or more additives, such as rennet. The starter is usually one of the lactic acid-forming bacteria type. The milk may be either fresh or reconstituted, and generally has a minimum of fat content. However, for particular types of cottage cheese, varying amounts of fat may be present.

FIGURE 1.4: ACIDIFICATION OF WHEY DURING COTTAGE CHEESE PRODUCTION

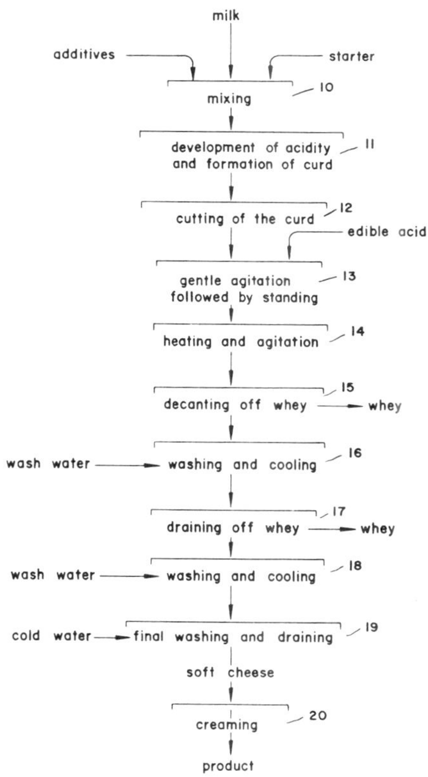

Source: J.L. Holder and D.R. Morgan; U.S. Patent 3,411,920; Nov. 19, 1968

In step (11) the inoculated milk is held under controlled temperature conditions, during which time lactic acid develops, and eventually a coagulating acidity is reached with resulting coagulation and formation of the desired curd. After coagulation of the milk, the curd is cut by conventional procedures and equipment as indicated in step (12). At this time, the mass consists of the cut curd together with whey. Assuming that the process is carried out batch fashion, the amount of free liquid whey is sufficient to cover the cut curd to a substantial depth. An amount of edible acid is now added in step (13) to reduce the pH of the whey to 4.3 to 4.4.

The acid employed may be citric, phosphoric, acetic, lactic or tartaric. Citric and phosphoric are preferable because the presence of residual amounts of such acids in the finished product is compatible with flavor requirements. Preferably the acid is added as a water solution as, for example, a 50% solution of citric acid in water. A suitable procedure is to add the citric acid solution to the free whey at the top of the mass, and then cause it to be dispersed throughout the free whey of the mass by gentle agitation. Such agitation also causes some additional separation of whey from the curd.

After the addition of acid and gentle agitation, which in a typical instance will be carried out over a period of 2 to 7 minutes, the mass is permitted to stand without further agitation for a period of from 5 to 15 minutes to allow for further expulsion of whey from the curd, with firming of the curd. While subsequent heating in step (14) may be started at the end of the period just specified, it has been found that heating may commence about 2 minutes after the acid is added, provided the curd is not agitated for a period of from 5 to 15 minutes after the acid has been dispersed in the whey.

Heating in step (14) serves to harden the curd. It has been found that the curd can be heated more rapidly than usual thus shortening the cooking period. It is preferred to cook to substantially higher levels, for example, 130° to 135°F. Thus heat can be applied to the mass in step (14) to elevate the temperature from ambient to 126° to 140°F. over a period of 50 to 115 minutes, and then held at such temperature level for 5 to 15 minutes to effect cooking and firming to the desired degree. The overall period of time required for step (14) may be 75 to 120 minutes, in contrast with time periods of 105 to 165 minutes used in other processes.

Following heat treatment in step (14), the mass is subjected to further operations for the removal of whey and for washing the curd. This can be carried out by removing free whey in step (15), as by decantation, after which wash water is applied in step (16), and whey diluted with wash water is removed in step (17) as by drainage. A further application of wash water and subsequent drainage of diluted whey is indicated in steps (18) and (19). Application of wash water in step (16) serves to reduce the temperature of the curd to 80° to 90°F. In step (18), the temperature is reduced by a second wash water to 60° to 70°F., and finally in step (19), to 38° to 45°F. by addition of a cold wash water. After step (19) the washed curd can be used as such or can be subjected to creaming in step (20) to produce the final cottage cheese product.

Example: 950 gallons of pasteurized skim milk at 93°F. were placed in a suitable vat. A lactic acid starter was prepared described in U.S. Patent 3,017,274, and 550 gallons of such a starter at 85°F., having a noncoagulating acidity of 0.35% were added to the skim milk. To this mixture at 90°F. were added 10 ml. of rennet.

The mixture was permitted to stand for a period of 3 hours and 15 minutes, during which time the acidity developed to 0.50%, corresponding to a pH of 4.7. The curd was then cut with conventional 5/16 inch curd knives. A citric acid solution was prepared by dissolving 8 lbs. of food grade citric acid in 2 gallons of water. The curd was gently agitated after cutting to begin the expulsion of whey from the curd, and at that time the citric acid was added by pouring it on the surface of the whey overlying the curd, the gentle agitation serving to distribute the added acid throughout the vat. The mass was then allowed to stand for 15 minutes during which time there was some further expulsion of whey from the curd, and some firming of the curd. The curd and whey at that time were at a temperature of about 90°F.

Heating was then started using a jet type cooker which heated the whey by injection of steam. Five minutes after the heating was started, mechanical agitators were turned on slow speed and the heating continued at a steady rate, whereby the temperature of 132°F. was reached in a period of 90 minutes. The whey overlying the curd was then drained off down to the level of the top of the curd, and then wash water at about 72°F. was added to cause the temperature of the curd to be lowered to about 80°F. Thereafter whey diluted with wash water was drained from the curd. Two additional wash waters were then applied to the curd whereby the curd temperature was reduced to 60°F. and 40°F. respectively. After final washing, the curd was creamed and packaged for sale.

Disodium Phosphate Treatment

G.D. Flickinger and E.G. Stimpson; U.S. Patent 3,591,390; July 6, 1971; assigned to National Dairy Products Corporation describe a method for treating cottage or similar cheese to provide the desired texture in the finished product. In the method, an agent is added to the cottage cheese curd to reduce the firmness. The agent may also be used to provide desired texture of the curd when acidic materials are added to the cottage cheese curd. The agents used to texture the cottage cheese curd are selected from phosphate or citrate compounds or mixtures. A particularly preferred compound for this purpose is disodium phosphate.

Example 1: 200 gallons of skim milk containing 8.8% by weight solids is used to make cottage cheese. The skim milk is first pasteurized at a temperature of 165°F. for 30 minutes. The skim milk is introduced into a vat at 90°F. to provide a suitable environment for the starter. When an acidity of 0.50% equivalent lactic acid has been obtained a coagulum is formed. The coagulum is cut with conventional cottage cheese knives to provide quarter inch curd cubes in whey. The curd is then cooked in the whey until the curd reaches a temperature of 150°F. The whey is then drained and the curd is washed with two successive washings of 60°F. water. It is determined at this time that the curd is too firm. The curd is then creamed with a 40°F. cottage cheese creaming mixture.

A solution of disodium phosphate is then prepared. The solution contains about 10 weight percent disodium phosphate. The solution of disodium phosphate is then added to the creamed cottage cheese to provide a level of 0.4% by weight of disodium phosphate per pound of drained cottage cheese curd. The blended mixture of creamed cottage cheese and disodium phosphate is then stored at a temperature of 40°F. overnight. It is then determined that the curd particles have been restored to a desired texture and the curd is softened. Other curd from the same batch is creamed and stored without addition of disodium phosphate. It is determined that after storage the curd firmness is still too high.

Example 2: Cottage cheese is prepared as described with the exception that the curd is cooked to a temperature of 125°F. It is determined after preparation of the creamed cottage cheese that the curd firmness is at a desirable level. Disodium phosphate is then added to pineapple syrup to provide a level of 7.2 weight percent of disodium phosphate. The disodium phosphate solution is then added to the creamed cottage cheese at a level sufficient to provide 0.25% of disodium phosphate per pound of drained cottage cheese curd. Pineapple chunks are then added to the creamed cottage cheese at a level sufficient to provide 15 weight percent based on the creamed cottage cheese. The cottage cheese with added pineapple is then stored at a temperature of 40°F. for a period of 1 day. The cottage cheese is tested and it is found that no firming of the cottage cheese has been effected and that the cottage cheese curd has a desirable texture.

A control sample of creamed cottage cheese is also prepared to which no disodium phosphate is added. After addition of the pineapple chunks and storage at 40°F. for 1 day the curd firmness has increased to a point where the cottage cheese is rated unacceptable. The method may be used to provide desirable curd body and texture in cottage cheese curd when it has been determined that curd firmness is too high. The method may also be used to provide desirable texture and body in cottage cheese curd when acidic materials are added to the curd.

CREAMING PROCESSES

Citric Acid Fermenting Organisms

F.J. Babel and D.W. Mather; U.S. Patent 2,971,847; February 14, 1961; assigned to Research Corporation found that by creaming cottage cheese with a culture of citric acid-fermenting organisms in milk or a milk fraction containing added citric acid, the resulting creamed cottage cheese had a uniform, pleasing and stable flavor. It was also found that the creamed cottage cheese thus produced had greatly improved keeping qualities. The improved flavor is largely due to the biacetyl content introduced by the cultures. The improved keeping quality of the creamed cottage cheese is due in part to the volatile acid content, largely acetic and propionic acid of the cultures, which are known to be inhibitors of mold and slime formation.

The citric acid-fermenting organisms which have been found to be particularly useful in the method are Leuconostoc citrovorum and Leuconostoc dextranicum. Pure cultures of the citric acid-fermenting organisms should be used. Absence of lactic acid-producing organisms such as Streptococcus lactis or Streptococcus cremoris is particularly important as excessive lactic acid is objectionable in creamed cottage cheese. The citric acid-fermenting organisms grow best at about 25°C. After sufficient growth of the citric acid-fermenting organisms has taken place, citric acid is added and the fermentation allowed to proceed at a temperature of 21° to 25°C.

The substrate for the citric acid-fermenting bacteria may be skim milk, milk, or mixtures of skim milk and milk, skim milk and cream, or milk and cream. The growth medium is heated to 200°F. (93.3°C.) and held at that temperature for 1 hour, or it may be heated under 15 pounds steam pressure of 15 to 20 minutes. After heating, the growth medium is cooled to 21° to 25°C. for 24 to 48 hours to permit sufficient growth of the organisms. Following incubation, the medium is adjusted to about 4.3 with a sterile citric acid solution containing 15% citric acid. The medium is again incubated at 21° to 25°C. for about 24 hours to permit fermentation of the citric acid. The medium then is ready for use.

The citric acid-fermenting organisms may be grown in a substrate consisting of whole milk, skim milk plus added cream, or whole milk plus added cream, which is heated before inoculation with the pure culture of the citric acid-fermenting organism to a temperature sufficient to destroy most of the original bacterial flora. However, when cream is heated to a high temperature, the fat has a tendency to separate as an oily layer. It is, therefore, preferable to grow the organism in a substrate of sterilized skim milk and after the fermentation of the citric acid is substantially complete mix the culture with cream in the desired amount to produce a creaming composition.

One of the advantages of this method is that the flavor of creamed cottage cheese can be standardized in several ways. For example, if the substrate or growth medium is skim milk, it could be used in the following amounts in preparing a creaming mixture containing 12% fat.

12 parts 20% fat cream plus 8 parts culture
12 parts 30% fat cream plus 18 parts culture
12 parts 40% fat cream plus 28 parts culture

If the substrate or growth medium is milk containing 4% fat, it can be used as follows:

8 parts 20% fat cream plus 8 parts culture
8 parts 30% fat cream plus 18 parts culture
8 parts 40% fat cream plus 28 parts culture

The creaming mixture used for cottage cheese generally contains about 12% fat. Cottage cheese is mixed with a creaming mixture (12%) in the proportion of 2 parts by weight of cottage cheese to 1 part by weight of 12% creaming mixture. This gives creamed cottage cheese with 4% fat. If a creamed cottage cheese is desired that will show little free cream, the fat content of the creaming mixture can be increased. For example, if a 16% fat creaming mixture were used, it would be added at the rate of 1 part by weight of creaming mixture to 3 parts by weight of cheese.

The amount of culture in the creaming mixture can be varied according to the amount of flavor compounds present. It has been found that a creaming mixture containing 12% fat, made up of 12 parts of 20% cream plus 8 parts culture (skim milk) is suitable. To provide a creaming mixture that is smooth in body and texture, the mixture of culture and cream can be passed through a homogenizer using about 500 lbs. pressure per sq. in.

Example: Sterile skim milk at a temperature of 70°F. (21.1°C.) was inoculated with 2% of a pure culture of *Leuconostoc citrovorum* and the culture medium was incubated at 70°F. for one day. The culture medium was then acidified to pH 4.3 with a sterile aqueous solution of citric acid and further incubated at 70°F. for about 17 hours. The culture was then mixed with cream containing 20% milk fat in the proportion of 12 parts of cream to 8 parts of skim milk to obtain a creaming composition containing 12% milk fat.

The creaming mixture was added to cottage cheese in sufficient quantity to bring the creamed cottage cheese to a milk fat content of 4%. The creamed cottage cheese was salted at the rate of 1% by weight. The following table gives the pH and the biacetyl (ac^2) and acetylmethylcarbinol (amc) contents of the uncreamed curd, regular creamed curd and curd creamed by the method of this process.

	Uncreamed curd			Regular creamed curd (salted)			Curd creamed by process		
Days stored at 45°F.	pH	ac^2 (ppm)	amc + ac^2 (ppm)	pH	ac^2 (ppm)	amc + ac^2 (ppm)	pH	ac^2 (ppm)	amc + ac^2 (ppm)
0	4.68	0.50	9.1	4.99	0.30	6.5	4.94	1.6	41.2
4	4.66	0.58	10.6	5.05	0.34	6.7	5.00	2.2	54.4
6	4.68	0.50	8.9	5.05	0.34	7.4	4.98	2.2	45.6
8	4.69	0.52	9.5	5.02	0.38	6.8	4.98	2.4	53.6
11	4.65	0.52	9.1	5.03	0.34	6.3	4.92	2.8	53.4

The table shows that the uncreamed curd and the regular creamed curd contained very little biacetyl. The curd creamed with the creaming mixture of the process was much higher in biacetyl initially, and the biacetyl content increased with storage for an 11 day period. The cottage cheese creamed by the method had much better keeping quality than cottage cheese creamed in the regular manner.

Lactic Acid Forming Organisms

W.K. Moseley, P.R. Elliker and W.E. Sadine; U.S. Patent 3,323,921; June 6, 1967 describe a method of preparing a flavor and stabilizing additive for dairy products which incubates a substantially pure culture of a lactic acid bacterium in a milk substrate for 12 to 24 hours at about 21°C. The resulting culture can then be added to other milk materials to furnish the final flavor and stabilization additive.

The lactic acid-producing bacteria are rapid growers and can be inoculated from dry or liquid seed directly into pasteurized or sterilized milk. This inoculum is incubated at relatively low temperatures for 12 to 24 hours. The resulting culture is in turn inoculated directly into the final bulk tank of culture milk or cream, incubation of which for 12 to 24 hours, also at relatively low temperature, gives the final desired flavor additive, for example, a creaming mixture additive or creaming mixture base for creaming cottage cheese curds.

A variation of this procedure, particularly useful in making cottage cheese since it shortens the process, involves inoculation of the creaming mixture with a culture derived from the first transfer from dry or liquid seed or from the succeeding bulk culture, followed by incubation of the inoculated culture for 6 to 12 hours at 21°C. The resulting cultured creaming mixture can be salted and added to the cottage cheese curd without homogenization.

The lactic acid bacteria useful in this process are grown or incubated at about 21°C.; temperatures of 20° to 22°C. are operative. It is important that the incubation temperature be kept below 22°C. since, at higher temperatures, the bacteria produce undesirably large amounts of lactic acid and the amount of biacetyl present actually decreases, perhaps because of bacterial destruction.

The length of time required to provide an adequate growth of a lactic acid Streptococcus such as S. diacetilactis can vary from 6 to 24 hours depending upon the growth characteristics of the strain of bacteria employed, the size of the inoculum, the culture medium employed, the temperature of incubation, etc. Actually, in many instances, an incubation period of 6 to 12 hours will suffice. The length of time is not critical, the critical factor being the attainment of a substantial growth of the lactic acid bacteria prior to using the incubated culture medium as a flavor and stabilizing additive. Useful culture media incudes various milk-based substrates such as homogenized whole milk, skim milk, cream, etc.

Example 1: A portion of pasteurized nonfat milk is cooled to 21°C. and inoculated with a culture of S. diacetilactis (ATCC No. 15346) in sufficient amount to provide a luxurious growth after 16 hours of incubation at this temperature. The culture is then cooled to about 2°C. and mixed with pasteurized, homogenized, salted sweet cream of approximately 20% fat content at the rate of 2 parts of nonfat culture to 3 parts of sweet cream. The resulting mixture is added directly to cottage cheese curd.

Example 2: The process of Example 1 is repeated except that pasteurized homogenized milk or creams of various fat contents are cultured with suitable strains of S. diacetilactis (ATCC No. 15346) in place of the pasteurized nonfat milk of that example. The final culture is incorporated in creaming mixtures with desired final fat content. One advantage of this process in the manufacture of cottage cheese is that the direct seed method of inoculation can be employed for producing our creaming mixture additive, thereby eliminating the necessity for cottage cheese manufacturing plants lacking facilities for propagating organisms to carry the culture over extended periods of time.

Cottage cheese creamed in the manner set forth above not only has superior aroma and taste because of the presence of appreciable amounts of biacetyl, but also has a greatly extended shelf life over that of conventionally prepared cottage cheese. Both the persistence of biacetyl and lack of bacterial contamination apparently are due to the presence of large numbers of cells of the lactic acid bacteria, specifically S. diacetilactis, in the cottage cheese. The reason for the ability of S. diacetilactis to reduce contamination apparently is due in part to the ability of this rapidly growing lactic acid bacterium to competitively inhibit other bacteria in the milk substrate.

Stabilizing Agent in Creaming Mixture

B.J. Flick and J.P. Pisani; U.S. Patent 3,506,456; April 14, 1970; assigned to Kraftco Corporation provide a creaming mixture which includes a stabilizing agent compatible with and derived from dairy products, and which provides desirably increased viscosity and adhesion to cottage cheese particles. The stabilizing agent comprises particles of cottage cheese curd which may be obtained from any substantially solid particles formed in the manufacture of cottage cheese.

In particular, the particles of cottage cheese curd may be cottage cheese curd which has been ground, comminuted or otherwise reduced in particle size, or fines produced in the manufacture of cottage cheese. The cottage cheese curd or fines are of a reduced particle size so they may be uniformly suspended in one or more of the fluid milk products used in preparing the creaming mixture. In general, the particle size should be of such reduced size that there is no grainy mouth feel encountered in the creaming mixture into which the fines are distributed. Cottage cheese curd fines recovered from either the whey or the wash water of a cottage cheese make procedure may also be used as the stabilizing agent. The particle size of such fines is generally too large to permit their effective use by merely blending with the creaming mixture. It is preferable to further reduce the particle size of the fines by blending the fines with the creaming mixture then homogenizing the resultant blend. It is important that the comminuted material, whether obtained from whole curd or curd fines, be thoroughly blended with at least a portion of the creaming mixture prior to combining that portion with cottage cheese curd to prepare creamed cottage cheese.

The stabilizing agent is used at levels from 5 to 60% by weight of the creaming mixture. At levels below 5% no substantial amount of stabilization is observed. The level of stabilizing agent which may be provided in the creaming mixture without causing undesirably high viscosity is dependent on the level of fat in the creaming mixture. For example, creamed cottage cheese comprising fat in excess of about 4% cannot have creaming mixture having more than about 30% stabilizing agent based upon the weight of the creaming mixture without becoming too viscous. Higher levels of comminuted cottage cheese curd within the indicated range may be used in creaming mixtures containing less fat. Since cottage cheese curd usually has about 15 to 25% by weight of solids, the total cottage cheese curd solids present in the creaming mixture will be from 1 to 15% by weight of the creaming mixture.

Example: A vat of skim milk having 9.0% solids and 0.2% fat was made into large curd cottage cheese by conventional make procedure which included addition of starter to the milk, addition of rennet at a level of 1 cc of commercial single strength rennet to the milk, and culturing the milk to an acidity of 0.50% equivalent lactic acid until the milk had set to form a coagulum. The resultant coagulum was then cut, cooked, and drained, following which it was washed with two successive cold water washes and again allowed to drain. The resultant cottage cheese curd was then creamed with a dressing having about 11.5% milk fat, 10% nonfat milk solids and 2.7% salt.

After the dressing had been uniformly distributed throughout the curd, approximately 18% by weight of the blend was separated from the remaining material and was homogenized by conducting it through a single stage homogenizer at a pressure of 1,000 psi. The resultant homogenized creaming mixture was then added back to the remaining blend of cottage cheese curd and dressing, and was uniformly distributed throughout. The resultant creamed cottage cheese was packaged in conventional packaging apparatus in which the creamed cottage cheese was metered and dispensed into containers. The viscosity of the creaming mixture was sufficient to provide uniformity of composition in the containers throughout the packaging cycle. About 3 weeks after packaging, the containers were opened and evaluated for storage stability. The creamed cottage cheese was still satisfactory, with no change in body or consistency of the curd or of the creaming mixture.

PRESERVATION PROCESSES

Sorbic Acid

The process described by G.A. Perry and R.L. Lawrence; U.S. Patent 2,974,046; March 7, 1961; assigned to Corn Products Company is predicated upon the fact that sorbic acid, or any of its edible salts such as calcium, potassium and sodium sorbate, which on acidification yield sorbic acid, will, when added to a cultured milk product in small and apparently critical amounts depending upon the pH of the cultured milk product promote growth of useful bacteria, and almost totally inhibit the growth of spoilage bacteria, for periods of time exceeding the maximum periods during which it is likely that such a product would be expected to have a shelf life, or to be kept in the home.

In a series of tests using 31.1 pound samples of creamed cottage cheese taken from regular production in a dairy plant, the control samples were free of sorbic acid and the test samples contained 0.05% sorbic acid, based on the total weight of the creamed cottage cheese which had been uniformly distributed through the cottage cheese by prior dissolution. At the beginning of these tests, samples of cottage cheese from both the control and the test masses were made the subject of bacterial plate counts and microscopic examination, and the number of bacteria present was calculated on a per gram basis.

Since useful bacteria and spoilage bacteria could not, at the beginning of the test, be visually distinguished, the count of 5,000 per gram established at the beginning of the test was assumed arbitrarily to be one-half of each type. The samples were then stored at 50°F. for varying periods of time, as indicated in the table which follows. At the end of each period of time a representative interior portion of a control and of a test sample were subjected to bacterial plate counts and microscopic examination of the bacteria contained therein. Each time that a count was made, both on the starting day and on each subsequent day, a gas test was run on an extracted sample, this being the usual gas test employing formate ricinoleate broth.

In the table which follows, the bacterial count is given on a per gram basis for each of the control and test samples for the periods of time set forth in the table, as well as the results of the corresponding gas test.

	Control Samples (no sorbic acid)			
Days	Useful Bacteria	Spoilage Bacteria	Gas Test[a]	pH
0	2,500[b]	2,500[b]	neg.	4.9
6	(c)	500,000	pos.	4.9
10	(c)	2,000,000[e]	pos.	4.8
14	(c)	1,200,000[f]	pos.	4.4
23	(c)	600,000[f]	pos.	4.4
	Test Sample (0.05% sorbic acid)			
Days	Useful Bacteria	Spoilage Bacteria	Gas Test[a]	pH
0	2,500[b]	2,500[b]	neg.	4.9
6	15,000	(d)	neg.	4.9
10	105,000	(d)	neg.	4.9
14	240,000	(d)	neg.	4.8
23	4,000,000	(d)	neg.	4.8

(a) Gas Test — Formate ricinoleate broth.
(b) At day 0 the bacterial count was 5,000 per g. for each of the control and test samples. One-half of these were assumed to be useful bacteria, and the other half spoilage bacteria.
(c) No detectable increase.
(d) No detectable increase in bacterial, yeast or mold counts.
(e) Inedible due to excessive spoilage bacteria coupled with yeast and mold growth.
(f) Decrease in spoilage bacteria after 10 days due to competitive action of yeast and mold growth with attendant lowering of pH.

This demonstrates that sorbic acid has a selective capacity for controlling the growth of useful and spoilage bacteria. It appears to promote the growth of useful bacteria, and to inhibit the growth of spoilage bacteria. In each case, a control mass of creamed cottage cheese, and a mass containing 0.05% of sorbic acid, each of which had been held at 50°F. for the number of days indicated, was then subjected to organoleptic examination by a panel of judges. In the case of each test sample, the panel did not report any significant increase in spoilage factors and reported favorably on odor and on flavor retention. On the other hand, each of the control samples, after 10 days of incubation, were found to be objectionable in odor, taste and appearance, and to be inedible.

Selective control of bacterial growth in creamed cottage cheese is effected by adding to the milk product from 0.03 to 0.15% by weight on the total milk product, of sorbic acid or its equivalent in the form of the calcium, potassium or sodium salts of sorbic acid. The lower limit in the stated range is determined by loss of effectiveness. For example, when the sorbic acid is as little as 0.01% there seems to be no discrimination between useful bacteria and spoilage bacteria.

The upper limit of the range is determined by the appearance of an organoleptically noticeable sour taste characteristic of sorbic acid. This sour taste begins to appear in cultured milk products having a pH close to neutrality when sorbic acid is above 0.15%. For example, when the curd in creamed cottage cheese is made by enzymatic treatment, such as rennet, the curd and hence the resulting creamed cottage cheese has a pH of about 6.4, and the permissible upper limit of sorbic acid is about 0.15%. However, in the case of cultured milk products having a lower pH, such as a pH of from about 4.6 to about 5.1, which is characteristic of creamed cottage cheese made with washed cultured curd, the sour taste begins to appear when sorbic acid exceeds about 0.08%. It appears, therefore, that the permissible upper limit of sorbic acid declines when the cultured milk product is slightly acid, and becomes greater as the cultured milk product approaches neutrality.

In adding sorbic acid to a mixed cultured milk product such as creamed cottage cheese, it is preferred that the sorbic acid be mixed first with the cream dressing where it is relatively highly soluble in the fats present and wherein its inhibiting effect on spoilage bacteria improves the storage characteristics of the cream dressing prior to being mixed with the curd. The thus-treated cream dressing is then mixed with curd in the usual way. Ordinarily, creamed cottage cheese consists of about 4 parts curd and 1 part cream dressing. The amount of sorbic acid added to the cream dressing is calculated upon the total weight desired in the mixed creamed cottage cheese.

The preferred range for sorbic acid is 0.05 to 0.07% based on the total weight of the cultured milk product treated. Where calcium, potassium or sodium sorbate is used, an amount is chosen which will release the desired amount of sorbic acid in the final product.

Example 1: To 25 pounds of washed cultured cottage cheese curd having a pH of about 4.5, there were added 6 pounds of cream dressing containing 0.31 pounds of salt to which 0.016 pound of sorbic acid had been added following pasteurization and homogenization while the cream dressing was at a temperature of about 80°F. The cream dressing was cooled to 40°F. and the curd and dressing were thoroughly mixed together. In this mixture sorbic acid constituted about 0.05% by weight.

Samples of this creamed cottage cheese, preserved at 50°F. for 3 weeks, showed no significant spoilage characteristics, had acceptable odor and flavor, and showed a negative gas test.

Example 2: A batch of cultured cottage cheese was made up exactly as in Example 1, except that the 0.016 pound of sorbic acid was replaced by 0.02 of calcium sorbate. The effective quantity of sorbic acid in the final product remained the same as in Example 1, namely, 0.05% by weight, and the results were the same.

Bacteria and Sorbic Acid

R.L. Lawrence and G.A. Perry; U.S. Patent 3,022,176; February 20, 1962; assigned to Corn Products Company describe a process by which a cottage cheese having a prolonged flavor and shelf life is produced. According to the process a cultured milk product containing large numbers of added microorganisms of the family Lactobacteriaceae and a sorbic acid component in an amount equivalent to 0.025 to 0.15% by weight of sorbic acid and having a pH of 4.3 to 5.5 is produced.

This process produces a cultured milk product by inoculating a milk substrate with large numbers of flavor-forming microorganisms of the family Lactobacteriaceae, incubating the inoculated milk substrate for from 4 to 48 hours or until total counts reach 10 million to 1 billion per gram at temperatures of from 50° to 90°F., and adding salt and a sorbic acid component in an amount such that the final cultured milk product has from 0.025 to 0.15% sorbic acid.

In carrying out the process, a culture of microorganisms capable of growing at low temperatures such as 45°F. and lower, and possessing the unique characteristic of multiplying in the presence of chemical preservatives, such as sorbic acid and its salts is introduced into the milk substrate. The family Lactobacteriaceae and especially those microorganisms belonging to the genus Leuconostoc have this characteristic. These flavor-forming groups of microorganisms possess the characteristic of being tolerant to sorbic acid and of being able to multiply at low temperatures, while contaminants of a psychrophilic nature are effectively restrained by the sorbic acid component which functions as a preservative. These spoilage microorganisms which grow at low temperatures destroy the desirable products of flavor-forming bacteria.

Thus, by introducing into the milk substrate helpful flavor-forming microorganisms and a sorbic acid component, not only is the growth of spoilage microorganisms prevented, but due to the selectivity on the part of the sorbic acid and its edible salts, the flavor of the cultured milk product is improved by encouraging the growth of these flavor-forming microorganisms.

To secure the most intense flavor formation, the flavor-forming microorganisms should be in the range corresponding to that of a fully cultured milk. This is usually obtained by introducing a heavy, actively growing inoculum of 8 to 10% by weight of the inoculum into a milk substrate. The amount of sorbic acid introduced into the finished cultured milk product is 0.025 to 0.15% by weight of the cultured milk product. At the time of inoculation, a small amount of sorbic acid component such that it does not restrict the growth of aroma-formers, i.e., 0.05 to 0.15% sorbic acid preferably is added to the milk substrate. The remainder of the sorbic acid required to maintain a desirable concentration in the finished product is added after processing. For example, creamed cottage cheese may be prepared by performing a preliminary fermentation of cream, containing 0.02 to 0.15% sorbic acid, by inoculation of the cream with large numbers of flavor-forming microorganisms of the family Lactobacteriaceae, preferably of the genus Leuconostoc and then incubating the inoculated cream for up to 48 hours at temperatures of 50° to 90°F. At the end of the incubation period, salt is added and sorbic acid component is added in the amounts required to maintain 0.08 to 0.3% sorbic acid in the finished product. The fermented cream is then chilled to a temperature of 35° to 45°F. prior to mixing it with curd.

Example 1: To 6 pounds of dry curd of pH 4.9 was added 2 pounds of cream dressing containing 16% butterfat and 4% salt. After blending, the mixture was packaged in 12 ounce waxed containers and capped with transparent plastic lids. The tubs of creamed cottage cheese were stored at 45°F. and examined periodically for flavor, odor, and microbial growth. After 6 days, the total microbial count on nutrient agar exceeded 11,000,000 per gram consisting mainly of large colonies of gram-negative spoilage bacteria and yeasts. The fresh flavor had disappeared and sourness developed and after 9 days visible growth on the surface was clearly noticeable.

Example 2: To 6 pounds of dry curd of pH 4.9 was added 2 pounds of cream dressing of 15% butterfat content. A pure culture of flavor-forming bacteria of the genus Leuconostoc was added to the cream dressing and the inoculated cream was held at 72°F. for 16 hours. After this time 4% salt was dissolved in the cultured cream which was then blended with curd. The product was packaged as Example 1 and held at 45°F. After 7 days a slight "off" odor and flavor characterized by sourness was noticeable and total microbial counts were in excess of 1×10^8 per gram, about equally divided between contaminants and flavor-forming bacteria. Spoilage was not prevented and after 9 days surface growth was evident.

Example 3: To 6 pounds of curd of pH 4.9 was added 2 pounds of cream dressing of 16% butterfat content and which also contained 0.20% sorbic acid and 4% salt. This product was packaged as in Example 1. After 15 days at 45°F. there was a mild, clean flavor-development, which was completely free of sourness. Plate counts were 1,000,000 per gram, mostly of the nonspoilage aroma type microorganisms.

Example 4: To 6 pounds of curd of pH 4.9 was added 2 pounds of cream dressing of 16% butterfat content which had been inoculated with a culture of flavor-forming bacteria of the genus Leuconostoc and these bacteria were permitted to act upon the cream at 72°F. for 18 hours. After this time, 0.2% sorbic acid and 4% salt were added. A good, flavorful, mildly-aromatic fresh cheese (not sour) resulted and after 15 days the plate counts were 1.5×10^8 and no contaminants were found.

Example 5: A batch of creamed cottage cheese was prepared exactly as Example 4 except that 0.07% sorbic acid was added to the cream prior to inoculation with Leuconostoc, the remainder being added after incubation. The flavor of this product improved over its original freshness and held at a high level for more than 15 days. The counts of aroma-formers were 1.7×10^8 and no contaminants were found.

As shown in Example 1, the experimental cheese did not possess good keeping quality and was poorest in flavor. Neither was the product of Example 2 markedly different despite the fact that aroma-formers were present in large numbers. Growth of contaminants was the cause of gradual deterioration and loss of flavor in the cheeses of Examples 1 and 2.

Example 3 illustrates the effectiveness of sorbic acid towards inhibiting the growth of spoilage micororganisms. The cheese of Example 3 is also lacking in flavor developments due to the small numbers of Leuconostoc present. The beneficial effect of the combination of large numbers of Leuconostoc and sorbic acid is clearly demonstrated in Example 4 and 5 and the beneficial effect of controlling the growth of spoilage microorganisms is shown.

Stabilized Chlorine Dioxide

J.F. Wentworth; U.S. Patent 3,147,124; September 1, 1964; assigned to Chemical Research Laboratories of America, Inc. developed a process for retarding spoilage in cheese by adding stabilized chlorine dioxide to cheese. It may be added to the milk before it is subjected to acid, or it may be added to the wash water which is used to separate the whey from the curds.

Chlorine dioxide is customarily prepared in an aqueous solution since it is unstable and, under certain conditions, explosive. When in solution it is subject to rapid decomposition. Stabilizing agents, however, may be added to the chlorine dioxide solution to stabilize against decomposition so that the solution may be stored for relatively long periods of time.

Chlorine dioxide has the characteristics of an effective spoilage retarding agent for cheeses. It has germicidal properties which increase as the acidity of the solution increases. It is nontoxic in dilute solutions. An additional desirable characteristic is that chlorine dioxide aids the coagulation of milk when heated. Although this process may be used to retard spoilage in all cheeses, including cream cheese, it will be described specifically as it is used to make cottage cheese resistant to spoilage. The addition of stabilized chlorine dioxide to cottage cheese is more effective than hypochlorite for increasing the shelf life of cottage cheese because it is not so readily inactivated by organic matter, such as nonfat milk solids. There are advantages in adding chlorine dioxide to the skim milk rather than to the wash water.

Since stabilized chlorine dioxide is nontoxic to animals, it would be safe to add a small amount to milk used to make cottage cheese. At the pH 6.4 to 6.8 of fresh milk, stabilized chlorine dioxide would retard growth of microorganisms to a lesser extent than at a low pH. This is particularly advantageous in making cottage cheese. It would give the acid-forming organism a chance to get started in the chlorine dioxide treated skim milk. As acid is developed, chlorine dioxide would automatically check the growth of the starter organisms. The addition of stabilized chlorine dioxide to milk would slow down initial growth of the acid starter organism, but this may be overcome by adding a larger amount of the organism or by increasing incubation time.

There is a tendency for milk plants to pasteurize milk at temperatures above that required by public health officials, so that the time required for pasteurization may be reduced. These higher temperatures are not satisfactory for skim milk intended for cottage cheese manufacture, because the milk will not coagulate as firmly as when lower temperatures are used. When lower pasteurization temperatures are used, however, there is a danger of inadequate bacterial destruction. The addition of stabilized chlorine dioxide to the milk would insure adequate bacterial destruction for the lower pasteurization temperatures which are necessary to obtain cottage cheese of the desired physical properties.

Example: Skim milk is pasteurized and a solution of stabilized chlorine dioxide is added to the pasteurized milk to form a solution containing 200 ppm of stabilized chlorine dioxide. 0.5 to 5.0% of S. Lactis starter organism and 1 ml. of rennet per 1,000 pounds of milk are then added to the skim milk. It is then heated to a temperature of 72° to 88°F. for 5 to 16 hours. When coagulation of the curd is complete, it is cut, cooked for 1 1/2 to 2 1/2 hours at 115° to 130°F., whey drained, and then washed with cold water to remove the remaining whey.

Chlorine Dioxide and Mold Inhibitor

In the process described by W.H. van Wieren and J.C. Langberg; U.S. Patent 3,232,768; February 1, 1966; assigned to The Borden Company a source of chlorine dioxide is added to the wash water and a yeast and mold inhibitor added to the washed curds to prevent the growth of any contaminating organisms in cottage cheese and prevent the further growth of the organisms in the starter culture. The conventional soaking time of the curd in the wash water is not altered. The result is a marked increase in the keeping quality of the cottage cheese.

The use of a source of chlorine dioxide in the wash water without the subsequent addition of the yeast and mold inhibitor to the curds does not give satisfactory results. In like manner, the use of only the yeast and mold inhibitor is also unsatisfactory. It is necessary that both be used to obtain the increased keeping quality and uniform flavor.

The source of chlorine dioxide preferably is added to all of the wash waters, although if several wash waters are to be used, the first wash water only need contain chlorine dioxide. Addition of chlorine dioxide to the first wash water appears to be more effective than addition to subsequent wash waters because of the higher temperature and lower pH. After the last wash water has been drained from the curd, the yeast and mold inhibitor is added to the curds. In preparing creamed cottage cheese, the inhibitor is preferably added to the cream mix to be admixed with the curds. The curd may then be salted, or left unsalted, and packaged for use. In some instances the curd may be refrigerated prior to packaging.

It has also been found that the storage stability of the creamed cottage cheese can be greatly improved if the cream mix is heated to 180°F. for 20 minutes prior to mixing with the cheese curd. It is preferred to add a food antioxidant to the cream mix in addition to the heat-treatment to obtain the prolonged keeping quality. It is also desirable to pack the creamed cottage cheese so that oxygen is excluded either by the use of gas-packing in an atmosphere of nitrogen or carbon dioxide, vacuum packing, or by filling the container with cheese to permit no head space.

The chlorine dioxide can be any commercially available source of chlorine dioxide, preferably a stabilized form that is safe for use in foods. The yeast and mold inhibitor can be sorbic acid, sodium sorbate, sodium propionate, sodium benzoate and other commercially available inhibitors commonly used in foods. The antioxidant used in the cream mix could be, for example, nordihydroguaiaretic acid or a propylene glycol solution of butylated hydroxyanisole with propyl gallate and citric acid.

Satisfactory proportions of the chlorine dioxide are 15 to 200 ppm in the wash water. The proportion of yeast and mold inhibitor added to the curd corresponds, for 100 parts by weight of the curd, to 0.05 to 0.075 part by weight. The amount of antioxidant added to the cream mix is about 10 to 50 ppm, based on 100 parts by weight of the cream mix.

Example: Cottage cheese curd was made in the usual manner and divided into two equal portions. One portion was given three washings in water containing 100 ppm chlorine dioxide, and the other three washings in water containing 15 ppm chlorine dioxide. Each of these portions, when drained, was in turn divided into two equal parts, all of which were creamed. Two portions of the cream mix contained 0.075% sorbic acid. The cheeses were packaged, stored at 47° to 50°F., and periodically tested for spoilage. The results are set forth in the table below.

	100 ppm ClO_2		15 ppm ClO_2	
Weeks	No Sorbic Acid	0.075% Sorbic Acid	No Sorbic Acid	0.075% Sorbic Acid
0	1,500	1,600	12,000	9,000
1	230	230	9,800	6,600
2	140	200	3,000,400	4,600
3	80	170	5,200,000	2,100
4	200	160	15,000,000**	3,100
5	xx*	210	-----	12,000
6	-----	<100	-----	16,000
7	-----	270	-----	190,000

*No count possible; surface completely covered with mold.
**Surface completely covered with mold.

Creamed cottage cheeses made by washing the curd with 200 ppm chlorine dioxide and adding 0.075% sorbic acid to the cream mix were found to be edible after 13 weeks with no slimy surface growth or mold contamination. There was a development of rancidity in the samples that had no antioxidant added, or in which the cream had not been given high heat treatment.

Heating to Extend Shelf Life

K.R. Breidenstein; U.S. Patent 3,697,293; October 10, 1972; assigned to Kraftco Corporation described a method for manufacturing a cottage cheese having an extended shelf life which relates to heating the cheese while it is being transported to a packaging station. During such transport the cheese curd is heated to a temperature of at least 130°F.

In this method for making cottage cheese, the skim milk, whether having been heat treated by pasteurization or high temperature heat treatment, is acidified. Acid is established in the skim milk at a pH sufficient to cause the formation of a coagulum which can be cut. The curd is then heat conditioned in the whey. The temperature of the whey is gradually increased over a period of about 2 hours as the curd is carefully stirred in the whey. The heating is restricted so that the temperature of the curd does not exceed 130°F.

The heat conditioning of the curd must be sufficient to establish curd firmness sufficient to enable the curd to be handled during the subsequent washing steps. However, the firming of the curd should be restricted to a level which is less than about 60% of the final curd firmness desired. Additional curd firmness is subsequently developed during washing of the curd and by subjecting the curd to heating while being transported, as will be described more fully hereinafter. It is important to note that this heat conditioning is less than the usual cooking of curd in the normal cottage cheese make procedure.

The firmness of the curd may be determined by a firmness test. The firmness of the curd is measured by placing a curd particle at the bottom of a cylindrical tube and by dropping weights onto the particle from known heights. The relationship between the weight and the height required to shatter the curd particle may be used to measure the firmness of the particle, which may be expressed in dynes. The firmness of the curd is adjusted by subsequent steps of this process.

Additional acid may be established while the curd particles are maintained in the whey during the heat conditioning step. In general, the longer the time in which the curd particles remain in the whey after additional acid is provided, the easier the separation of the whey will be and the more rapid will be the development of curd firmness during subsequent steps. For skim milk, it is preferred to establish a whey acidity of from 0.51 to 0.56% equivalent lactic acid.

The whey may then be partially or substantially completely separated by a suitable procedure, such as by transporting the curd and whey to a drain table provided with a screened outlet which retains the curd while permitting the whey to be drained. It is desirable to maintain a blanket of whey over the curd particles. Therefore, the whey is drained only to that point where the level of whey is at or above the level of curd particles.

After draining the whey, washing of the curd may then be effected by adding wash water to it. The wash water may be heated prior to its addition to the curd. It is preferred that the temperature of the wash water be less than 130°F. If the wash water is heated to above 130°F., firming of the curd occurs to an undesirable extent, and the subsequent heat treatment step of this process provides a curd which is too firm. If the wash water is at a temperature of less than 90°F., the heat conditioning of the curd in the whey should be prolonged to provide additional firmness. Regardless of the firming effected by the addition of wash water, i.e., whether the wash water is heated or nonheated, the firmness of the curd after the washing step is preferably from 50×10^{-3} to 80×10^{-3} dynes and must be less than about 80% of the firmness desired in the final curd product.

Multiple washings of the curd may be used at the discretion of the manufacturer of the cottage cheese. After the final washing, the wash water is drained from the curd and the curd may be creamed with a suitable creaming mixture. The creaming mixture may be added at various temperatures, but should preferably provide an equilibrium temperature of curd and creaming mixture of less than 140°F. The curd is then creamed and heat treated to provide a heated curd which may be packaged. Heating of the curd is effected while the creamed curd is being transported from the drain table to packaging equipment which may be effected by any suitable means, such as by pumping the curd through a pipe. Preferably, the heating of the creamed curd is effected by injecting steam into a flowing stream of the curd. The temperature of the creamed curd is raised by the heating step to at least 130° up and to 170°F.

Heating substantially above 170°F. gives undesired effects such as over firming and off-flavors. As indicated, steam injection is the preferred method of heating the curd because of the rapidity with which the heating of the curd may be effected. It is preferred to provide a nozzle for injection of steam into the curd in which the full force of the steam is not directly impinged upon the curd particles to prevent shattering of the curd particles. This may be done by introducing the steam into a nozzle which has a plurality of outlets. If the desired increase in temperature is relatively small, i.e., no more than about 10°, the steam may be low pressure steam of 15 psig or less. If the temperature increase of the curd is more than about 10°, high pressure steam of greater than 15 psig is preferred. In general, the higher the degree of superheat of the steam, the less the condensate that occurs in the curd.

After increasing the temperature of the creamed curd, now at a temperature of at least 130°F., it is packaged into suitable containers. The increase in temperature of the curd effects desired final firming of the curd and provides a heated curd which may be packaged while hot and which provides extended shelf life. In this connection, the time at which the curd remains at the elevated temperature effects both the final curd firmness and the shelf life properties of the creamed curd. It is preferred that the final curd firmness be in the range of 90 to 110 x 10^{-3} dynes. Firmness within this range and desirable shelf life properties are obtained when the temperature of the curd in the package is maintained within the range of 130° to 170°F. for from 30 to 3 minutes, the time inversely proportional to the temperature. Thereafter, the curd is cooled. No particular cooling cycle is required, however, it is preferred to reduce the temperature of the curd to less than 120°F. within 60 minutes after holding at the desired temperature.

Example: 200 gallons of natural skim milk, containing 8.8% solids were used to make cottage cheese. The skim milk was first heat treated at a temperature of 238°F. for 34 seconds, which effected about 73% denaturation of the serum protein. The skim milk was introduced in a vat at 95°F. and a Streptococcus lactis starter was added. The skim milk was maintained at a temperature of 90°F. to provide a suitable environment for the starter. When an acidity of 0.48% equivalent lactic acid, corresponding to a pH of 4.6, had been obtained, a coagulum had formed, and cutting of the coagulum was started. The coagulum was cut into quarter-inch cubes with conventional cottage cheese knives. The curd thus formed was permitted to remain in the whey at a temperature of 90°F. until an acidity of 0.54% equivalent lactic acid had been obtained. The whey was then drained from the curd until the whey reached a level of 1 inch above the top of the curd particles. This was equivalent to draining about 60% of the total free whey.

Water which had been heated to a temperature of 130°F. was then added to the curd in an amount sufficient to provide a mixture of curd and water substantially equivalent to the level of skim milk originally present. The curd was maintained in suspension in the water for a period of about 30 minutes, at which time the temperature of the curd was 121°F. The firmness of the curd particles at this time was 69 x 10^{-3} dynes, as determined by the test previously described and was less than 70% of the final curd firmness.

The water was then drained until the level of the liquid in the vat was about 1 inch above the level of the curd particles. Water, which had been heated to 130°F. was again added at a level sufficient to provide a volume half that of the original level of skim milk in the tank. The curd particles remained in this second wash water for a period of 30 minutes. At this time the curd particles had reached a temperature of 128°F. and the firmness was 87 x 10^{-3} dynes. The water was then fully drained from the curd. A creaming mixture at a temperature of 145°F. was then added to the curd to provide a creamed cottage cheese product. The equilibrium temperature of the product, after the creaming mixture was added and blended with the curd was 125°F. The firmness of the curd at this time was 121 x 10^{-3} dynes.

The curd, immediately after the creaming mixture had been added and blended, was pumped through a line containing a steam injection nozzle. The steam injection nozzle contained four outlet points for the injection of steam into the flowing steam of creamed cottage cheese. The steam was added at a level sufficient to increase the temperature of creamed cottage cheese to 145°F. The hot cottage cheese was then packaged into containers. After 10 minutes, the containers were transferred to a cooler and were cooled and maintained at a temperature of 40°F. The temperature of the creamed curd was below 50°F. in less than 24 hours. The firmness of the creamed curd was 104 x 10^{-3} dynes 24 hours after the packaging of the creamed curd.

The total time required to produce the creamed cottage cheese from the time that the skim milk was introduced into the vat until the creamed cottage cheese was packaged was 9 hours. The cottage cheese produced by the above method was judged to be of excellent quality. It remained of good quality for a period of 8 weeks while stored at a temperature of 45°F.

STORAGE AND PACKAGING

Chlorinated Brine Storage

H. Leber and L.A. Perini; U.S. Patent 2,962,379; November 29, 1960; assigned to Dairymen's League Co-Operative Association, Inc. describe a method of treating cottage cheese which permits it to be stored for a long period of time without deterioration. The method involves mixing cottage cheese curds with salt brine in a container which is chemically inert to the brine and curds. The salt brine is chlorinated to 10 ppm and acidulated to a pH of 6.0. The brine is supplied in a quantity and concentration to completely immerse the curds and depress the freezing point of the entire mass to a temperature at or below 22°F. The mixture is then stored above 32°F., for example, 40°F., for a period of time sufficient to permit the brine to diffuse and permeate the entire mass of curd.

After the initial storage period, for example, 18 hours, the mixture is transferred to a cold storage room and stored for the period desired at a temperature of 22° to 25°F. After the storage period, the brine is washed from the curd which is then ready for sale to the consumer. The product may be further treated by the use of additives, such as a mixture of cream and milk solids to improve the fat and solids content of the product.

The cottage cheese can be stored for long periods of time without deterioration. For example, cottage cheese stored for as long as three months showed no evidence of bacterial growth. Furthermore, no foreign odor could be detected and the product had the clean flavor of fresh cottage cheese. The stored product had a score of 97 and the shelf life in a home refrigerator was even greater than the 19 and 15 days of freshly made cottage cheese. The brine, in addition to permitting the product to be stored at a low temperature without freezing, also produces an inhibiting effect on bacteria and does not adversely affect the taste or texture of the product.

Example: Bags of Cryovac (a laminated film of a polyvinylidene chloride resin) are used which measure 20 inches by 36 inches. The plastic bags are used as liners in metal cans of 60 pounds' capacity. Cottage cheese curd manufactured by a conventional method is then placed in the bag in the container to provide a layer on the bottom of the bag. Two quarts of a 25.14% sodium chloride solution containing chlorine in the ratio of 10 ppm and acidulated to pH 6.0 by the addition of lactic, phosphoric, or citric acid, are then added to the top of the layer of curd. Curd is again added until the bag is about half full, after which two more quarts of the brine are added. Another layer of curd is added after which four quarts of the brine are poured on the curd in the top of the bag. The amount of curd is weighed as it is added in layers so that the resulting mixture in a bag comprises 45 pounds of cottage cheese curd and 19.9 pounds of brine which is sufficient to completely immerse the curd.

The open end of the plastic bag filled with curd and brine is then gathered to expel air from the bag, and twisted into a rope to close the open end of the bag. The twisted, closed end is fastened by applying an aluminum clip to the lower closed end of the rope and the loose end of the rope is tucked down between the bag and inner wall of the can. A cover is applied to the top of the can and a dating tag applied.

The cans of packaged curd and brine mixture are placed in a refrigerated room held at 40°F. for 18 hours to permit the brine to diffuse and thoroughly permeate the entire mass of curd. After the initial storage, the cans are transferred to a cold storage room maintained at 22° and 25°F. for the period of storage required. The cans should be stacked in the cold storage room so that none of the cans contact each other or the walls of the room to allow circulation of air and avoid localized temperature differentials. The curds retain their original form and texture without freezing at this temperature, and the bacterial count decreases during periods of storage which may be as long as 6 months or more.

When the cottage cheese is to be used after a period of storage the clips are removed from the plastic bags. The twisted rope end of the bags are unwound and the loose ends of the bags pulled down over the exterior of the can in which it is contained. The curd and brine mixture of a number of cans is emptied into a cheese vat and fresh water at a temperature of 40°F. is added in the ratio of 6 volumes of water to 1 volume of mixed brine and curd. This dilute mixture is gently stirred in the vat for 15 minutes. At the end of the stirring operation the fluid is drained from the curd. The washing step is repeated, using the same quantity of wash water at a temperature of 40°F. to which chlorine is added in an amount to provide 10 ppm and acidulated to a pH of 6.0 by the addition of lactic, phosphoric, or citric acid. The mixture is again gently stirred for a period of 15 minutes, after which the liquid is drained from the curd. The resulting curd is trenched and allowed to drain for 45 to 60 minutes.

This washing step leaves a residual amount of salt in the cottage cheese amounting to 0.5% by weight so that the salt content may be varied for the particular market where it is to be sold. The amount of fresh water used in the washing step may be varied to give the required salt content to the finished product. A prepared mixture of cream and/or milk solids together with salt, if required, may be added to the curd to give the desired fat and solid content to the cottage cheese. The stored cottage cheese is then ready for distribution and consumption.

Canned Cottage Cheese

The process of J.F. Shaver; U.S. Patent 3,075,842; January 29, 1963; assigned to Gerber Products Company is designed to provide a palatable cheese product packed in a can or glass which is sterile and stable. The flavor, texture, and color are all excellent and the product has a very useful market as a prepared food for infants. Preferably for infants' diets, a cheese having a soft curd, such as cottage cheese, and which is low in solid content is employed. Any creamed or uncreamed, salted or unsalted cottage cheese may be used whether frozen or fresh. Any combination of rennin and lactic acid coagulated cottage cheese is usable and the curd may be large or small, or the cottage cheese may be in homogenized form as a starting material. The pH of commercial cheese suitable for use in this process may be in the range of from 4.2 to 5.6.

Two processes are available for the production of the cheese: the batch process and the line process. The preferred process is the line process. The essence of the line process is to mill or grind fresh or frozen cottage cheese so that it will pass through a 0.031 screen and then mix the cheese with water and other ingredients, including, for example, flavored materials such as fruit, sugar, whole milk solids and thickening agents. After the slurry is formed the pH is adjusted to 4.35 to 4.55. The slurry is then homogenized at a pressure of 500 to 4,000 psi.

The aqueous slurry is then conveyed to a heating chamber where the product is heated to 245° to 270°F. To insure sterility of the product, the heated material is held for 10 seconds to a minute or so, and is then conveyed to a conventional flash evaporator where it is cooled to between 200° and 250°F., and then filled in containers and held for a period of 3 to 15 minutes at 190° to 212°F. to insure sterilization.

A typical formulation for line processing of the product might involve the following:

Ingredients	Pounds/100 Gallons (840 lbs.)
Frozen cottage cheese	337
Pineapple	120 - 150
Sucrose	70
Whole milk solids	40
Thickening agent	30 - 45
Lemon concentrate	1.25 - 1.50
Water	250 - 350

The range of total solids is about 26 to 29% on a dry basis. The optimum heating temperature is about 260°F. and the holding time 20 seconds. A suitable thickening agent is W-13, a modified waxy corn starch (American Maize Company). Any fruits may be used such as pineapple. The pH is generally modified downwardly from the normal pH of cottage cheese to the desirable range by means of lemon juice or other acidifier.

The essential steps for producing the desired product with the line procedure embodies mixing the cheese and other ingredients including the thickening agent, and milling this to a fine particle size, placing the material in an aqueous slurry, adjusting the pH, and then homogenizing the mixture to present a very finely divided particle size before subjecting to the relatively high heat treatment. A preferred procedure is to mill the cheese alone and then add the milled cheese to the other ingredients with water, and then pass to the homogenization step after adjusting the pH. It is theorized that when the material in the finely divided state is subjected to high heat, there is a denaturation of the casein protein content in the presence of a protective colloid system formed by the starch thickening agent, thus tending to retain the protein particles in the desired particle size and prevent them from agglomerating and forming large, rather than minute, discrete particles which produce a smooth texture to the end product.

If a cheese product is desired which does not use a fruit or other acid-containing flavor which reduces the pH below 4.5, then it is necessary, from a standpoint of providing sterility, to subject the canned product to retort heat treatment which might take the form of 240°F. for 50 minutes or so. That is, when the pH of the product is above 4.55, it is generally necessary to subject the canned product to cooking temperatures after canning while this is not so when the pH is below 4.55 and where the product may be hot filled.

Air and Moisture Impermeable Container

The object of the packaging process for cottage cheese described by H. Rode; U.S. Patent 3,111,410; November 19, 1963; assigned to W.R. Grace & Co. is to overcome the disadvantages of yeast and mold formation by virtue of the fact that the cottage cheese is tightly packed into an air and moisture impermeable container, while at the same time insuring the prevention of ballooning of the container by gas-forming bacteria.

The method involves the conversion of greater amount of the lactose in the starting milk material into lactic acid than is normal by the addition of more than 1% of a pure strain of Streptococcus lactis to the milk. The pH of the curd is carefully maintained within a range of 4.9 to 5.2 by flushing the curd with 20 to 35% by volume distilled water (pH approximately 7.0) either before, during or after the addition of rennet extract. Up to 1% or more sodium chloride and/or buffering salts depending upon the flavor desired are then added. Finally, after drawing off the whey, the curd is pressed, preferably through a tubular nozzle, into an evacuated moisture and gas impermeable thermoplastic film tube, the ends of which are subsequently twisted and sealed tight with a metal clamp for example.

Example 1: Skim milk was heated to 20° to 22°C. in a curd producing vat or trough and 1% lactic acid culture was added. Rennet was added at the rate of 1 ml. of the commercial extract per 1,000 pounds of skim milk. The commercial extract of rennet is diluted in 40 times its volume of cold water to assist in distributing the rennet uniformly throughout the vat of milk. After 5 to 6 hours, the milk became a curd and had a pH of 5.0. The curd was subjected to the usual ripening, the partial conversion of the lactose into acid occurring in about 12 hours. The whey was then drawn off in the curd producer, or by working in a trough, and the curd was filled into curd sacks.

Drawing off of the whey was discontinued after the whey reached a pH of 4.9 to 5.2. The water content amounted to 70 to 80% depending upon the fat content. The fat content was adjusted accordingly in several samples when straining the curd in the usual manner. Several other samples were worked into creamed cottage cheese by the addition of cream. Several samples were then taken and each was packaged in a different manner. In Sample A, the cottage cheese was packaged in polyvinyl chloride cups having a capacity of 250 grams, and the cups were closed manually by placing a lid on the top. In Sample B, the cottage cheese was packaged according to the method of this process by forcing it through a pipe into a tube of gas and moisture impermeable thermoplastic film material, twisting the ends of the tube and sealing the twisted portions with metal clamps.

A suitable film material may be a polyvinylidene chloride-polyvinylchloride film (Cryovac). The Cryovac film tube had inner diameters of 35 mm. and 45 mm. for 100 grams and 250 grams packages, respectively. The stability of the packaged cheese was determined organoleptically. The stability measured in days is shown in the following table.

	Fat Content, 10%	Fat Content, 30%
Sample A:		
Polyvinyl Chloride Cups — days	4.5	6.0
Sample B:		
Cryoval Film Tube		
(without additional vacuum) — days	14.8	15.0
(with additional vacuum) — days	18.0	17.5

Example 2: In the following example, the lactose is substantially converted to lactic acid, and the excess lactic acid is flushed out with distilled water. 20 to 30% distilled water (pH approximately 7.0) was added to the starting milk material and the milk-water mixture was heated to approximately 20° to 22°C. The addition of rennet was calculated on the basis of the total amount of liquid (milk plus water) whereas the addition of the lactic acid culture of starter was calculated on the basis of the amount of milk alone and was equal to at least twice the amount of starter standardly employed, i.e., 2%. Processing of the cheese curd was then carried out as in Example 1, i.e., the whey was separated and the cottage cheese was packaged in a substantially air-free, gas and moisture impermeable film tube.

The portions of cheese thus packaged showed no alteration in flavor or separation of whey, no yeast or mold formation and the quality, texture and body of the curd was maintained even after a three-week storage period. There was in addition no evidence of ballooning of the film tube.

Example 3: The skim milk was treated as in Example 2, adding the rennet and twice the amount of *Streptococcus lactis*, as in Example 1, and flushing out the excess acid with 20% distilled water (approximately pH 7.0) after formation of the curd. The whey was then separated in the usual manner and when the curd was strained, the sodium chloride content was increased to about 1%. The portions of cheese packaged in the film tubes displayed a stability of more than 24 days.

SOFT UNPROCESSED CHEESES

CREAM CHEESE

Continuous Draining of Curd

In the continuous manufacture of cream cheese, a very important operation which determines the quality of the final product, consists of draining the curds. Capillary action and other inherent effects play a fundamental part in this operation in that the direct contact of a filter cloth absorbs the moisture in an almost instantaneous manner.

When the texture of the filter cloth has effected the absorption corresponding to its absorbing capacity, it becomes impossible for it to suck up any supplementary moisture. It can only begin to function once more if care is taken to evacuate all moisture which would prevent supplementary absorption or to obtain a super-saturation of the filter.

If a compression force is applied to the mass of curds to compel its residual moisture to super-saturate the filter, this force acts indiscriminately on the moisture and the dry material to be preserved, which results in losses. To avoid these losses, it becomes necessary to resort to a slow filtration which has an adverse effect on the quality of the product. In addition, the intimate and prolonged contact between the residual serum and the dry material gives rise to digestion phenomena which also result in losses.

The best method is still the progressive evacuation of the moisture contained in the filter cloth. If the texture of the filter could continuously retain the power of absorption which it possesses in the dry state, the absorption of moisture would take place in an automatic and continuous manner and with considerably increased rapidity. Any compression would in that case be unnecessary.

The method described by J.M.A.C. Berge; U.S. Patent 2,974,045; March 7, 1961; assigned to Ch. Gervias SA, France enables a solution of this kind to be obtained. This method is characterized in that it consists in effecting the draining operation by applying to the face of the cloth or other material constituting the filter for the curds, opposite to that in contact with the curds, a layer of supplementary material adapted to absorb by capillary action the moisture, which accumulates in the filter, this layer of material being subjected to periodic or continuous wringing to free it from the moisture absorbed. The supplementary layer referred to is displaced due to its contact with the material of the filter and is then passed through a wringing device.

The displacement of the supplementary absorbent layer is transmitted to the band forming the filter solely by contact and the resistance to friction between the two absorbent materials. The material of the filter is thus automatically and continuously freed from an excess of moisture, and it becomes possible to carry out a complete draining of the curds.

An apparatus which enables this above-described method to be carried into effect is characterized in that it comprises a filter formed by a layer of filter cloth of an absorbent nature, one face of which receives the curds, and a second supplementary layer of sheet also constituted by an absorbent material applied against the face of the filter cloth opposite to that which is in contact with the curds, a suitable device being provided to ensure continuous or periodic wringing of this latter sheet. Accompanying drawings are given in the patent.

Texturizing Process

Since it is desirable to prepare soft, uncured cheese having the long shelf life of hot packed cheeses, while also having the more open body and texture of cold packed cheeses J.C. Hurlburt and G.H. Kraft; U.S. Patent 3,072,489; January 8, 1963; assigned to National Dairy Products Corporation describe a method of texturizing soft, uncured cheese to provide these characteristics in the finished cheese. In accordance with the process, soft, uncured cheese prepared by any suitable method is subjected to a method of texturizing which imparts to the cheese a more open texture and body.

The method involves incorporating an inert gas into fluidized soft, uncured cheese and forming a uniform dispersion therewith, the concentration of gas being in excess of that required for saturation of the cheese and sufficient to provide a more open body and texture in the cheese. The dispersion is maintained while the cheese is packaged so that an improved product is obtained which not only has extended shelf life, but also has improved more open texture and body.

Cream cheese, Neufchatel cheese or other soft, uncured cheese is fluidized prior to final packaging. Fluidizing can be accomplished by heating to about 135° to 180°F. Alternatively, the cheese can be fluidized by beating or whipping solid or plastic cheese until it is flowable, even though the temperature may be below the above-indicated temperature. The cheese, when fluidized, is in a condition for suitable mixing with inert gas. Moreover, the cheese can be pumped and various ingredients can be added while in the fluidized state.

The inert gas is dispersed in the fluidized cheese in a suitable mixing apparatus at a controlled level and preferably under a pressure of about 50 psi or more. The pressure assures formation of the dispersion, with the inert gas in sufficiently dispersed form to achieve desired texturizing without production of a cheese which has a crumbly texture. Thus, a sufficient concentration of the inert gas should be intimately dispersed with the cheese to open the texture of the cheese.

It has been found that the concentration of inert gas necessary for such purposes is one which is sufficient to saturate the fluidized cheese and to provide at least a 1%, by volume of the cheese, overrun. That is, the total volume of the cheese after the inert gas introduction should be at least 1% greater than the total volume of the cheese before the inert gas introduction into the cheese. The relative volumes, however, are measured at atmospheric pressure.

It has been found that the overrun should usually not be more than about 5%, by volume, of the cheese, inasmuch as larger overruns and consequent greater concentrations of inert gas in the cheese may result in finished cheese which has a crumbly, generally undesirable texture. It has been found that a 3% overrun usually provides the best results. The substantial criticality of the level of inert gas in the cheese is a unique feature with soft, uncured, cheeses.

Inert gas should be maintained in dispersion in the fluidized cheese up to and including setting of the cheese to a solid condition. However, in many instances, the dispersion during processing will undergo substantial working before setting of the cheese. As an example, movement of the dispersion to the filling and packaging equipment normally results in substantial working of the dispersion. So also does passage of the dispersion through a line strainer. Such substantial working of the dispersion tends to promote loss of inert gas with consequent reduction of the percentage overrun and reduction in the texturizing of the cheese.

Accordingly, it is desirable and/or necessary to provide an increased pressure in the system containing the dispersion, at least from the point of formation of the gas-cheese dispersion to the point of passage of the fluidized cheese dispersion to filling equipment. The increased pressure aids in retaining the gas in finely dispersed form in the cheese. However, once the substantial working of the cheese has ended, the dispersion can be subjected to atmospheric pressure without serious loss of inert gas. Thus, the cheese can be passed into a filling machine operating at atmospheric pressure so long as the filling machine does not substantially work the cheese.

Example: Cream cheese prepared with a conventional cream cheese procedure which included pasteurization of the cheese was fluidized by heating to 160° to 165°F. and the fluidized cream cheese was passed by a screw conveyor to a blending apparatus. Locust bean gum and salt were added to the cheese during passage through the conveyor. The cheese was held in the blending apparatus for a short time at about 160° to 165°F. and was then passed by a gear pump to a two-stage "mix-air" pump, the second stage having three times the capacity of the first stage.

In the "mix-air" pump, nitrogen gas was introduced in a metered amount to the hot cheese (still about 160°F.) to provide an overrun volume of 3% of the volume of the cheese. The hot cheese and nitrogen were intimately mixed under pressure in the "mix-air" pump to provide a stable uniform dispersion of the gas with the cheese, which exited the pump at about 95 psig and was then passed while still fluid to a 30 mesh line strainer to keep the gas uniformly distributed in the cheese and remove any lumps from the cheese.

A back pressure valve was provided in the line downstream from the line strainer to maintain the fluid gas-cheese dispersion under a pressure of about 70 psig until the dispersion was ready to be packaged. The fluid gas-cheese dispersion was passed downstream of the back pressure valve and into the open hopper of a cheese filling machine, which machine controlled the flow of the cheese into one-half pound packages. The machine operated at atmospheric pressure without substantial working of the cheese. The cream cheese in the filled packages was wrapped, allowed to cool to a fully set, solid condition and then several such one-half pound packages were opened and the solidified cheese samples were removed, broken open and evaluated visually and in the mouth for texture. Each of the samples exhibited a more open texture and desirable body, than control samples subjected to the identical procedure, except that no gas was introduced into the control samples. Moreover, all samples had an extended shelf life (about 4 weeks).

Addition of Monoglycerides and Hydrophilic Colloid

L.L. Little; U.S. Patent 3,359,116; December 19, 1967; assigned to The Battelle Development Corporation describes a process in which cream cheese is made by adding an edible acid to animal milk to which has been added 0.2 to 3.0% high monoglycerides (ratio of mono- to diglyceride of at least 5:3) and a hydrophilic colloid. The milk containing the high monoglyceride and colloid is pasteurized and homogenized prior to being acidified. Lipoids in water can be substituted for the milk. Cream cheese prepared by this process requires milk containing 20 to 40% fat.

One of the advantages of the method is that it is adaptable to modern automated processing techniques. The ingredients may be added to a batch tank to effect precise standardization of the composition of the finished product. The mixture may be pasteurized by any desired means including ultra high temperature sterilization. It may be homogenized at any desired stage in the process, and may be rehomogenized to bring out a desired body characteristic. Preferably the product is cooled before acidification, but acidification can be accomplished in the hot product at any stage of the heating cycle. The variations that can be devised in the sequence of processing steps is almost without limit. A number of these variations are given below.

(1) Make up batch in pasteurizing vat. Heat to 160° to 195°F. for 30 to 0 minutes, as desired. Homogenize. Cool to 80° to 40°F. Add food grade acid solution to pH of approximately 4.75. Mix thoroughly. Package. Store under refrigeration.

(2) Make up batch in pasteurizing vat. Heat to 160° to 195°F. for 30 to 0 minutes, as desired. Cool to 80° to 40°F. Add food grade acid solution to pH of approximately 4.75. Mix thoroughly. Homogenize. Package. Store under refrigeration.

(3) Add ingredients to batch tank. Pasteurize at 170° to 300°F. for 60 to 0 seconds. Homogenize. Cool to 80° to 40°F. Add food grade acid solution to pH of approximately 4.75 and mix thoroughly. Package. Store under refrigeration. If product was sterilized and food acid added aseptically, product need not be stored under refrigeration.

(4) Add ingredients to batch tank. Pasteurize at 170° to 300°F. for 60 to 0 seconds. Homogenize. Cool to 80° to 40°F. Add food grade acid solution to pH of approximately 4.75. Homogenize. Package. If product was not sterilized and aseptically acidified, it must be stored under refrigeration.

(5) Add ingredients to batch tank. Pasteurize at 170° to 300°F. for 60 to 0 seconds. Cool to 80° to 40°F. Add food grade acid solution to pH of about 4.75. Homogenize. Package. If product was not sterilized and aseptically acidified, it must be

stored under refrigeration.

(6) Add ingredients to batch tank. Pasteurize at 170° to 300°F. for 60 to 0 seconds. Add food grade acid solution at temperature of 170°F. or above. Homogenize. Package. If product was sterilized (heated to a temperature of 250° to 300°F.) it need not be stored under refrigeration.

(7) Add ingredients to batch tank. Pasteurize at 170° to 300°F. for 60 to 0 seconds. Add food grade acid solution to pH of about 4.75. Homogenize. Cool to 80° to 40°F. Homogenize. Package. If product was not sterilized and acid solution added aseptically, the product must be stored under refrigeration.

(8) Add ingredients to batch tank. Pasteurize at 170° to 300°F. for 60 to 0 seconds. Homogenize. Add food grade acid solution to pH of about 4.75. Cool to 80° to 40°F. Homogenize. Package. If product was not sterilized and acidified aseptically, the product must be stored under refrigeration.

One of the major uses of cream cheese is in making cheese cake. Many experimental cream cheeses made by this process have been tested in making cheese cake and produced cheese cake comparable in every way to the very best cheese cakes available. Also, cheese cakes made with this cream cheese withstood freezing and prolonged frozen storage as well as those made with conventional cream cheese.

Combining Ripened and Unripened Mix

D.R. Carswell and J.C. Hurlburt; U.S. Patent 3,492,129; January 27, 1970; assigned to National Dairy Products Corporation describe a process designed to provide a cream cheese which is subject to a minimal amount of wheying off without requiring the use of nondairy stabilizers.

In the process, a cream cheese mix is ripened to a predetermined acidity (calculated as percent lactic acid) and to this mix is added a substantially unripened second cream cheese mix to reduce the acidity of the mixture to less than about 0.7%. The mixture may then be separated into cream cheese and whey according to conventional procedures.

A first mix is prepared from cream and milk and/or skim milk. The first mix should contain about 10 to 14% butterfat by weight of the mix, although butterfat concentrations of 6 to 20% may be used to produce either cream cheese or Neufchatel cheese.

The first and second mixes may be heat treated prior to use in making cream cheese at an elevated temperature for a period of time sufficient to denature the serum protein. As a result of denaturation heat treatment a physico-chemical change is effected in the serum protein which causes such protein to become sensitive to acid and to coprecipitate with casein upon acidification to the iso-electric pH. Denaturation of serum protein may be effected at 185° to 230°F. for 6 minutes to 30 seconds, respectively.

The mixes, if not heat treated to effect denaturation of serum protein, may be pasteurized and homogenized in the usual manner. The mixes are then cooled to 60° to 90°F. The first mix is then inoculated with a lactic acid starter culture and ripening is effected. A small amount of rennet may be added, according to conventional procedures. The ripening of the first mix may be effected by a long or short set as desired. When a long set is used, about 0.75 to 1% of starter by weight of mix is added, while in a short set about a 5% level of starter is used.

Ripening is performed at about 60° to 90°F. in accordance with conventional procedures. The first mix is allowed to ripen until the acidity of the coagulum reaches a level between 0.80 and 0.85%. The higher acidities in the ripened mix result in a better flavored final product and are therefore preferred.

The second mix, or "unripened mix", is pasteurized, if not previously heat treated to effect denaturation of serum protein, and homogenized. It is then added to the ripened first mix until the acidity of the mixture has been lowered to less than 0.7%, preferably from 0.50 to 0.65%.

The amount of unripened mix required to reduce the acidity of the mixture within the desired range is, of course, dependent upon the level of acidity which has been reached in the ripened first mix and the total acidity of the unripened second mix. The mixture of ripened mix and unripened mix may be effected by any suitable means.

The resultant coagulum, which now has an acidity between 0.50 and 0.70% may then be heated to 130° to 180°F., under agitation, to break the viscosity of the mix. The whey is then separated from the curd by centrifugal separation or the use of draining bags.

A preferred method of separating the whey from the curd is by centrifugal separation while the curd and whey are in a heated condition. After separation, it is generally preferred to pack the cream cheese product at an elevated temperature to increase shelf life. However, further processing of the curd by cold working of the curd to improve texture, such as on a chill roll, may be effected. When such a process is employed the heated curd is deposited on a refrigerated roll in a relatively thin layer to chill it rapidly, after which it is removed from the roll. The chilling process has been found to provide a desired curd texture.

Example: A cream cheese mix containing 12% butterfat was prepared. The cream cheese mix was heat treated at 230°F. for 30 seconds in a tubular heat exchanger to denature the serum protein. After being heat treated, the cream cheese mix was homogenized at 2,000 to 3,000 psi in a single stage homogenizer.

Half of the cream cheese mix was then pumped into a holding tank to provide a first mix and was then cooled to 70°F. To the first mix there is added 0.75% by weight of lactic acid cream cheese starter, and one milliliter of rennet per thousand pounds of mix, and ripening of the first mix was then effected at about 75°F. When an acidity of 0.85% had been reached, the unripened half of the cream cheese mix previously prepared was added to the ripened mix while the mixture was gently stirred until the resultant acidity was 0.6%.

The resultant mixture of unripened mix and ripened mix was then heated to 175° to 180°F. in a swept surface heat exchanger, i.e., and a heat exchanger in which heating is effected in a thin film unit with repeated scraping of the heat exchange surface. The mixture was then passed through a centrifuge in which the whey was separated from the curd to provide a cream cheese having a fat content of 33.4%.

At the time of separation of the whey from the curd the mixture had a pH of 4.8 which is substantially higher pH than is considered suitable in conventional practice. The cream cheese was then blended with salt but no additional nondairy stabilizers were added. The cream cheese was then packed in conventional packages at 165° to 175°F. The cream cheese was fully acceptable in appearance and flavor and exhibited minimal wheying off when the cheese was stored for six weeks.

PROCESSES SUITABLE FOR ANY SOFT CURD CHEESE

Low Temperature Centrifugal Separation of Whey

C.L. Angerer and G.W. Wilson; U.S. Patent 3,012,889; December 12, 1961; assigned to The Borden Company provide a process by which both cream cheese and Neufchatel curd may be separated from the whey by centrifuging without exposing the ripened mix to the usual high temperatures.

In the process a sweet mix is first prepared in the conventional procedure, that is, sweet milk and cream are blended to obtain a mixture having an appropriate amount of butterfat which for cream cheese is between 11 and 20%. For Neufchatel cheese, a butterfat content between 6.5 and 11.0% is used.

The next step is to add to the sweet mix a small amount of gum used in the hot packing of finished cream cheese for preventing excessive leakage of moisture. Suitable gum for this purpose is locust bean gum. The proportion of gum solids added should be between 0.01 and 0.2% by weight of the sweet mix.

After adding the gum, the sweet mix is stirred and heated to 165°F. over a period of about half an hour. This heating serves not only to pasteurize the sweet mix but also to activate the gum with the mix, so that the gum addition will cause the unusual effects in the mix after ripening which are to be described. The temperature required to activate the gum with the sweet mix varies somewhat with the heating time and the particular gum used. A temperature about 165°F. is usually about optimum.

The pasteurized gum reacted sweet mix is next pumped while hot through a homogenizer at a pressure about 2,500 pounds per square inch. Homogenizing the sweet mix (either before or after the pasteurizing of it) is a part of the conventional procedure for making cream and Neufchatel cheese. Homogenizing the special sweet mix of this process performs the additional function of perfecting the distribution and activity of the gum.

After homogenizing, the sweet mix is cooled to a setting temperature about 72°F., and is run into a setting tank, for example, a conventional coil pasteurizer provided with stirring means. The sweet mix is then inoculated as in the conventional process with a conventional amount and kind of starter. A little rennet is also added. The proportion of rennet should not be less than 0.1 cc nor more than 3 ounces per 1,000 pounds of mix. About 14 cc (1/2 oz.) of rennet ot each 1,000 pounds of mix, is the preferred proportion.

The stirrer is advantageously used for a few minutes to mix in the starter and rennet. The batch is then allowed to set and ripen usually overnight at a temperature around 70° to 74°F. until the mix acidity of 0.7 to 0.9% is reached.

After the mix has set and ripened to the proper acidity, the stirrer is operated for a sufficient time to break up the ripened mix at the setting temperature (e.g., about 70° to 74°F.). Such stirring soon results in breaking the ripened mix into small curd particles floating around in substantially clear whey. This requires no heating to accomplish. Violent or prolonged stirring is not needed and should be avoided. About 2 to 5 minutes' stirring in a conventional coil pasteurizer or in a vertical tank having equivalent stirring means, is usually enough.

The broken up ripened mix is then pumped continuously through a tubular heater or other suitable warming device to a small gravity feed surge tank for feeding the mix to the centrifuge. The temperature of the mix fed to the centrifuge is kept within the range of 108° to 115°F. In practice, 110°F. is a very satisfactory temperature.

The curd separated by the process may be "cold packed" as cream cheese or Neufchatel cheese in accordance with conventional procedure. Such "cold packed" cheese has a flavor and texture at least as good as can be produced by the conventional bagging method and superior to that obtained when a ripened mix is heated to high temperatures for centrifuging, and after centrifuging is cooled down again for so-called cold packing.

The shelf life of "cold packed" cream cheese or "cold packed" Neufchatel cheese made by this method is generally somewhat longer than that of such cold packed cheese made by the conventional bagging process, because the centrifuging procedure reduces the amount of handling and other opportunities for contamination involved in the manufacturing procedure.

Acid Precipitation

Z.D. Roundy; U.S. Patent 2,963,370; December 6, 1960; assigned to Armour and Company describes a rapid process for the manufacture of soft-type cheeses. It is an acid precipitation process in which the fermentation processes and accompanying use of starter cultures and rennet is eliminated and one in which a curd is produced which is amenable to separation by centrifugation with prior heat treatment. Also, the method produces a curd which has little or no tendency to "weep" or leak whey when held in storage.

It is customary, but not essential, to adjust the fat content of the milk or cream and milk mixture. In the preparation of baker's cheese, the milk mixture generally contains either no fat or merely a trace of fat. In the production of Neufchatel-type cheese the milk and cream mixture is adjusted to contain approximately 5 to 8% fat. In the production of cream cheeses the milk and cream mixture may contain from 7 to 20% fat.

Since this process requires only a short period of time, pasteurization is not usually essential since insufficient time is allowed for undesirable organisms to develop, and although it is desirable, it is not essential that the milk and cream mixture be homogenized. Homogenization will generally reduce the loss of fat in the whey and aid in the production of curd with a uniform fat content.

It is desirable to acidify the milk and cream mixture to about its isoelectric point, that is to a pH of 4.6 to 4.7 while at 50° to 120°F. This acidification is accomplished by the addition of 0.3 to 0.4% by weight of an edible acid dissolved in water.

It has been found desirable to heat the water in which the edible acid is to be dissolved. In the preparation of baker's cheese it has also been found desirable to acidify the mixture to a point slightly beyond the isoelectric point or to about a pH of 4.5. The precipitation of the curd is accomplished in a very short time, as soon as the mixture is completely acidified.

To accomplish uniform acidification it has been found desirable to agitate the milk and cream mixture during the addition of the aqueous solution of acid. The preferred temperature for the acidification step is from about 80° to 95°F. The curd and whey may be separated by drainage through unbleached muslin, however, when precipitated by this method the curd and whey may be separated by means of centrifuge without further treatment.

Low Acid Culture

R. Cuomo; U.S. Patent 3,047,400; July 31, 1962 describes a method of making sweet curd cheese known as ricotta which uses fresh nonexcess acid milk in a quantity which constitutes 80% of the beginning mixture. The other 20% constitutes pure whole milk having a 0.6% butterfat content plus proteins, lactose and mineral salts and a solids content of 15 pounds per 100 pounds of milk.

The milk may be substituted with reconstituted spray dehydrated skim milk to which is added a quantity of butter to increase the butterfat content. The dried skim milk may be substituted for the fresh milk so that it constitutes up to 50% of the beginning mixture.

The beginning milk mixture is added to a suitable vat. A low acid starter culture is then added to the mixture. The starter employed is a curdled milk precipitated by means of strictly controlled lactic acid bacteria. The titratable acidity of the mix is controlled to within 0.19 to 0.22%. By using a low acid culture, the titratable acidity is held down, thereby preventing any appreciable decrease in the compressibility of the curds.

Once the milk mixture is in the vat and the necessary starter has been added, the resulting mixture is pasteurized by either a continuous or batch system. The minimum temperature of both systems must be 176°F. During cold winter months, it is advisable to employ somewhat higher temperatures during pasteurization.

If the substituted starting materials are used, it is necessary that this milk mixture with the added starter be homogenized, or emulsified to insure a smooth texture of the mixture and the resultant cheese produced. Under these circumstances the best results for this homogenizing or emulsifying are obtained at 142°F., and then the pasteurization is continued. This added step also prevents the butter oil which has a light body from floating to the surface.

After the milk has been properly pasteurized, it is cooled to the setting temperature of 95°F. during the summer and 96.8°F. in the winter months. Rennet is added for a 30 minute coagulation. This fairly short setting time also, along with the use of a low acid culture, results in a better maximum compressibility of the curds after completion of the chemical reactions, probably because there is less gassing or aeration of the curds during the shorter period. Once the mixture becomes firm, the mass or batch is cut into small cubes. The cubed curd is permitted to settle for 5 to 10 min.

The resulting whey is drained off and the proper quantity of salt may then be added. The curd is molded into tubes of approximately 2 1/4 inches diameter and 8 inches long. The molds are permitted to cool on the draining table and are then refrigerated for about 12 hours at a temperature of 34° to 40°F. The molded cheese is then removed from the molds and cut by guillotine knives into units of 1 inch lengths. These units are then wrapped in moisture impermeable paper or cellophane and are ready for distribution and consumption.

The resultant product may be held in stores or house refrigerators for a week to 10 days or more while still remaining sweet. This cheese when it is a product of wholly fresh milk may be produced with or without homogenizing or emulsifying and if produced with either one of these processes it gives a finer grain cheese. The sweet curd cheese is a food which is rich in proteins of high biological value which is easy to digest due to its low or almost zero acid content, which is unlike cottage cheese.

Acidogen and Proteolytic Enzyme

E.A. Corbin, Jr.; U.S. Patent 3,620,768; November 16, 1971; assigned to Diamond Shamrock Corporation describes a process in which cottage cheese, baker's cheese, cream cheese and Neufchatel cheese can be produced in standard dairy equipment without using bacterial fermentation or other time-consuming steps.

The method involves first acidifying cooled milk, i.e., milk maintained at temperatures of 32° to 45°F., by addition of free acid until a pH of 4.88 to 5.20 is reached. The cooled, acidified milk is next heated to just in excess of 60° to 100°F. for 30 to 90 minutes. Acidogen and proteolytic enzyme are then added to the heated, acidified milk which is then allowed to remain in a quiescent state to just in excess of 60° to 100°F. for 30 to 90 minutes so that a cuttable cheese curd is formed.

The resulting curd is finally cut, cooked, drained and washed to obtain the desired cheese. The process is particularly suited to making cheese curd at the lower pH ranges, i.e., below pH 4.88, which includes the cheeses mentioned.

Example 1: This example is directed to the preparation of cottage cheese curd. To one gallon of fresh, cold, pasteurized skim milk having a temperature of 45°F. and a pH of 6.80 was added, with stirring, 9.5 ml. of 85% food grade phosphoric acid diluted with 20 ml. of water. 9.5 ml. of 85% phosphoric acid weighs 16.2 g. and represents 0.41% by weight of 85% phosphoric acid in one gallon of skim milk (3,900 grams). pH of the acidified milk after phosphoric acid addition and mixing was 5.17. The skim milk was an unfortified skim milk containing about 9% by weight total solids and about 0.01% by weight butterfat.

The acidified milk was heated for about 30 minutes, with stirring, to 80°F. in a water bath. Temperature of the water bath was regulated so that it did not exceed the milk temperature by more than 20° to 30°F. While heating the milk, 2 ml. of Starter Flavor (Commercial Starter Distillate) and 0.6 g. of potassium sorbate, (Mold Inhibitor) were added. When the temperature of the milk reached 80°F., the pH was 5.06.

8 g. D-glucono-delta-lactone, an acidogen, was dissolved in 50 ml. of water; 15 ml. of 1/100 dilution of Hansen's Rennet Extract, Single Strength (an enzyme) was

added to the solution and thoroughly mixed. 8 g. D-glucono-delta-lactone per gallon of skim milk represents 0.205% by weight lactone. The resulting solution of acidogen and enzyme was immediately added to the milk heated to 80°F. and uniformly mixed. pH of the heated, acidified milk containing acidogen and enzyme was 5.06.

The heated, acidified milk was allowed to stand undisturbed in a quiescent state for 1 hour at 80°F. A cuttable cheese curd was formed. The curd was cut into 3/8-inch size cubes and allowed to stand for 15 minutes at 80°F. Shrinkage of the curd occurred and the whey was expressed. pH of the whey was 4.85.

The cut curd was stirred gently and cooked until the whey temperature was 110°F. after 30 minutes of cooking. A cooking rate of 1°F. per minute was obtained by regulating the water bath temperature so that it did not exceed the whey temperature by more than 20°F. When the whey temperature reached 110°F.,,pH of the whey was 4.70 and the curd was firm and shrunken in a normal manner without matting of the curd. Cooking and stirring was continued for a total of 50 minutes. The final cooking temperature was 130°F. Whey drained from the curd after cooking had a pH of 4.68.

The curd was washed and drained twice, first with 90°F. water and then with 70°F. water. Thereafter, the curd was washed with 66°F. water, allowed to stand for 15 minutes in the wash water and drained. Weight of the drained curd was 478 g. and the yield of curd was 12.2%. The product was a satisfactory cottage cheese curd.

Example 2: This example is directed to preparation of baker's cheese. To 1 gallon of commercial skim milk having a temperature of 42°F. and a pH of 6.78 was added, with stirring, 9.5 ml. of 85% phosphoric acid. The milk had the same composition as the milk used in Example 1. pH of the acidified milk after phosphoric acid addition and mixing was 5.05. The acidified milk was heated about 30 minutes, with stirring, to 80°F. in a water bath using the procedure described in Example 1 above.

10 g. of D-glucono-delta-lactone, an acidogen and 0.15 ml. of pure Rennet Extract (Hansen's Rennet Extract, Single Strength), were added to the acidified milk at 80°F. and thoroughly mixed. pH of the acidified milk containing the acidogen and enzyme was 5.05.

The milk was allowed to stand undisturbed in a quiescent state for 1 hour at 80°F. A normal cheese curd formed and was broken up mechanically. pH of the whey was 4.75. The curd was allowed to set and the whey was expressed for 50 minutes. pH of the whey was 4.66. The curd was placed in a cloth bag and the bag was hung to drain for 1 hour. Intermittent pressure was used to help expel the whey. Weight of the drained curd was 450 g.; moisture content of the curd was 73.5% and the yield of curd was 11.5%. The product was satisfactory for use as a baker's cheese curd.

Example 3: This example is directed to preparation of cream cheese curd. To 4 quarts of half milk and half cream having a temperature of 44°F. was added, with stirring, 8.5 ml. of 85% phosphoric acid diluted with 40 ml. of water. The half milk and half cream had a total solids of 19.5%, butterfat content of 10.5%, a pH of 6.52 before acidification and a pH of 5.00 after acidification. The acidified half milk and half cream was heated for 30 minutes, with stirring, to 80°F. pH of the heated, acidified cream-milk mixture was 5.03. 10 g. of D-glucono-delta-lactone dissolved in 60 ml. of water was mixed with 1.2 ml. of Hansen's Cottage Cheese Coagulator and the resulting mixture was stirred into the heated, acidified half milk and half cream. Then the half milk and half cream was allowed to stand undistrubed in a quiescent state at 80°F. for 60 minutes. A heavy, firm curd formed.

The curd was cut and allowed to stand 10 minutes at 80°F. pH of the cut curd and whey was 4.78. The cut curd was stirred and cooked until the whey temperature reached 130°F. pH of the whey was 4.60. Then the cut curd was cooked at 130°F. for 15 minutes. The curd was cooled to 80°F. pH of the whey was 4.40.

Thirty-six g. of salt was added to the curd at 80°F. The curd was cooled to 60°F., poured into a cloth bag and drained overnight at 38°F. in a refrigerator. The curd was emptied from the cloth bag the next day. pH of the curd was 4.80. The curd was dry, soft and very slightly grainy. It was mixed in a Hobart blender with sufficient lactic acid to reduce pH of the curd to 4.60 and to blend the curd into a smooth homogenous mass.

The homogenized curd was packaged and refrigerated to obtain a product which had the same appearance, taste and texture as a cultured cream cheese. Neufchatel cheese can be produced by this process by using a cream-milk mixture containing about 5 to 6% fat instead of the 10.5 to 11.5% used.

DRIED BAKER'S CHEESE

Milk Condensation

Z.D. Roundy and N.R.H. Ormond; U.S. Patent 2,956,885; October 18, 1960; assigned to Armour and Company describe a dried baker's-type cheese product and its preparation. In the process it is customary to remove the butterfat from the milk starting material. The milk mixture generally contains either no fat or merely a trace of fat. The milk may be pasteurized to improve its flavor and keeping qualities.

Since the process requires only a very short period of time, pasterurization may not be necessary since insufficient time is allowed during the manufacturing process for undesirable organisms to develop. The milk is also generally heated in the "hot well" during the condensing process thereby destroying viable organisms and inactivating enzymes which may be present in the milk.

Instead of using rennet or starter culture as in the usual procedures, it is desirable to acidify the milk and cream mixture to a pH of 4.5 to 4.7. The temperature of the milk during the acidification step should be from 35° to 120°F. It is preferred to perform the acidification with the milk at a temperature of about 50°F.

The acidification step is accomplished by addition of about 0.3 to 0.4% by weight of an edible acid dissolved in water. It has been found desirable to acidify the mixture to a point slightly beyond the isoelectric point or to about a pH of 4.5 to 4.6. The precipitation of the curd is accomplished in a very short time, as soon as the mixture is completely and uniformly acidified. Even if the process is being carried out in a batchwise process, it has been found desirable to agitate the milk mixture during the addition of the aqueous solution of the acid.

The temperature at which the best yields are obtained so far as particle size is concerned is about 50°F. When the milk and acid mixture is at a temperature below 45°F., the reaction of the acid on the casein of the milk is retarded. At a temperature of over 50°F. the acid reaction is fairly rapid, thereby creating particles of increased size. The faster the agitation during the acidification step, the smaller is the size of the particles produced.

The percentage of total solids in the acidified milk mixture also has an effect upon the particle size of the precipitated curd. A total solids content of about 20 to 30% appears to be optimum. It is desirable to use as high a solids concentration as possible since it is less costly to remove moisture from the product by condensation procedures prior to acidification than by drying procedures after the curd is precipitated. The upper limit of concentration is determined by the viscosity of the solution since it is necessary to have a milk solution which can be agitated sufficiently to insure uniform admixture of the milk and acid solutions.

Rather than precipitate the milk in a batch process it has also been found desirable to meter the acid solution into the milk solution in a continuous process. After the acid is metered into the milk solution, the mixture of milk and acid solution is kept under constant agitation to insure uniform admixture of acid in milk and thereby to produce particles of uniform size. After precipitation of the curd, the curd-whey solution is at a pH within the range of from 4.0 to about 4.9. This solution is then dehydrated by conventional spray drying equipment.

The spray dried baker's-type cheese product is improved in its baking characteristics when either natural or synthetic stabilizers are added. A desirable stabilizer can be formed by mixing into the cheese approximately 3.3% of carboxymethyl cellulose in 6.7% of instant starch, thus making a final cheese product containing approximately 10% of stabilizer. This combination was found to produce the best cake volume as well as having many other excellent baking characteristics.

A baker's-type cheese blend containing the stabilizer (an instant starch and carboxymethylcellulose combination) in the amount of about 10% by weight of the blend was placed in sealed cans and stored at 40°F., 72°F., and 100°F. for storage tests. It was found that good flavored cheese cakes can be produced from powdered cheese product that had been in storage in excess of eight months.

Example: 5,480 pounds of skim milk which had been heat treated to produce a phosphatase-negative milk and condensed to 36% solids was introduced into a rectangular, coil vat. The milk was normal in acidity (pH 6.3). 2,500 pounds of water was added, lowering the solids content to approximately 25% and adjusting the temperature to 50°F. 650 pounds of 15% citric acid was added, making a total of about 8,630 pounds of liquid having 23% solids to be dried.

The acid was added to the coil vat until the pH reached 5.5, indicating that most of the acid can be added to the bulk milk supply if the isoelectric point is not reached. Thousand pound quantities were removed from the coil vat and placed into vertical vats equipped with a lightning agitator. Then the rest of the acid was added to bring the pH down to 4.5. Two such thousand pound vats were used so that one could be dried while the other was being prepared. The product was spray dried in a concurrent dryer with a 0.040 inch nozzle orifice using an inlet temperature of 260°F. and an outlet temperature of 160°F.

The 1,776 pounds of dry powder was then blended by a mechanical blender in 300 pound batches, using 270 pounds of dried cheese product, 9 pounds 14 1/2 ounces of carboxymethylcellulose (CMC), and 20 pounds 1 1/2 ounces of instant starch. This blend was packaged in polyethylene lined drums for storage. Upon subsequent usage in commercial bakeries, this product was found to be very stable and uniformly high in quality.

Heat Denatured Serum Protein

D.G. Vakaleris; U.S. Patent 3,128,192; April 7, 1964; assigned to National Dairy Products Corporation developed a process for producing a dry product. The process includes at least 10% heat-denatured serum protein in the product. ("Serum protein" refers to milk protein other than casein.) This is an important feature, and is responsible in part for improved water absorbing characteristics. Upon reconstitution with as much as four times its weight of water, the product provides a cheese having texture and body well suited for baking applications.

The product may be prepared by at least two procedures. In the first procedure, skim milk is subjected to high-temperature, short-time heat treatment, by which denaturation of serum protein is effected. Denaturation of serum protein is effected under conditions under which about two-thirds of the heat-denaturable serum protein is denatured.

Denaturation is carried out at temperatures substantially higher than normal pasteurization temperatures. The extent of denaturation at a particular temperature is functionally related to the duration of treatment. To minimize undesirable flavor formation, denaturation may be effected above 200°F. for relatively short periods of time. To obtain maxiumum yields of denatured serum protein, the skim milk is subjected to 220°F. for about 30 seconds.

This treatment effects denaturation of substantially all of the heat-denaturable serum protein. The heat treated milk is inoculated with starter and rennet, in accordance with conventional methods. The milk is allowed to coagulate to form a curd which is cut, and the curd is cooked to expel whey. The whey is drained off, the curd is washed with water, and suspended in water to provide a slurry.

The slurry is homogenized and spray dried to provide the dry product. The product comprises about 80% milk protein, less than about 10% lactose, and, for desirable keeping quality, preferably comprises less than about 5% moisture. When prepared according to the preferred procedure, the milk protein of the product comprises about 15% heat-denatured serum protein.

In the second procedure, a cheese curd is prepared from skim milk by a conventional cottage cheese-making procedure and is suspended in water to provide a slurry. The slurry is homogenized and spray dried to provide a dry powder. Furthermore, denatured serum protein is prepared from whey which is subjected to heat treatment to effect denaturation of the serum protein.

Subjecting the whey to the conditions of time-temperature set forth in connection with the first procedure may be utilized to effect denaturation of the serum protein. The denatured serum protein is separated from the remainder of the whey, washed with water, and suspended in water to provide a slurry. The slurry is dried to provide a dry powder comprising denatured serum protein. Also, in the second procedure, the dried powders are blended at proportions in which the protein in the resultant dry product includes about 15% of denatured serum protein. The product readily reconstitutes in water to provide a smooth, firm paste having a desirable baker's cheese flavor.

If the dried powders are blended in proportions in which the protein comprises substantially less than 10% denatured serum protein, the resultant dry product may have insufficient capacity to absorb water, and baking performance is not as satisfactory. If the protein comprises substantially more than about 20% of denatured serum protein, the reconstituted product acquires a somewhat grainy texture, but this coarse texture may be avoided by suitable adjustment of the mineral balance.

Example: 3,000 pounds of skim milk was heat-treated by passing the skim milk through a Votator unit. The surface of the Votator cylinder was maintained at about 220°F. and the retention time of the skim milk in the cylinder was about 30 seconds. This high-temperature, short-time heat treatment effected denaturation of about 75% of the nitrogenous materials in the serum, which was substantially all the heat-denaturable serum protein.

The heat-treated skim milk was introduced into a conventional cottage cheese vat, and was set with starter and rennet. The acidity was allowed to develop until the pH of the curd was about 4.9, providing a whey acidity of about 0.39%. At this acidity the curd was cut, and the curd was then cooked to 130°F. Whey was drained from the curd, and the curd was washed three times with water. 0.384 pound of cheese solids per pound of skim milk solids were obtained.

Water was added to the curd, and the mixture was milled to provide a slurry comprising about 12% solids. The slurry had a pH of 4.9. The slurry was homogenized at 2,000 psi and spray-dried to provide the product. The dried product comprised about 80% milk protein, about 8% lactose, and about 4.1% moisture. The milk protein of the product included about 15% denatured serum protein.

The product was stable for about 4 months when held at room temperature, and was stable for more than a year at refrigerator temperatures. Upon reconstitution with water a smooth firm cheese was obtained which had a desirable baker's cheese flavor. In this connection, the paste had a relatively firm body upon the addition of up to four parts of water per part of the dry baker's-type cheese.

The product was used in preparing a cheese cake, and the cake had a desirably smooth texture, high volume and rigid structure. In comparison, a control cheese cake was prepared from a commercially-available dry baker's-type cheese product, which product included vegetable gum for water absorption purposes. The cheese cake made from the product of this example had substantially higher volume, firmer structure, and better flavor than the control cheese cake. Pumping and handling properties of the homogenized slurry can be improved by pH adjustments. The denatured serum protein could also be added separately to a product pasteurized at lower temperatures.

Heating Curd and Whey

E.T. Biasi and J.G. Skibiak; U.S. Patent 3,431,115; March 4, 1969; assigned to

Pennsalt Chemicals Corporation describe a process for making baker's cheese comprising the steps forming the cheese curd, heating it to a selected temperature and delivering it to a centrifuge having a boundary surface and a conveyor surface, and moving the solids with the conveyor surface to a solids discharge opening once the solids have settled on the boundary surface. The moisture content of the final cheese product can be accurately controlled by adjusting the temperature to which the cheese is raised prior to centrifuging.

Skim milk having a fat level content between 0.1 and 0.5% is delivered to the cheese-making tank. Also delivered is an appropriate quantity of a bacteria culture for use as a starter and an appropriate enzyme such as rennet. The tank is maintained at room temperature, 72°F., for 8 to 12 hours. When the cheese curd is properly formed, and this is detected by the mixture reaching an appropriate pH level such as 4.5 to 5.0, the cheese curd with its whey is pumped or delivered by gravity in a stream to a heat exchanger.

In the heat exchanger the temperature of the mix is uniformly and rapidly raised to a selected temperature depending on a desired moisture content of the cheese product. Preferably the heat exchanger may be heated by steam. For the heating to be uniform, the heat exchanger can lead the flow of ripened mix for a considerable distance as through zig-zag passages or simply over a sizeable length of delivery tube.

The curd and whey are heated in the range of 95° to 160°F. depending upon the desired moisture content of the final cheese product. The exact reason why temperature before centrifuging is critical in the amount of moisture in the final cheese product is not known. However, it is believed that the protein solids coalesce and bind together to a greater extent as the temperature is raised. This binding makes the centrifuging more efficient.

From the heat exchanger the stream of the heated cheese curd and its whey is delivered to a centrifuge of the type having a solid wall and an internal screw conveyor and having separate liquid and solid discharge ports in which the heavier cheese solids settle against the boundary wall and the conveying surface of the conveyor screw which rotates at a different speed from the speed of the bowl. This action moves the settled solids to the solids discharge port from which they are thrown outward and collected. These solids comprise the final cheese product. The whey is discharged through a separate liquid discharge port at the opposite end of the centrifuge and is for disposal or further processing.

Example: The cheese curd at a temperature of 72°F. was externally heated to 148°F. by steam. At this temperature the stream was delivered to a P-2000 Super-D-Canter centrifuge operating at approximately 4,800 rpm and generating a centrifugal force of approximately 4,500 times gravity. The differential speed of the conveyor was 3 rpm. The cheese which discharged was totally acceptable and of uniform quality, having a moisture content of 68% and a pH of 4.8. Subsequently the temperature of the mix was reduced so that a thermometer on the discharge end of the heat exchanger indicated 122°F. The cheese product which discharged under these conditions had a moisture content of 74%, approximately the maximum moisture content which bakers using such cheese are able to accept. It can be seen that the process offers at least an accurate control of the moisture content of the discharged final cheese product. At the same time it affords a means for effectively replacing draining the whey from the cheese in drain bags.

PROCESSES APPLICABLE TO ALL CHEESES

CURD FORMATION

Without Starter Culture

E.G. Hammond and D.D. Deane; U.S. Patent 2,982,654; May 2, 1961; assigned to Iowa State College Research Foundation, Inc. discovered that a suitable curd for cheese production can be formed by directly adding to milk a neutral substance which will slowly react with water to form an acid capable of coagulating casein. The agent must be one which will form a nontoxic acid and which can be completely dissolved or dispersed in the milk. The agent must hydrolyze relatively slowly compared to the rate at which it can be dissolved in the milk. If the acid development is too fast with respect to the rate of solution, local precipitation of the curd may occur near particles of the acidogen or acid-forming agent.

In cases where enough acidogen is added to lower the overall pH of the milk below with respect to the rate of solution and the coagulum begins to form before the acidogen is entirely dispersed, an uneven coagulum may result, or if an even coagulum is obtained, the acidity may be difficult to control and may interfere in subsequent steps in making the cheese.

If enough of the acidogen is added to milk to drop the pH from its normal value of 6.5 to 6.8 to a final pH of 4.6 to 4.7 after complete hydrolysis, the acidogen must be dissolved before the pH reaches the point where coagulation begins, which is about 4.9 to 5.3. In addition, the acidogen should, in order to produce a nutritive and digestible cheese product, be capable, upon hydrolysis, of lowering the pH or coagulating the milk without also reacting with and modifying the protein of that milk.

These results are best obtained with water-soluble agents capable of slowly hydrolyzing to form carboxylic acids. Cyclic esters such as lactones and lactides have been found particularly effective as acidogens for this purpose. For instance, excellent results have been obtained with the low melting lactide (41° to 42°C.) of lactic acid and with D-glucono-delta-lactone. Both of these materials dissolve or mix readily with milk before coagulum formation begins and both form acids which are believed to be nontoxic and which do not produce undesirable flavors or odors in the final product.

The low melting lactide of lactic acid may be prepared by heating lactic acid under reduced pressure. After the lactic acid is dehydrated and polymerized, the temperature is raised and the lactide is distilled. The crude distillate is a mixture of high and low melting lactides. These may be separated by crystallization from suitable solvents followed by filtration. The more soluble form melts at 41° to 42°C. in contrast to the less soluble isomer which melts at 128°C.

To make a cheese product by this method, the acidogen is added directly to milk and the reactants are then agitated until they are thoroughly mixed. By varying or adjusting the composition of the milk as the starting material, the character of the resulting cheese may be controlled to a considerable extent. For example, Cheddar cheese is ordinarily made from whole milk, or milk in which the fat content has been adjusted to approximately 3.5%, while cottage cheese is usually made from skim milk, or from reconstituted, concentrated skim milk or nonfat dry milk solids.

The amount of acidogen added to the milk depends principally upon the buffering capacity of the milk. The buffering capacity is highly correlated with the total solids of the milk, and the amount of acidogen needed for a given level of total solids is found empirically. After this, the amount of acidogen which will give the desired pH upon complete hydrolysis may be calculated from the total solids. When the acidogen is to be the only agent used for coagulation of the milk, its amount should be calculated so that upon complete hydrolysis the liquid will have a pH between 4.5 to 4.8, with an optimum pH of 4.7.

Rennet or proteolytic enzymes may be added to the milk with the acidogen to increase the rate of coagulation or to alter the properties of the curd. Also to improve the cooking properties of the curd, $CaCl_2$ may be added to the milk with the acidogen. If the cheese is to be ripened, microorganisms may be added with the acidogen to alter the flavor and physical properties of the curd.

After thorough distribution of the acidogen in the milk, agitation is discontinued and the coagulum is allowed to form. Specifically, agitation should be discontinued before coagulation has begun, that is, before the pH of the liquid has dropped to a value of 5.1, or at least to a pH of 4.9 to 5.3. If agitation is continued beyond the point when coagulation begins, then the curd will not form as a solid mass but will instead come down as a finely divided precipitate.

After the acidogen is dispersed, the hydrolysis is allowed to proceed with or without the action of enzymes until the hydrolysis and the action of the enzyme, if any, has proceeded to the desired point. If the product is to be cottage cheese the pH is dropped to a pH of 4.6 to 4.7. The formed coagulum is then cut with vertical and horizontal wire curd knives to reduce the coagulum to small cubes. After allowing some time for the expression of whey to begin, the curd and whey is heated or cooked by slowly raising the temperature to 120° to 140°F. When the cooking is judged to be complete the whey is drained, and the curd is washed, drained, and refrigerated.

Example: To 2,000 g. of reconstituted skim milk containing 8.73% solids, having a pH of 6.70 and temperature of 25°C., 17.46 g. of powdered low melting lactide of lactic acid were added and stirred into the milk with a mechanical stirrer. Stirring continued for 20 minutes. Fifteen minutes after the lactide was added, 2 ml. of a 1 to 1,000 dilution of commercial rennet was added. The milk was allowed to stand after the lactide and rennet had been stirred in. Eighty minutes after the addition of the lactide, the coagulum was cut with 1/4 inch knives and cooked by a standard procedure for cottage cheese. The yield of curd was 16%, and the curd had a solids content of 20.1%. The curd had the normal appearance and texture of cottage cheese curd and had a bland flavor. No off-flavors were detected.

Cheese Curd and Casein

N.J. Berridge; U.S. Patent 2,997,395; August 22, 1961; assigned to National Research Development Corporation, England describes a process for the manufacture of cheese curd or casein which comprises holding the curd-forming mixture, consisting of milk, rennet and if required acidifying substances, at a temperature of not more than 15°C. for a period of time such that the enzymic action of the rennet is nearing completion or is complete; subsequently quickly heating the mixture to a temperature of not less than 30°C., and holding the mixture at that temperature so that rapid coagulation of curd and rapid exudation of whey can occur.

Where acidifying substances are added, they may be the usual starter or other microorganisms or an acid, e.g., hydrochloric acid, or a combination of both starter and acid. Suitable starters have been found to be Streptococcus thermophilus and Streptococcus cremoris. Other acidifying substances may be used if required, for example, lactone or an ester and esterase. The acidifying substances are added in sufficient quantities to give the milk a pH of 5.6 to 6.0. In a preferred process, a small quantity of starter (2%) is used in combination with hydrochloric acid. This small amount of starter produces acid after the curd has been formed and assists in the ripening of the curd.

The second stage of the process, i.e., the rapid heating stage, is carried out by allowing the curd-forming mixture to flow as a thin layer over a heated surface, e.g., the surface of a plate or trough, or by depositing the curd-forming mixture on a heated surface from which it is subsequently removed after a suitable period of time has elapsed (e.g., 20 to 100 sec.). This process can be used in the continuous production of cheese. Moreover, the curd can easily be obtained in an acceptable form after curdling.

In further work N.J. Berridge; U.S. Patent 3,120,443; February 4, 1964; assigned to National Research Development Corporation, England found that the temperature restrictions of less than 15°C. for the stage of the process at which the enzymic action takes place and above 30°C. for the stage at which coagulation occurs are not essential to the process.

An acceptable curd can be obtained provided that the coagulation stage is carried out on a curd-forming mixture in which the enzymic action of the rennet is nearing completion or is complete, but in which coagulation has not commenced regardless of the process by which this mixture has been obtained, provided also that the curd-forming mixture is heated in a form in which the surface area is large compared with the volume as opposed to being heated in bulk.

The process provides for the manufacture of cheese curd or casein, which comprises heating a curd-forming mixture, in which the enzymic action of the rennet is nearing completion or is complete but in which coagulation has not visibly commenced, in a form in which its surface area is large compared with its volume to cause rapid coagulation of the curd-forming mixture followed by rapid exudation of whey.

The curd-forming mixture can be obtained in the state in which the enzymic action of the rennet is nearing completion or is complete but in which coagulation has not visibly commenced by allowing the curd-forming mixture to stand for a period of time, "the rennetting time", which depends upon the quantity of rennet in the mixture, the pH of the mixture, the temperature at which the mixture is allowed to stand, and also on the biochemical properties of the milk. Thus, the greater the proportion of rennet in the mixture the shorter the rennetting time. Similarly the lower the pH of the mixture, within limits, the shorter the rennetting time.

The pH can be adjusted by the addition of acidifying substances which may comprise the usual starter alone, for example, Streptococcus cremoris or other micro-organisms, or an acid, e.g., hydrochloric acid, or a combination of starter and acid. The acid is added to shorten the rennetting time, but the pH should not in general be lowered below 5.6 as, if this is done, subsequent coagulation of the mixture gives rise to a curd which is soft and sticky when hot. A pH range of 5.6 to 6.0 has been found to be convenient.

The rennetting action can be allowed to take place at even lower pH values, however, if the pH of the mixture is subsequently adjusted to above 5.6 and the mixture allowed to stand for some time at this higher pH before coagulation. Even using this method, however, the lower limit of pH cannot in general be reduced below 5.0, as, for instance casein is completely precipitated by acid at pH 4.7, precipitation beginning at a pH of about 4.8 to 4.9.

Moreover, it has been found that certain starters shorten the time for which the curd-forming mixture can be stored at a low temperature without coagulation. When using these starters, therefore, it is sometimes advantageous to add the rennet to the milk and allow the enzymic action of the rennet to take place before adding the starter. The starter is then added just prior to the coagulation step.

The higher the temperature at which the rennetting action is allowed to take place, the shorter the time required and also the greater the tendency to coagulate. The temperature must be such, therefore, that the rennetting action is complete or is nearing completion before coagulation takes place. Such a temperature is preferably below 20°C.

In cases where no acidifying substances are used in the curd-forming mixture, larger quantities of rennet (e.g., 20 to 30 times) or longer holding periods at low temperatures (e.g., 5 hours) or a combination of the two (e.g., 5 times the rennet and one hour holding period) are desirable. If two short a holding period is used, the milk will subsequently take too long to coagulate on being quickly heated, and the curd formed will be too soft with a result that the exuding whey will contain a surplus of fat. If too long a holding period is used, the curd-forming mixture will thicken in the cold, with the result that once again large losses of fat in the whey will occur.

The conditions under which the rennet action takes place have been found to affect the form of the curd produced by subsequent coagulation and, in general, a rennetting time of about one hour at about 10°C. has been found to produce the most acceptable curd upon coagulation. If such a rennetting time and temperature are to be used, the curd-forming mixture can be first tested to determine whether or not such a treatment will be satisfactory in completing the enzymic action and, if necessary, the contents of the mixture adjusted so that the treatment will be satisfactory.

The treatment of the curd-forming mixture to effect completion or near completion of the enzymic action of the rennet can if desired be performed continuously. Thus milk, rennet and, if required, acidifying substances can be continuously introduced into any form of delay line such as a long pipe or a number of vessels connected in series or a tank which is divided into a series of compartments by a number of transverse partitions provided with holes allowing flow of fluid between the compartments.

The rate of inflow of milk, rennet, etc. can be adjusted so that the average time taken to pass from one end of the tank to the other, at which the mixture is removed, via the intervening compartments is a satisfactory rennetting time. Test samples can from time to time be removed from different sections of the tank and checked to ensure that rennetting is in fact proceeding satisfactorily.

When a tank with five sections is used, the condition of the curd-forming mixture is generally satisfactory if a sample from the second or third section (counting in the direction of milk flow) clots when allowed to flow very gently one drop at a time on to the surface of water at 60°C. The drops sink but coagulate before they mix. If this result is obtained only with samples from from the fourth or fifth section the curd will generally not be satisfactory. If a sample from the first section coagulates there is danger of premature coagulation. The milk may be too acid. The situation may be adjusted by alterations in the inflow rates of the milk, rennet and/or acidifying substances according to requirements.

The form of the curd-forming mixture, and the temperature to which and the time for which the curd-forming mixture is heated to effect rapid coagulation of curd vary with the conditions of the process. The larger the surface area of the mixture as compared with its volume and the higher the temperature to which it is heated, the more rapid will be the coagulation. The curd should not in general be heated above 70°C., however, as such a hot curd tends to stick and is difficult to handle. A preferred temperature for carrying out the process is in the range of about 30° to 70°C., and temperatures from 50° to 70°C. have been found especially suitable.

The maximum temperature is usually limited by the starter. Each strain has a different sensitivity. When, as is usually the case, an active starter is required in the curd it is necessary to find by experiment the effect of different time-and-temperature combinations on the starters to be used to ensure that the starter used is capable of remaining active under the conditions of the process. Coagulation should occur very quickly if an acceptable curd is to be obtained, but the heating is often continued after coagulation for a short time (e.g., 30 sec.) as this does not harm the curd and it assists the expulsion of whey.

The heating of the curd-forming mixture is preferably carried out by contacting it in the form of a thin film with a heated metal surface. A thin film is required in order that all the curd-forming mixture can be quickly heated, and thus rapidly coagulated and a film less than 1 mm. thick has been found to be convenient. If it can be arranged that both sides of the film be heated, then thicker films may be used.

In one particularly convenient method, the curd-forming mixture is delivered as a thin layer on the heated metal surface of a rotating cylinder and is allowed to remain in contact with the heated surface until coagulation of curd has occurred. A layer, thickness of about 0.2 mm. has been found suitable using this method. A suitable apparatus for implementing this process is described in the patent.

Freezing and Thawing Curd

T. Chikuma; U.S. Patent 3,091,539; May 28, 1963; assigned to Goro Chikuma, Japan describes a method for making cheese which involves freezing the curd and then melting it in water and pressing to dehydrate the curds. Curds are prepared by a known process, and the curds are instantly cooled and frozen at an exceedingly low temperature. After that the curds are fed into a filter such as a cloth bag. They are then subjected to melting by putting them into water or adding water at a suitable temperature, and the remainder in the filter is collected and compressed in a press box. The resulting material is inserted into a salt solution and ripened as usual.

In the process of freezing the curd, a continuous or a semi-continuous system may be employed and the freezing temperature must be suitable for freezing the curd at about -20° to -30°C. The process of compressing is employed before the material is subjected to the process of salting. Equipment such as a jacketed cheese-vat, a curd knife, a stirrer, etc., are not needed in this process, and it is also not necessary to apply skill and experience, such as for stirring without breaking soft curds during the heating up period. The product obtained is superior, and soft cheese, such as cottage cheese which is obtained in this process contains less lactic acid and is suitable for storage.

Example: 30 kg. of sterilized milk containing 3.2% of fat by weight of total composition is transferred into a cylindrical jacket equipped with a stirrer and made of stainless steel. Some kind of starter, such as *S. lactis* or lactic acid bacilli in skimmed milk, are added to the milk which has a lactic acid content of about 0.160% by weight of total milk. 1.5 g. of rennet is added. The mixture is then fed into a vat and maintained at room temperature for 15 min., and a curd is obtained. This curd is instantly frozen in the freezing room at about -30°C., and subjected to melting by setting it into warm water of a temperature of about 40°C. Small lumps which remain in the filter are transferred to a rectangular case 25 x 12.5 x 20 cm., and compressed there under a weight of 5 kg. for 15 hours. This pressed lump is set into a saturated aqueous solution of salt, and a cheese product is obtained. The yield is 2.9 kg.

Use of Milk with Desired Whey Protein Nitrogen Value

P.P. Noznick and R.H. Bundus; U.S. Patent 3,316,098; April 25, 1967; assigned to Beatrice Foods Company obtained higher yields by subjecting the fluid milk to a high heat treatment. With milk having a W-P-N (Whey-protein value of 1 to 3 mg. per cubic centimeter of nitrogen) of 2 mg. per cubic centimeter, a cheese was obtained with a higher yield of cheese per pound of solids. This increase of pounds of curd per pound of solid amounts to at least 1.5 to 3.0 pounds and better, subject to the total protein composition of milk, i.e., the higher the protein the higher the yield. In brief, there is a substantial increase in the pounds of curd per pound of solids, e.g., a mean increase of 10 to 20% of yield over the same milk using conventional processes. By high heat treatment is meant heating at temperatures and/or hold times substantially above pasteurization methods which are 161°F. for 15 seconds or 145°F. for 30 minutes. The times and temperatures referred to for this process are a temperature of 185°F. for a hold time of at least 900 seconds, or 300° to 305°F. and a hold time of 1 second or less, i.e., substantially instantaneous. While normal yields from 9% total solids skim milk are about 12 to 13 pounds of curd per 100 pounds of skim milk at a curd total solids content of 20%, the yields of this process are 16 to 18 pounds and even higher per 100 pounds of skim milk at a curd total solids content of 20%. This is a marked increase in yield.

Example 1: Raw sweet skim milk in the normal range of total solids content of about 8.5% is heated up to 262°F. and is subjected to this temperature of 262°F. for 100 seconds to heat the skim milk. Cheese made from this liquid skim milk has the advantage that apparently, the heating alters the whey proteins so that they remain in the curd and are not lost in the whey. Thereby a greater yield of curd is obtained from the skim milk than usual. The cottage cheese is then prepared as usual.

Example 2: 32 gallons of whole milk testing 3.5% butterfat was heated at 255°F. for 15 seconds. The milk was tube cooled in a heat exchanger to 92°F. for optimum growth of the organisms and 0.1% calcium chloride, 0.5% *Streptococcus lactis* culture, cheese color at the rate of 2 ounces per 1,000 pounds of milk and rennet at the rate of 3 ounces per 1,000 pounds of milk and 2% normal Cheddar cheese whey at 70°F. from a batch of Cheddar cheese made by usual time and temperature treatment of the raw milk of 165°F. for 15 seconds, were added and stirred into the milk.

The increase in acidity was observed to be much faster in this milk than in the heat treated conventional control milk. After development of the body at pH 6.02, the curd was cut, cooked to 106°F. for about 1/2 hour (above 110°F. desirable organisms were destroyed), drained, cheddared, cubed, salted and pressed into wheels.

These conditions maintain the *Streptococcus lactis* culture through cheddaring and provide optimum growth conditions for the other cheese curing bacteria. The control cheese yield was 9.45 lbs. of cheese per 100 lbs. of milk, whereas cheese made by this example yielded 10.9 lbs. of cheese per 100 lbs. of the same milk. The control product was not prepared by initially heating the milk or adding whey as described above. This showed that the initial time and temperature treatment of the milk and the whey addition are critical factors. While the pH does not appear critical, it was evident that coagulation of the curd was obtained at a higher pH than by conventional processes, e.g., as evidenced by observation and comparison of the example and control sample.

Two methods of improving the flavor of the final cheese product made by this process are offered by I.B. Eggen and R.H. Bundus; U.S. Patent 3,297,451; January 10, 1967; assigned to Beatrice Foods Company. One method is to homogenize the skim milk prior to the high heat treatment, the second being to add citric acid or a nontoxic citrate, such as sodium citrate, potassium citrate, or ammonium citrate, prior to the fermentation step. The citrate can be employed whether or not the skim milk is homogenized prior to the high heat treatment. When a citrate is added to the milk before heat treatment, it is used in small amounts, for example, 0.1 to 0.2% by weight.

Example 1: Raw sweet skim milk having 8.5% solid content was homogenzied at 8,000 psi at 100°F. and then was heated up to 262°F. for 100 seconds. The product was cooled to 90°F. and 10% of Streptococcus lactis added. The mixture was allowed to incubate for 4 hours. Thus curd was then firm enough to cut. The curd was cut and then heated at 130°F. for 1.5 hours. The whey was then subjected to the usual draining. The hot drained curd was cooled and rinsed with sterile cold water at 35°F. The product had improved flavor over conventionally prepared cottage cheese.

Example 2: The procedure of Example 1 was repeated except 0.2% of sodium citrate was added prior to the homogenization of the raw skim milk. The homogenization prior to the high heat treatment can be carried out at 5,000 to 10,000 psi at a temperature of 90° to 180°F.

Heat Treatment and Semipermeable Membrane

A process for the continuous production of curd according to N.J. Berridge; U.S. Patent 3,520,697; July 14, 1970; assigned to National Research Development Corporation, England comprises the steps of renneting milk at a temperature such that rennet action can occur without curd formation, holding the milk at that temperature at least until rennet action is nearing completion or is complete and contacting the renneted milk with one surface of a semi-permeable membrane, the other surface of which is in contact with water at elevated temperature so that the milk is heated to a temperature at which curd formation can occur.

The milk should be contacted with the membrane in such a form that the milk is rapidly heated throughout its bulk to the required temperature in a form in which its surface area is large compared with its volume. For example, the membrane can be in the form of a thin tube through which the milk is caused to flow. Alternatively, two sheets of membrane placed close together to form effectively a flat tube can be used or the milk can be flowed as a thin layer over a sloping sheet membrane. Regenerated cellulose (Visking dialysis tubing) and collodion (cellulose nitrate) are suitable materials for the membrane.

A plurality of membranes arranged in parallel and fed from a common source can be used to increase the throughput of the apparatus. When this arrangement is used means should be provided to ensure that the rate of feed to all the membranes is the same. For example, when the membranes are in the form of tubes having small diameter the connection from the common source to each membrane tube may be via a capillary tube which limits the rate of flow of the milk to the required degree.

The milk can first be inoculated with a suitable starter culture such as Streptococcus thermophilus and its pH adjusted to between the natural pH and 5.7, for example, between 6.1 and 6.3. For cheddar cheese the preferred pH is between 6.2 and 6.3. The milk is then cooled to between 0° and 10°C., treated with rennet, and held at this temperature at least until the specific rennet action is nearing completion or is complete as shown by coagulation when a drop of milk is allowed to fall from a height of 2 mm. into water at 50° to 60°C. The milk can also be held at the holding temperature for times considerably in excess of the minimum time needed for the completion of rennet action. A long holding time may modify the properties of the curd.

The milk is then brought into contact with the membrane, preferably by flowing it over one surface of the membrane, while hot water is flowed over the other surface of the membrane. During this stage the milk should be heated to a temperature sufficient to cause clotting but not so high as to kill the starter culture or to cause protein to be deposited from the whey (between 30° to 50°C.). In some circumstances, e.g., when making soft cheeses, the starter can be added at a later stage in the process and then the milk can be heated to temperatures above those at which cultures are killed.

The required temperature of the water on the other side of the membrane will depend on the length of time that the milk is in contact with the heated membrane, on the initial temperature of the milk, the nature and thickness of the membrane, and rate at which the water flows over the membrane.

Preferably the membrane should not be heated at the point where the milk first comes into contact with it, so that the milk flows over the surface of the membrane for a short distance before coming in contact with the heated part of the membrane where curd formation occurs and the tube or ducts leading to the membrane do not become heated. This ensures that curd formation does not occur in the tube or ducts leading to the membrane.

The amount of curd sticking to the membrane surface is decreased particularly greatly if the water is acidified. Phosphoric acid or hydrochloric acid may be used for acidifying the water. Even when acidified water is used a small amount of protein may stick to the membrane, and it may be necessary periodically (say every 8 hours) to interrupt the milk flow and flush the membrane with water followed by an acidic or alkaline cleansing solution. If any further particles of protein remain stuck to the membrane treatment with pepsin solution may be required. The hot water may be recirculated, but should be changed from time to time.

Apparatus is designed for this continuous production of curd which comprises means for supporting a semi-permeable membrane, means for causing milk to flow over one surface of the semi-permeable membrane, means for contacting the other surface of the semi-permeable membrane with water at elevated temperature and means for cooling the milk before it comes into contact with the semi-permeable membrane.

Bubbling CO_2 to Reduce Coagulation Time

The cheese making process described by C.O. Claesson and E.M.H. Claesson; U.S. Patent 3,560,224; February 2, 1971; assigned to Alfa-Laval AB, Sweden relates to a method in which the coagulation time is reduced and a well developed curd is obtained consistently despite variations in the coagulation properties of different batches of milk.

This can be achieved by adding carbon dioxide (CO_2) to the milk from which the cheese is to be made. In this way, a substantial reduction is obtained in the differences in milk coagulation properties attributable to different specimens of cows or different stocks and to different location periods, and also the coagulation time is reduced and a well developed curd is obtained. Preferably, the pH value of the milk is adjusted to about 6.0 to 6.5 by this addition of carbon dioxide.

The process is further explained with reference to Figure 3.1 which is a diagram showing the time of coagulation of milk from different cows at different pH values of the milk. The curves there indicate the coagulation time for different milk samples after rennet addition at different pH values of the milk, where the pH value has been adjusted by bubbling carbon dioxide through the milk. It appears from the diagram that the coagulation time diminishes rapidly as the pH value of the milk is lowered. At a pH value of 6.5, e.g., the variation in coagulation time between different milk samples is only 6 minutes, which means both a favorable standardization and an actual reduction of the coagulation time. It also appears from the diagram that milk which is noncoagulating (i.e., the coagulation time is infinite) can be made to coagulate in about 35 minutes after a slight change of the pH value (0.05 unit).

FIGURE 3.1: BUBBLING CO_2 TO REDUCE COAGULATION TIME

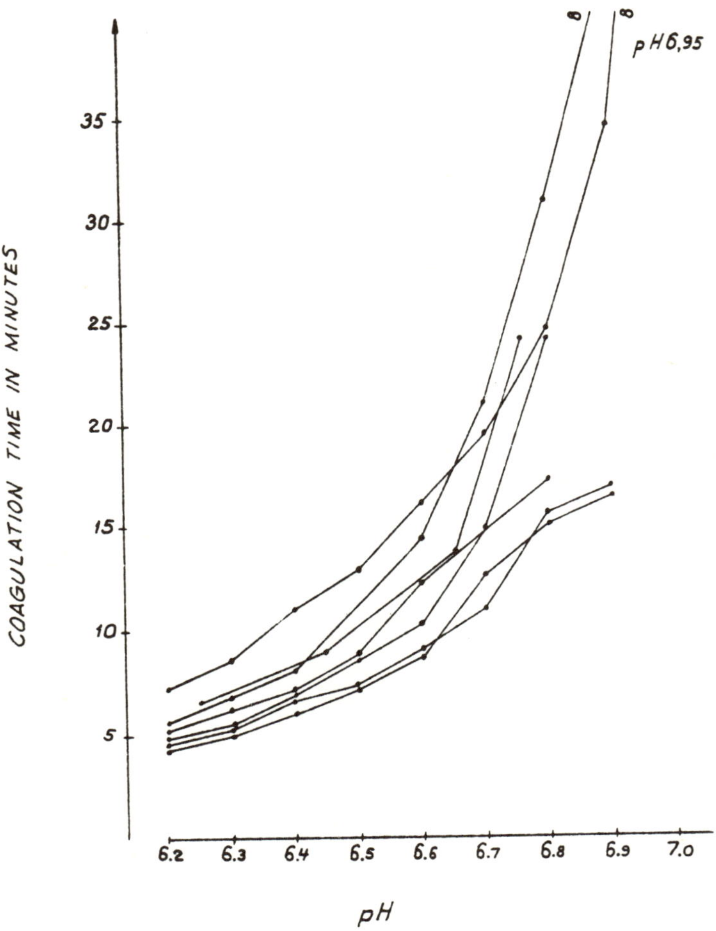

Source: C.O. Claesson and E.M.H. Claesson; U.S. Patent 3,560,224; February 2, 1971

An adjustment of the pH value by bubbling carbon dioxide through the milk thus has a marked effect upon the rennet coagulation time. This effect is caused by the more rapid course of the enzymatic phase, but apart from this an anion effect is achieved which brings about changes in the ion environment of the calcium-caseinate and calcium-phosphate complex. These changes are probably caused by the carbonic and hydrocarbonic ions, which influence the course and result of the coagulation phase.

The addition of carbon-dioxide brings about the liberation of substances that directly influence the aggregation of the casein during the coagulation by which the curd structure is influenced in a desired direction. The following table shows the reduction of the coagulation time in sec./0.1 pH for different milk samples after addition of different pH-lowering substances.

Milk Sample No.	HCl, sec./0.1 pH	Lactic Acid, sec./0.1 pH	CO_2, sec./0.1 pH
1	158	140	254
2	181	215	242
3	489	-	550
4	424	-	472

It appears from the table that the addition of CO_2, in addition to the pH effect, has a specific influence on the coagulation course. Probably the casein aggregation is influenced by the hydrocarbonic ion. When the properties and syneresis of the curd were determined according to Koestler, the curd obtained from milk treated with CO_2 had better values throughout.

Example: The equipment used for this example comprises a vat having a capacity of 4,000 to 6,000 liters and adapted to heat its content by means of heating coils in the jacket of the vat. As is conventional for cheese-making, the vat is also equipped with an agitator for stirring. For the purpose of this example, the agitator is provided with a conduit connected to a tube for carbon dioxide gas, and the gas supply is controlled by a pH meter through a solenoid valve. The cutting of the curd is effected by a device mounted on the agitator.

In the production of round-eyed cheese, 3,000 liters of pasteurized milk are introduced into the vat and the temperature is adjusted to 32°C. The pH value of the milk is adjusted with carbon dioxide to 6.5 to 6.4 while stirring. Thereafter, lactic acid and aroma bacteria and 0.5 liter of rennet are added while stirring. After 30 minutes, the curd is cut and the temperature is raised to 42°C., and thereafter the whey runs off while the curd under working is tapped into molds.

Use of Concentrated Milk

The production of cheese curd is discussed by C. Giddey and G. Dove; U.S. Patent 3,653,919; April 4, 1972; assigned to Alfa-Laval AB, Switzerland. In the process concentrated milk with a dry substance content of about 30% is renneted at 15° to 30°C. for 5 to 15 minutes to insolubilize the casein and obtain a homogeneous non-gelled viscous mass. This is then cooled below 10°C., after which parallel streams of the renneted milk are heated rapidly to 25° to 45°C. to polymerize the casein and obtain curd in the streams. The curd is then washed in an aqueous bath, separated from the bath, and finally recollected to form a homogeneous mass.

Example: Pasteurized milk with a fat content of 3 1/2% and a pH value of 6.6 is concentrated to one-third of its volume by means of evaporation at 30°C. The concentrated milk is then cooled to 25°C. and renneted by the addition of 0.03% rennet of the concentration 1 per 10,000 (calculated on the volume of the unconcentrated milk). The milk thus renneted is kept at the latter temperature for about 12 minutes until a viscosity is obtained and qualitatively estimated. That viscosity should correspond to a quantitative value of 90 centipoise at 25°C.

Now the milk is quickly cooled to 7°C. and then injected into a water solution of calcium chloride through a spinning nozzle. The spinning nozzle has 200 holes with a diameter of 0.2 mm. The aqueous calcium chloride solution should contain about 500 mg. calcium chloride per liter and have a temperature of 35°C., and the water solution also contains a culture of acidophilic bacteria of the type *Bacillus casei*. The fibers or threads thus formed are separated from the solution by means of straining, and they are then washed in double their volume of water and pressed into a cheese mold, in which the resulting cheese mass is left to ripen in the ordinary manner.

When heating the viscous mass by injecting it in the shape of threads or lamella-shaped jets into a bath at a temperature of 25° to 45°C., it is particularly advantageous to divide the viscous flow also perpendicularly to the direction of the flow in addition to its being divided into parallel flows in the spinning nozzle. This should be done immediately when the viscous mass is injected into the bath. The division perpendicularly to the direction of flow can easily be effected by any suitable means, such as a rotating knife mounted close to the spinning nozzle at its outlet side.

In order to obtain the absolutely maximal effect as regards heat exchange and washing action, when heating by injection into a bath at 25° to 45°C., the bath should be vigorously stirred during the injection. Constant stirring is preferably performed in a way such that the pieces or curd are entrained into a circular movement that will eventually lead to a gathering and separation of the curd particles from the bath.

The curd fibers obtained at the moment when the viscous milk is pressed through the spinning nozzle, or the cut pieces of fibers obtained when dividing the flow perpendicularly as well, provide the curd with an extremely suitable structure, because they make it easy to extract quickly and effectively the unwanted constituents of the curd by means of the water solution into which the fibers are "spun." This is largely due to the fact that the so-called syneresis takes place extremely quickly and effectively after the curd formation when done according to this process.

In addition, the ion exchanges and the inoculation of the bacterial culture that is intended to give the cheese its taste and final aroma are greatly facilitated. The same applies to all the subsequent operations that are traditionally performed, whether during or after the separation of the curd from the whey, and this is by virtue of the fibrous structure of the curd.

The ripening of the cheese made from a curd obtained by this process is facilitated and quickly results in a product of highest quality. The curd is very suitable for forming, and the fibers or pieces of fibers, which correspond to the conventionally obtained grains of curd, join well together and make a perfectly homogeneous mass. The ability of the curd to keep the wanted constituents of the milk, particularly the fatty constituents, is considerably greater than in the case of curd obtained by the continuous processes of cheese-making. The method is well suited for introduction of automation, and this is particularly so due to the fact that the transformation of the casein into the insoluble form, during the first state of the nonenzymatic phase of the renneting, takes place in a homogeneous, flowing mass without polymerized micelles, i.e., without separation of whey. Therefore, the viscous mass which is first obtained can easily be cooled continuously by means of a heat exchanger of ordinary type.

The viscous mass obtained when the casein transforms into the insoluble state can be kept for an unlimited time by means of simple cooling; and this, of course, will also greatly simplify the process. The curd obtained by this method is sufficiently firm, after the formation of the fibers, to be subjected to all kinds of mechanical operations which might be desired, such as drying by means of centrifuging, washing, etc.; and these operations can be carried out without any special precautionary measures.

Concentrated Milk and Acidified Serum

M.P. Goujard; U.S. Patent 3,698,918; October 17, 1972; assigned to Compagnie Gervais Danone, France describes a process for the manufacture of milk curd by mixing previously acidified lactic whey with concentrated milk, the acidity of the whey being between 0.5 and 2.5 and the pH of the mixture being between 3 to 5. The proportions of dry extract of the milk and the conditions of separation of the curd are so selected as to recover substantially as much of the whey as was used to produce curdling of the milk.

According to this process for manufacturing milk curd by mixing milk with an acidified serum derived from a previous curdling operation, the acidification of the serum is performed by means which do not cause degradation of the lactose. Such means may be chemical, and include the addition to the serum of a suitable mineral or organic acid.

The means may also be physio-chemical, the acidification of the serum then being obtained by electrodialysis. Advantageously the acidification of the serum is continued until the pH of the serum falls to a value of 0.5 to 2.5, preferably from 0.9 to 2, these conditions being variable according to the nature of the acidification used.

The serum is then mixed with the milk to bring the pH of the mixture to a value between 3 and 5. It has been observed in fact that at a pH above 5.2 the curdling no longer takes place, and that at pH values lower than 2.8 the curdling results in very fine particles which are difficult to separate from the serum. The figures which have just been indicated correspond to an addition to the milk of about 60 to 20% of serum by volume. The operation is best performed at ambient temperature.

When the acidification is performed by chemical means, mineral acids may be employed, preferably acids such as hydrochloric acid and orthophosphoric acid, with which there is no risk of harming the quality of the product. However, other acids may be employed provided that suitable precautions are taken. Among the organic acids, there may be used acetic acid, citric acid and, preferably, lactic acid, the choice of these acids being determined by their dissociation constants, their organoleptic character and especially their nontoxic character.

So far as the carrying out of the process by electrodialysis is concerned, it may be carried out in any device of known type which directs the electrodialysis towards an acidification of the serum. Preferably, the initial serum is prepared by an acidifying electrodialysis of milk and separation of the curd obtained. Later, the process may be carried out continuously by acidifying a part of the serum collected after separation of the curd and mixing it with fresh milk.

When the process is carried out by electrodialysis, it is preferable to use selective anionic and cationic membranes and to dispose these membranes judiciously so as to achieve acidification of the serum without impoverishment of cations.

According to a particularly advantageous embodiment, the serum to be acidified is placed between two selective membranes of opposite signs disposed so as to retain the ions in the serum, the acidification resulting from the combination of hydrogen ions and the anions yielded by the solutions adjacent to the membranes. The solution generating the hydrogen ions is an acid solution, preferably a strong acid.

As the source of anions, preferably there is used a solution of the same nature as the serum to be acidified, with the object of not introducing foreign substances. This solution is advantageously constituted by a milk product, preferably surplus serum derived from a previous curdling operation.

This process lends itself to continuous operation without the hazards associated with curdling by means of rennet or by fermentation. The acidities both of the serum as well as of the serum-milk mixture may be easily controlled and reproduced. The weak pH value of the acidified serum ensures that the latter is not subject to any fermentative process.

As regards the curd, it is produced with a very fine, nongranular structure and it possesses excellent keeping qualities. In addition, the curd is obtained in yields which may be 50% better than those of other processes. This may be explained by the fact that the serum, according to this process, is strongly retained by the molecular products formed at the time of curdling.

It is thus possible to obtain yields of curd of about 350 to 600 g. per liter of milk treated with more than 100 g. of dry substance. These yields depend, to a certain extent, on the conditions of the centrifuging. Accelerations of from 250 to 1,000 g. and centrifuging times of from 1 to 10 minutes give excellent results.

In this process for the manufacture of milk curd by mixing with previously acidified lactic whey, concentrated milk is used and the acidity of the whey, the pH value to which the milk is brought, the proportion of dry extract of the latter and the conditions of separation of the curd are determined in such a manner as to recover substantially as much of the whey as was used, to produce curdling of the milk.

Thus, the various parameters involved are the following: (a) the degree of concentration of the milk represented by its dry extract percentage; (b) the degree of acidity of the whey, indicated by its pH value; (c) the pH value of the mixture of milk and acidified whey; (d) the conditions under which the curd and the whey produced are separated, indicated by the dry extract percentage of the curd. The process results in a considerable increase in the yield of milk curd, which increase is all the more as the proportion of dry extract of the milk increases.

The concentration does not have any appreciable effect on the quality of the product obtained. It results, however, in important advantages which show themselves, on the one hand, in a reduction in the quantity of acidified whey required to coagulate the milk and thus in a reduction in the volumes of liquid in circulation.

On the other hand, by suitably choosing the various factors entering into consideration, namely the acidity of the whey, the pH to which the acidified milk is brought, the conditions of separation of the curd and the proportion of dry extract of the latter, it is possible to determine a proportion of dry extract of the milk for which the quantity of whey recovered after separation of the curd is equal to the quantity of acidified whey employed.

This equality is very important when carrying out the process continuously, since it suffices to acidify the recovered whey and to return it in contact with a quantity of milk equal to that from which it derives in order to suppress the constraints which cause excesses of whey.

By employing a whey acidified to pH 2 to acidify the milk to a pH of the order of 4.35 to 4.45, then by separating the curd from the whey by centrifuging, the proportion of dry extract for which equality is achieved between the volumes of whey employed and whey recovered is of the order of 25%.

Thus, in one aspect the process comprises acidifying whey to a pH of about 2, concentrating milk to a dry extract content of at least 20% by weight, adding the acidified whey to the concentrated milk until the pH of the mixture thereof is from 4.35 to 4.45 and then separating the curd from the whey by centrifuging.

The curd and whey produced are preferably separated by centrifugation until the dry extract percentage of the curd is 15 to 18 weight percent for skimmed milk and from 19 to 22 weight percent for whole milk to obtain the best results. This gives a dry extract percentage of 22 to 25 weight percent for the concentrated milk, either skimmed or whole milk.

The quality of the products obtained and the yield of the process can also be improved by submitting the milk, before the application of the process, to a sterilization, which advantageously is combined with a homogenization when it is a question of whole milk. The sterilization of the milk, before its coagulation by means of acidified whey, has for its result a coagulation of the casein into very fine particles. This gives a curd of a new type having a texture which is very smooth and a consistency similar to that of a jelly. Moreover, this curd does not exhibit any syneresis.

The decanting of the acidified milk is slower, but this can be easily remedied by increasing either the duration of the centrifuging, or the speed of this operation. On the other hand, the yield of curd is higher. The homogenization permits the curd to be separated from the whey while avoiding migration of fatty globules outside the mass of curd.

Example 1: To skimmed milk concentrated to a dry extract content of 27.4% by weight there is added a whey acidified by electrodialysis to a pH value of 2 to give to the mixture a pH value of 4.40. For each liter of skimmed milk used, 3.870 l. of whey so acidified were added. Per liter of concentrated skimmed milk there is recovered 1,340 g. of curd having a dry extract content of 15%, which corresponds to 465 g. of curd having a dry extract content of 15% per liter of nonconcentrated skimmed milk having a dry extract content of 9.5%. The increase in yield compared with processes employing rennet is of the order of 40%. Centrifuging the above mixture at 2,000 rpm for one minute produced 1,130 g. of curd having a dry extract content of 17.8%; 3.760 l. of whey being recovered. This corresponds to 1,340 g. of curd with said dry extract content of 15% which can be obtained by shortening the centrifugation time, 3.550 l. of whey being recovered.

Thus, in order to obtain equality of the quantities of acidified whey and recovered whey, the dry extract of the milk should have been chosen at about 25% for the curd at 17.8% and about 22% for the curd at 15%. Without concentration, a liter of skimmed milk having a dry extract content of 9.5% mixed with 1.220 l. of whey acidified to a pH value of 2 would have given 350 g. of curd having a dry extract content of 15% with 1.900 l. of recovered whey.

Example 2: The starting material is skimmed milk having a dry extract content of 9.5%, which has been sterilized in an autoclave at 120°C. during a time of from 5 to 20 minutes. To each liter of this milk is added 1.220 l. of whey coming from a previous curdling operation and previously acidified to a pH value of 2 by electrodialysis. The whey is added in a quantity such that the pH value of the mixture is 4.4. The curd is separated by centrifuging at normal temperature.

There is obtained per liter of milk, 370 g. of curd having a dry extract content of 15%, 1.880 l. of whey being recovered. This represents an increase in yield of 15% compared with processes employing rennet. As has been indicated in Example 1, skimmed milk having a dry extract content of 9.5% but not sterilized gives per liter, 350 g. of curd having a dry extract content of 15%. The influence of sterilization on the amount of product obtained is not very important but the quality of the curd is better, the curd having a creamy aspect.

Example 3: The process is as described in Example 1, but starting this time with homogenized whole milk having a dry extract content of 12.5%, the fatty globules of which have a size less than 1 micron, and sterilized at 120°C. during 5 to 20 minutes. There is obtained, per liter of milk, 480 g. of curd having a dry extract content of 19%. This curd is homogeneous and of fine structure. This represents an increase in yield of 35 to 50% compared with known processes.

Whole milk having a dry extract content of 12.5%, but neither homogenized nor sterilized, gives per liter, 450 g. of coarser curd having a dry extract content of 19% after reincorporation, by heating, of the fatty material which had migrated in the whey. For 1 l. of homogenized whole milk at 12.5%, 1.170 l. of acidified whey were used and 1.710 l. of whey were recovered. With nonhomogenized whole milk, 1.150 l. of acidified whey were used and 1.720 l. of whey were removed. The following examples illustrate the combined application of concentration, sterilization, and if necessary, homogenization.

Example 4: Proceeding as described in the foregoing examples, 1 l. of skimmed milk concentrated to a dry extract content of 29.5% and sterilized gives 1,510 g. of curd having a dry extract content of 15%. This corresponds to 485 g. of curd having a dry extract content of 15% for 1 l. of skimmed milk having the usual proportion of 9.5% dry extract. The increase of yield is about 50% compared with known processes. The yield is greater than that given by simple concentration

(Example 1) or by simple sterilization (Example 2). To each liter of sterilized skimmed milk concentrated at 29.5%, 3.650 l. of whey acidified at pH 2 were added. 1,320 g. of curd at 17.2% were recovered together with 3.350 l. of whey. The curd has also a creamy aspect. This corresponds to 1,510 g. of curd at 15% and 3.160 l. of whey.

Example 5: Proceeding in the same way but starting with 1 l. of whole milk concentrated to a dry extract content of 28% and homogenized and sterilized, 1,350 g. of curd having a dry extract content of 19% is obtained. (2.510 l. of whey were added and 2.180 l. of whey were recovered.) This corresponds to 600 g. of curd having a dry extract content of 19% per liter of whole milk having the usual proportion of 12.5% dry extract. The improvement compared with Example 3 is very apparent.

AGING PROCESSES

Aging in Consumer Size Pieces

C.J. Berst; U.S. Patent 3,060,035; October 23, 1962; assigned to Armour and Company finds that green cheese or cheese curd can be packaged in consumer wrappers in small consumer size pieces prior to curing and curing will nevertheless proceed even if the wrapper is designed or treated to inhibit surface mold growth.

Using conventional hoops of rectangular, circular or other cross-section, the cheese curd is packed to make blocks of green cheese, which have been removed from the hoops. Instead of having a curing wrap applied at this point they will immediately be subdivided to produce any appropriate size and shape of consumer size piece such as the conventional wedge-shaped piece.

Each piece is individually wrapped in a mold inhibiting wrapping film. A variety of such films may be used. Under certain conditions of use in which the film is sealed about the surface of the consumer size cut, the film may be saran or cellophane to which metal foil has been laminated, with or without a coating of wax, or it may comprise a cellophane film coated on both sides with saran by extrusion or lamination. In some instances, such a composite film is further laminated or coated with polyethylene before being coated with wax.

Sorbic acid and propionates are two mold inhibitors used in the process, desirably not within the cheese but either incorporated in the wax coating on the inner face of the wrapper or added on the surface of the wax coating of the wrapper. Since the cheese has not yet been cured at the time the piece is wrapped, the wrapper is not ordinarily tightly sealed but may be left sufficiently open to permit the escape of the gas developed during curing. The mold inhibitor on the wrapper will contact only the surface of the piece of cheese and will not inhibit curing but will prevent mold growth on the surface.

If the wrapper is unsealed, it is particularly desirable that the wrapper be sufficiently large to have an ample lap of all margins so that any air moving into the wrapped package in response to atmospheric changes will have to pass for substantial distances in contact with sorbic acid or other mold inhibitor. A wrapping film such as Du Pont K202 which consists of cellophane with saran laminated to both faces by extrusion is coated with wax and sorbic acid is then coated on the wax surface on the side next to the cheese. The packed pieces are now placed in the merchandising box and in those wraps and containers in which they will ultimately be sold; the cheese is stacked in warehouses of proper temperature for curing for proper periods until desired aging is achieved.

When the wrapper is not sealed, only an occasional pieces will show surface growth mold and this occurs only seldom and is not sufficient to be objectionable from a merchandising standpoint. Of course, sealing would exclude the air entirely, and the spores carried by the air.

However, it is advantageous to permit the gas to escape from within the wrappers and accordingly the wrappers are desirably left unsealed, although in instances where sealing was practiced, the gas accumulations within the wrapper were not so large, ordinarily, as to be objectionable. It is only desired to note in this connection that, if curing takes place in a wrapper which has the necessary lap of surfaces treated with sorbic acid or the like, sealing is not essential.

Sealing sometimes occurs naturally because of the oily surface of the cheese. In packages sealed either purposely or intentionally, a brief exposure to vacuum during or following the curing period will withdraw the surplus gas. An ensuing steam treatment will reseal the wrapper.

In any event, the wrappers may be, and usually are, sealed prior to marketing the cheese. The sealing may be accomplished simply by natural oiling of the cheese causing the wrapping to adhere to the cheese itself as the result of the steam treatment. Alternatively, the wrapper may be heat sealed in a conventional way by causing the polyethylene material of lapping plies to adhere.

Citric Acid Addition and Vacuum Treatment

R.G. Wagner; U.S. Patent 3,105,764; October 1, 1963 describes a process for making natural aged cheddar cheese in which a small amount of citric acid or an inorganic salt of citric acid, such as sodium citrate, is added to the raw, heat treated or pasteurized milk along with the bacteria culture before the addition of the rennet. During the cheese making process and/or during the curing process, it is believed that the citric acid fermenters act upon the citrates in the milk to produce volatile acids which increase the flavor and aroma of the cheese.

The citric acid or salt is employed in the range of 0.00005% to 0.20% by weight of the milk. The bacteria culture used is a common type containing lactic acid producing bacteria and also containing citric acid fermenters. The citric acid fermenters, are *Streptococcus citrovorous* and/or *Streptococcus paracitrovorous*. It is thought that the citric acid fermenters act upon the citric acid to form volatile acids which contribute greatly to the flavor and aroma of the cheese. In addition, other changes which have not been clearly identified are also believed to occur in the cheese. The mixture of the milk, bacteria culture and the citric acid is processed by conventional cheese making methods.

According to the process, a vacuum treatment is used during the aging of the cheese which substantially reduces the aging period and produces improved body and texture for the cheese. In the aging process the cheese is initially maintained at about

45°F., for one-half month to 5 months, with a period of 2 months being preferred. After this initial aging period, the cheese is warmed to about 65°F. The cheese is maintained at this warm temperature for at least four days until the cheese is pliable and the gas production has substantially ceased. Generally this requires a period of 10 days to two weeks. The warming teatment serves to soften the curd and expand the gases so that they can be more readily expelled. In addition, the warmer temperature serves to accelerate the bacteria and enzyme action so that the bacteria has substantially "worked itself out" after the completion of the warming treatment.

The cheese is then transferred to a vacuum chamber and is generally at a temperature of 55°F. to 70°F. at this time. In the vacuum chamber, a vacuum is drawn on the cheese at a rate of less than 3 inches of mercury per minute until a vacuum of about 20 inches of mercury is obtained. Considering atmospheric pressure to be 29.92 inches of mercury at sea level, a vacuum of 20 inches of mercury would be equal to 7.92 inches of mercury absolute pressure or 3.87 psi. If this vacuum is drawn on the cheese at too fast a rate, the voids in the outer portion of the cheese will be sealed so that the gas in the inner portion of the cheese will not be able to escape.

After this initial vacuum period, the vacuum is preferably increased to approximately 27 to 28 1/2 inches of mercury (0.69 to 1.43 psi) for a period of at least 10 minutes with about 13 minutes being the usual procedure. This increased vacuum serves to accelerate the withdrawal of any gas remaining in the cheese.

On completion of the vacuum treatment, the air pressure on the cheese is restored to atmospheric at a very rapid rate so that the closed voids or slits in the cheese will not have the opportunity to expand. Generally, the pressure is brought up to atmospheric in a period of less than 10 seconds and generally in a period of two or three seconds.

The cheese is then returned to the aging chamber and aged at a temperature of 32° to 55°F. for an additional period of time which depends primarily upon the desired characteristics of the cheese, such as whether it is to be sharp or mild cheddar cheese.

The addition of the citric acid or other citric acid salt in combination with the vacuum treatment produces a cheese which has an improved flavor blend and the cheddar flavor is accentuated. The treatment substantially reduces the overall aging process and with this treatment, aging of six months provides a cheese having a flavor and texture comparable to that produced by aging for one year under other processes.

Aqueous Fluid System for Rapid Aging

P.P. Noznick, R.H. Bundus and A.J. Luksas; U.S. Patent 3,446,627; May 27, 1969; assigned to Beatrice Foods Company have found that they could accelerate the ripening process in the manufacture of cheese by ripening the cheese under conditions which comprise constantly agitating the substrate under growth conditions favorable for the organism. The cheese is ripened in a period of 2 to 10 days rather than 6 weeks to one year. The process is used to make Cheddar cheese, Swiss cheese, Neufchatel, brick, Limburger, Camembert, Edam, Muenster, provolone, Romano, Parmesan, Gouda or Liederkranz.

Any suitable protein-carbohydrate system with or without fat can be used as the growth medium. Such a system is provided by whole milk, skim milk, cream, or a mixture of vegetable proteins and a sugar with or without vegetable fat. Preferably simple hexose sugars such as glucose, galactose, lactose, fructose, or sorbose, i.e., sugars which will not form sucrose in the process are used.

The sugar selected, of course, must be one on which the bacteria, fungi or yeast can act. The vegetable protein can be soy protein or peanut protein. The glucose or other sugar must be employed in an amount sufficient to sustain the bacterial action. The system is sterilized, usually at 9 to 50% solids (in water) prior to adding the desired culture. It is also desirable to maintain a proper salt balance favorable to the growth of the organism.

After sterilization the desired culture, either bacteriological or mycological, are introduced with or without enzymes. Optimum growth conditions for the particular organism are then maintained. These conditions may be aerobic or anaerobic and can vary widely in temperature depending on the organism. Thus some cultures grow best at 124°F. (thermophilic organisms), others at 98°F. and still others at 71° or 35°F.

The conditions are chosen to favor the desired organism. The use of anaerobic or aerobic conditions or a condition between the two can be selected depending on the oxygen demand of the organism. The liquid system after flavor development, is formed into plastic consistency by the removal of moisture or is completely dehydrated by spray drying, roller drying or freeze drying. It is critical to agitate during the growth of the cultured organism on the substrate.

Example: 30 day old Cheddar cheese was macerated and fluidized with water to 30% solids. It was then introduced to a sterile incubation container and maintained under anaerobic conditions (a nitrogen atmosphere) at 85°F. and incubated for 2 days with constant agitation. This promoted the growth and reproduction of the anaerobic bacteria commonly found in Cheddar cheese and materially hastened the ripening process.

The thus cultured product was spray dried to give a highly flavored Cheddar cheese powder equivalent in flavor to a Cheddar cheese which has been aged 12 months. The procedure is based on a fluid ripening system. The procedure also requires vigorous agitation during the ripening process and only requires a relatively short period of time, e.g., 10 days or less.

MISCELLANEOUS PROCESSES

Recycle of Wash Effluent to Lower BOD

R.S. Timmins, B.S. Horton and R.L. Goldsmith; U.S. Patent 3,707,770; January 2, 1973; assigned to Abcor, Inc. describe a method which improves the management of water in a total cheese-manufacturing process, and which reduces the difficulties associated with water pollution due to high BOD waste effluent streams from a whey solution. The process involves the efficient and economical utilization of effluent water derived from a semipermeable membrane process, the use of this water to wash or rinse cheese curd and the recycling of this water back into the semipermeable membrane process after such use, thus closing the loop in a pollution cycle where a cheese solid is prepared and the waste whey solution is treated by membrane techniques. The process is described in detail in connection with the manufacture of a cottage cheese solid. However, it may also be used in any cheese manufacturing operation which requires the curd product to employ low BOD water or which incorporates a rinse or wash stage rather than merely a pressing of the curds.

The process is illustrated in conjunction with the flow sheet in Figure 4.1. Skim milk is employed to form a cheese curd in conventional fashion, producing a solid-curd product and a whey solution. The whey solution is directed for concentration of the protein fraction into an ultrafiltration section; for example, in an ultrafiltration section which is composed of cellulose acetate formed in a thin film directly on the inside of a plurality of porous polyolefin support tubes to provide a seamless membrane tube.

FIGURE 3.2: RECYCLE OF WASH EFFLUENT IN CHEESE PRODUCTION PROCESS

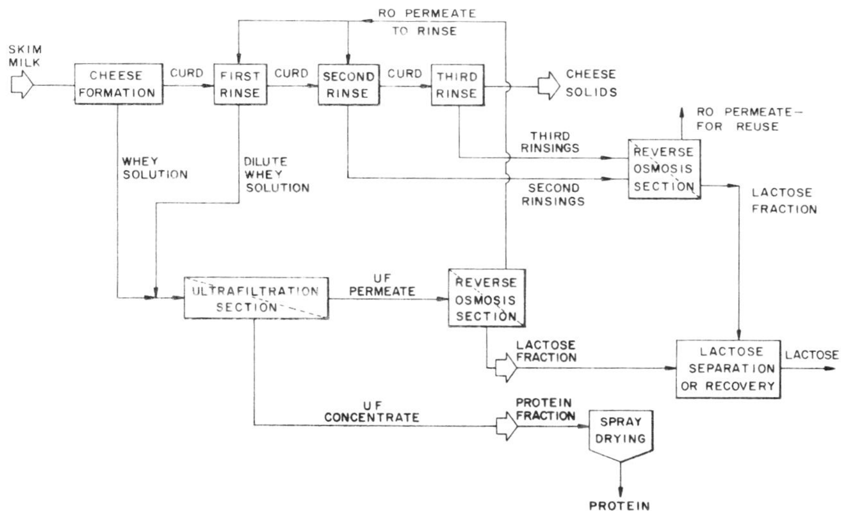

Source: R.S. Timmins, B.S. Horton and R.L. Goldsmith; U.S. Patent 3,707,770; January 2, 1973

The membrane is designed to provide over a 98% retention of an alpha lactalbumin and other high molecular weight soluble milk proteins. The ultrafiltration is operated at a pressure of 20 to 50 psi at a temperature of about 100° to 140°F. The concentrate of the ultrafiltration section comprises a protein fraction which may then be dried by evaporation, such as by a spray-drying technique to produce a solid protein fraction having unique lactose-free properties. The permeate from the ultrafiltration section is introduced into a reverse osmosis section operated at a pressure of 600 to

1,000 psi, at a temperature of 70° to 100°F. The reverse osmosis section may have a plurality of high-pressure tubes; that is, cellulose acetate membranes cast directly inside an epoxy-resin-reinforced braided glass fiber tube. The membranes are tailored to provide a retention of the lactose nonprotein nitrogen acids having a molecular weight over 50 or 100.

The concentrate from the reverse osmosis section comprises a lactose fraction which may be subsequently treated to remove the individual components of the fraction and the lactose present crystallized and rewashed to a high grade for use or the fraction recovered by drying without further separation. The permeate produced in the reverse osmosis section comprises a very low BOD, for example, less than 1,000 ppm, and very minor amounts of residual organic lactic acid and other products which may have been passed through the reverse osmosis membrane.

After formation of the cottage cheese curd, it is subject to a series of rinses as washings illustrated as a first rinse, second rinse and third rinse stages. The reverse osmosis low BOD permeate is recycled from the reverse osmosis section to the first rinse stage and used to rinse the curd. Typically, such a reverse osmosis permeate stream would have a composition of about 3% or less solids. The effluent from this first rinse stage is then added to the whey solution being introduced into the first stage ultrafiltration section.

The reverse osmosis permeate is also employed in a second rinse stage; however, due to the very low amount of solids in this stage, rather than adding the wash effluents from the second rinse stage back into the whey solution, the second rinse stage is discarded or otherwise treated. In the third rinse stage, as illustrated, regular wash water may be employed.

Where desired, the wash effluents from the second or third or subsequent rinse stages may then be subject to a separate reverse osmosis step or other treatment in order to provide a very low BOD water effluent stream which then may be employed within the plant, for example, as boiler water, additional rinsing, or other plant use, or if desired, discharged without significant water-pollution problems.

The process permits a very efficient and economical combination of a reverse osmosis treatment of a whey solution with the preparation of a solid cheese, while minimizing water-pollution problems and maximizing total water management within the cheese-manufacturing plant.

Reducing Moisture Content of Cheese

E.H. Cornwell and H.G. Foster, Jr.; U.S. Patent 3,482,999; December 9, 1969; assigned to Swift & Company describe a method for drying high-moisture cheese without causing the loss of normal fat content or case-hardening of the cheese surfaces.

Moist cheese is cut into chunks large enough that excessive mechanical working is avoided during the cutting process and small enough to present an adequate surface area during the subsequent drying process. After cutting the cheese into units of the proper size, it is dried in circulating air by heating it to a temperature which is not sufficiently high to produce the undesirable effects of exudation of the fat and case-hardening of the cheese surfaces. These deleterious effects may be entirely avoided by drying the cheese by a process which avoids excessive heating and mechanical working. Cutting the cheese into chunks having a length, width, and height, each measuring from about 1/4 to 4 inches, produces the best results. After the cheese mass has been cut into chunks of the proper size, it is ready for the actual drying operation. Efficient drying of large quantities of cheese requires agitation of the cheese together with heating. It is also important that circulating atmosphere be provided in order to carry away the moisture which evaporates.

Just as the avoidance of mechanical working is important in the cutting operation, it is also important in the drying operation. Thus, gentle tumbling is the preferred method of agitation. Tumbling will produce no appreciable mechanical agitation if the tumbling speed is maintained below about 200 inches per minute, expressed as the linear velocity of the cheese in relation to the dryer surface. There is no lower limit on the tumbling speed, as long as it is adequate to expose all of the surfaces of the cheese chunks to the atmosphere. The drying operation is carried out in a drum dryer because of the relative simplicity and ease with which the process may be carried out continuously in such equipment.

The cheese particles should also be heated and subjected to a current of warm, moving air. It is important that the temperature of the cheese and the surrounding air be kept relatively low since, when cheese is heated to a high temperature, it tends to become soft and pliable, and is thus more susceptible to mechanical working during agitation. Secondly, high temperatures tend to cause an undesirable tempering or case-hardening of the cheese surfaces, which prevents the efficient removal of moisture and harms the smooth texture of the finished product.

If the temperature of the cheese is not allowed to rise above about 95°F., while the temperature of the surrounding air is maintained at temperatures up to about 100°F., case-hardening and oil separation from excessive heating will be substantially avoided. Heating the cheese may be produced by any method which will not add moisture to the cheese. While the usual method involves heating the cheese with an atmosphere of warm air, other dry heat sources such as infrared lamps may also be employed.

Although overheating is the principal danger, the cheese should not be permitted to become too cold to permit efficient evaporation of the moisture. The minimum temperature is about 40°F. The humidity of the circulating air is not critical, except, it must be below the saturation level.

After drying to the desired moisture level, the product may be marketed or may be blended with dried cheese to produce a top quality cheese product of the proper moisture content. The dried product may also be converted into a high-quality process cheese by steam-injection cooking.

Example: Cheddar cheese with a moisture content of 41% was diced into 3/8 inch cubes and placed in a Link Belt Model 207-10 Roto-Louvre drum dryer, which has a drum diameter of 2.7 inches. The dryer was rotated at 1 rpm, giving a tumbling speed of about 97 inches per minute. The cheese cubes were fed continuously into the dryer at a rate of 313 pounds per hour.

Air was circulated through the dryer at a rate of 1,470 cubic feet per minute at an inlet temperature of about 100°F., emerging at an exhaust temperature of about 72°F. The cheese was fed into one end of the dryer drum at an average temperature of 51°F., giving an average discharge temperature of 66°F., emerging from the discharge end of the drum after remaining in the dryer for 43 minutes. The dried cheese cubes had a dull surface with no oil discernible and showed no evidence of case-hardening. The moisture content of the product was 34.1% while there was no appreciable fat loss during the drying process, giving a product conforming to the standards of a high-quality Cheddar cheese.

SPECIFIC CHEESE PROCESSES

SOFT RIPENED

Preparation with Sucresters

L. Nobile; U.S. Patent 3,164,477; January 5, 1965; assigned to Ledoga SpA, Italy found that by incorporating a small proportion of one or more sucrose esters of one or more edible aliphatic acids of 6 to 30 carbon atoms, both saturated and unsaturated, in the milk prior to formation of a soft cheese, a product is obtained which is more stable and has good keeping qualities.

The sucrose esters can be the individual esters or mixture of esters (sucresters) prepared by transesterification of sucrose and a lower aliphatic alcohol ester of an edible aliphatic acid or acids of 6 to 30 carbon atoms, or they may be contained in the complex mixtures obtained by the transesterification of sucrose with a natural triglyceride (glycerol triester), such products being known as sucroglyceride.

Both sucroglycerides and sucresters have good dispersing and emulsifying properties. Nevertheless, upon the addition of sucresters and sucroglycerides to the milk during the preparation of soft cheeses, a higher precipitation of substances contained in the milk is obtained, and the cheeses ultimately obtained in the usual processes of cheese manufacture have improved properties.

For practical purposes, an aqueous solution or suspension of 0.1 to 5% of the selected sucrester or sucroglyceride is prepared and added to the milk without modifying the usual process for soft cheese production. Such an amount of the solution or suspension (or solution-suspension) is employed that the content of the sucresters or sucroglycerides corresponds to 0.05 to 1% of the milk.

The sucresters and sucroglycerides used are the edible saturated and unsaturated monobasic fatty acid esters of sucrose (sucresters), with or without the partial glyceridic esters or such acids, and are preferably prepared in dimethylformamide (DMF) as the solvent, the amount of DMF being less than 100 ppm of such solvent.

The process is performed with heating. The temperature to which the milk is heated usually depends on the quality of the cheese to be produced, the conditions of milk (acidity and fat content), the coagulation power of the rennet and on the time it should take to coagulate. As a rule the temperature is 28° to 30°C. for soft cheeses. When the correct temperature has been reached, the necessary quantity of rennet is added to form a homogeneous mixture (the rennet is diluted with water and milk serum); the mass is thoroughly mixed and left to stand, heating is stopped and the pan covered to avoid surface dispersion of the heat.

The serum tends to gradually separate during the coagulation from the curd and to collect in a surface layer. However, as the mass gradually increases in consistency, separation becomes increasingly difficult and a certain quantity of serum remains in the coagulated mass. Because of its lactose, lactic acid and enzyme content, besides the presence of other substances, the serum easily alters in composition and this is the reason for its elimination, although the latter must not be complete, as small amounts of serum serve to start the fermentation of the caseinous mass, a phenomenon which is also known as ageing of the cheese.

If the removal occurs too quickly, the serum entrains a given and by no means negligible quantity of casein, and the normal yield of cheese is lessened. By contrast, if the coagulum is left standing too long, the contraction causes it to become too thick and makes removal of the serum more difficult, as well as the successive reduction of the curd. If all of the serum is eliminated immediately, the remaining caseinous mass is too dry and forming, as well as ageing, does not always take place regularly.

The addition of 0.5 to 1% of sugar esters and sugar glycerides of the milk shows that the curd forms much more regularly and shows a clear breakage without foam. The serum is generally clear without any lumps of curd remaining attached and the yield in weight of the cheese is good.

AMERICAN

Short Cure

J.E. Murphy; U.S. Patent 3,175,915; March 30, 1965; assigned to National Dairy Products Corporation provides a process for manufacturing American cheese having the characteristic flavor of aged Cheddar cheese that does not require extended curing of the curd, and provides a product having uniform flavor and body characteristics which is particularly suited for use as a raw material in the manufacture of process cheese and process cheese products.

In general, the method includes forming a cheese curd from raw or pasteurized milk by the addition of suitable amounts of a lactic acid starter and rennet, cutting and cooking the curd, salting the curd after the acidity rises to above about 0.40, reducing the moisture content of the curd to between 31 and 35.5%, placing the curd in an airtight enclosure which substantially prevents the formation of mold, and curing the curd at 65° to 80°F. until the pH of the cheese is within the range of 5.0 to 5.40.

Raw or pasteurized milk containing from 3.0 to 3.6% of milk fat is introduced into a cheese vat maintained at 86°F. and is inoculated with the usual amount of a lactic acid starter. When a suitable acidity is reached, rennet is added to set the curd. When the curd has attained a proper degree of firmness and the whey has an acidity of 0.10 to 0.12, the curd is cut into cubes one-fourth to one-half inch. The cut curd is agitated in the cheese vat to prevent agglomeration and is cooked to expel whey. The cooking is carried out at 100° to 103°F., the temperature of the vat being raised to within this temperature range in about 30 minutes. The cooking and agitation of the curd is continued until the acidity of the whey reaches about 0.15 to 0.16, at which time the curd and free whey are pumped onto a drain table maintained at about 100°F.

The curd is continuously agitated while on the drain table to prevent agglomeration of the curd granules. Free whey is drained from the curd until the level of whey just covers the curd. The curd is then allowed to remain in the whey as the acidity builds up. When the acidity reaches about 0.28, the remainder of the free whey is drained off leaving a moist granular curd which is retained on the drain table after the whey has been drained, and the acidity is allowed to continue to build up. During this period, the curd is sufficiently agitated to prevent agglomeration and matting of the curd. The increase in the acidity of the curd after the free whey has been drained causes the curd granules to shrink and additional whey is expelled from the curd.

When the acidity of the whey reaches about 0.50, the curd is salted with a suitable amount of salt which causes the curd to shrink further, expelling a further amount of whey. The curd is agitated during salting to insure an even distribution of salt throughout the curd and to prevent agglomeration of the curd. The salt is added to the curd in an amount so that the curd has a total salt content, including that salt initially present in the milk, of 1.2 ±0.1%. However, the total salt content may deviate slightly from this preferred value since a portion of the salt will be dissolved in the whey and drained from the curd with the whey. The deviation of the salt content of the curd may occur due to variances in the moisture content and degree of shrinkage of the curd in different cheese makes, which in turn causes a greater or lesser amount of whey to be expelled from the curd.

The granular curd obtained after salting has a moisture content of approximately 39 to 44%. To further reduce the moisture content of the curd to within a preferable range of 31 to 36%, a further amount of whey is removed. One method of removing additional amounts of whey is drum draining of the curd. After being allowed to drain for a suitable length of time, e.g., overnight, the moisture content of the curd is reduced to between about 34.5 to 35.5%, an average moisture content for curd produced according to the drum draining method being 35% moisture.

Air is then substantially displaced from the interior of the curd-containing barrels or drums. The displacement of the air from within the drum is believed to be partially responsible for the improved results obtained by the process. If desired, air may be displaced from the drums by the introduction into the drums of a gas which will not sustain mold growth, e.g., nitrogen, carbon dioxide, or mixtures of these. The introduction of a nonmold growth-sustaining gas further prevents the formation of mold and the production of off flavors by insuring that substantially all of the oxidizing gases are removed from the curd. It has been found that the nonmold growth-sustaining gas should be introduced into the drums in an amount so that any remaining oxygen in the drum will be consumed by the bacteria in the cheese curd before any visible mold growth occurs. Generally, it is sufficient to reduce the air content in the drums to less than about 3%.

The granular curd is then cured in the drum or barrel by storage in a suitable storage room having a controllable temperature. During curing, the temperature of the cheese is maintained within the range of 72° to 74°F. until the pH of the curd is about 5.0 to 5.40. At the end of one week the pH of the curd is about 4.8 to 4.85. The curd is considered fully cured when the pH of the curd is 5.0 to 5.40, a pH within this range being obtained in about 35 to 60 days. When this procedure for making stirred curd is employed, highly desirable cheese curd is obtained when the pH of the curd is 5.0 to 5.1, which is generally obtained in about 6 weeks.

Using Heat-Treated Milk

The process described by M.J. Roberts; U.S. Patent 3,650,768; March 21, 1972; assigned to Kraftco Corporation concerns a method for making American cheese and similar types of cheese from heat-treated milk to provide increased cheese flavor. In accordance with the method, a combination of a particular type of a proteolytic micrococcus and a particular type of lipase are added at specified levels to heat-treated milk which is to be made into American cheese. A particular lactobacillus microorganism may also be used in connection with the combination of the indicated proteolytic micrococcus and a specified lipase to provide improved body, flavor and/or texture to the cheese.

Example 1: 25,000 pounds of raw milk was heat-treated at a temperature of 162°F. for a period of 16 seconds to effect pasteurization. The milk was pumped to a vat and allowed to cool to 88°F. To the cooled, heat-treated milk was added 50 pounds of a milk starter culture including a proteolytic micrococcus. The culture contained 1.75×10^8 cells per gram of a proteolytic micrococcus. The proteolytic micrococcus culture was one which had been obtained from the University of Wisconsin, designated as T-3. To the milk was also added 6.25 grams of a lipase which had been obtained by extraction from the throat tissue of a calf. After addition of the above culture and the lipase enzyme, 250 pounds of a *Streptococcus lactis* starter culture was also added to the milk.

The milk was agitated and allowed to ripen for a period of one hour. 75 ounces of a rennet solution were then mixed into the ripened milk, and the stirring was stopped. After about 30 minutes the curd was sufficiently firm to cut and was cut with curd knives into 1/4" cubes. Stirring was then commenced and was continued until the whey was drained. Draining was completed in about 30 minutes, and thereafter the curd was packed in a layer about 7 or 8 inches deep on each side. When the curd was firm enough to be turned without breaking, it was cheddared. The curd was then run through a curd mill and was spread evenly over the bottom of a vat and stirred. 3 pounds of salt for each 1,000 pounds of milk were then mixed in and the curd was piled on each side of the vat while additional whey was drained.

After salting, the curd was transferred to cloth-lined metal hoops, and was pressed, dried and wrapped in a plastic film. After wrapping, the cheese was stored in a warehouse at a temperature of 45°F. to effect curing. The cheese was cured for a period of 6 months. At intervals of 2 months, a sample of the cheese was removed and inspected. The samples of cheese prepared with this procedure were compared with samples of cheese which had been prepared in a similar manner from the same milk with the exception that the combination of a proteolytic micrococcus and a lipase enzyme was not added to the heat-treated milk prior to setting. In all cases, the samples of cheese produced by this method were scored considerably higher in flavor than the control samples. The body and texture of the cheese produced by the above method was acceptable.

Example 2: Heat-treated milk was prepared for use in making Cheddar cheese in accordance with the procedure of Example 1, with the exception that 50 pounds of a liquid milk culture of *Lactobacillus lactis* was also added to the heat-treated milk prior to setting of the milk. The *L. lactis* milk culture was prepared by introducing a stock culture into skim milk and permitting growth to proceed until an acidity of 1.6% equivalent lactic acid was reached. The *L. lactis* stock culture was a strain obtained from the USDA and designated Lb7 (ATCC 7995). Cheddar cheese was then produced as in Example 1. Samples of Cheddar cheese produced were compared with Cheddar cheese which had been prepared in a similar manner from the same heat-treated milk but without the addition of the micrococcus, *L. lactis*, or the enzyme of this process.

In all cases, the samples of cheese produced by the above method were scored considerably higher in flavor than the control samples and were scored equal or better in body and texture.

BLUE

Addition of Blue Cheese Mold

T.I. Hedrick, E. Kondrup, and W.T. Williamson; U.S. Patent 3,365,303; Jan. 23, 1968; assigned to Research Corporation describe an accelerated method for making blue cheese which involves inoculating the milk with the blue cheese-forming mold prior to curd formation and ripening the cheese as small pieces or chunks. Incorporation of these steps into the conventional procedure for the preparation of blue cheese results in a shortening of the usual three-month curing period to one of 6 to 14 days. Such a shortened ripening period minimizes slime growth during the ripening period.

Other advantages of making blue cheese according to this method are that it results in a more uniform distribution of the mold throughout the cheese giving rise to a more uniform product. Instead of hooping and pressing the cheese into wheels weighing 5 or 6 pounds as was previously done, the curd is maintained in small pieces which expedites salting and helps accelerate the curing. The process avoids the practice in which the wheels, which were approximately 7.5" in diameter and 6" high, were punctured to provide and maintain small openings in the cheese in an attempt to promote ripening of the cheese.

The method includes the following steps: adding a cheese starter containing a lactic acid-forming organism, a blue cheese-forming mold and rennet to milk; dividing the resultant curd into small pieces; separating the divided curd from the whey; and periodically salting and stirring the divided curd until ripening of the cheese is complete.

Regularly pasteurized and homogenized milk is usually used. However, raw milk,

milk subject to subpasteurization conditions of time and/or temperature or milk subject to minimal pasteurization times and/or temperatures may also be employed in the making of a satisfactory blue cheese. Moreover, the cheese itself may be pasteurized during the course of its processing or in the course of manufacturing cheese products such as cheese spreads, salad dressings, etc.

A suitable pasteurization procedure is one in which the milk is heated at 130°F., homogenized in the usual way at 750 to 3,500 psig and then held for 30 minutes at 143°F. In an alternate variation, only the cream is homogenized and the milk reconstituted to a butterfat content of about 3.5% prior to the completion of pasteurization. Such pasteurization conditions do not completely destroy the milk lipases and the final flavor of the product cheese is not adversely affected. While not necessary under these conditions, additional lipase could be added to the milk. Under conditions in which more of the milk lipase is destroyed, the addition of further lipase is desirable.

The preferred order of addition of blue cheese-making ingredients to the milk is as follows: cheese starter, decolorizer, cheese-forming mold and rennet. This order of addition is not critical, however it is necessary that the milk be inoculated with the mold culture prior to curd formation. It may often be convenient to add these ingredients to the milk at the same time and even in admixture with each other.

The starter, which is generally added first, may be any conventional cheese starter used in the preparation of blue cheese. Such a starter contains a culture of a lactic acid-forming microorganism usually of the genus Streptococcus. A mixture of *S. lactis* with *S. diacetylactis*, *S. cremoris* and *Leuconostoc cremoris* is preferred. A suitable starter solution is customarily prepared by incubating pasteurized skim milk with a culture of the starter organism. The use of a decolorizer, such as chlorophyll in the form of its dilute aqueous solution, while not essential to the preparation of a satisfactory cheese, is preferred.

The milk may be inoculated with the mold-forming organism in any conventional manner. However, it is most convenient to use a commercially available mold powder. The particular mold employed is *Penicillium roquefortii*. It should be noted that there are many strains of this microorganism employed in blue cheese-making. Any satisfactory rennet may be utilized. The commercially available rennet solution is further diluted with water prior to its addition to the milk. Rennet contains the enzyme primarily responsible for curd formation.

When the acidity of the whey reaches about 0.11 to 0.12%, the curd which has formed is cut into smaller pieces about 1/4 to 1 inch in size. The cut curd is then gently stirred with the whey for a period of time usually about 60 to 90 minutes. The curd is then drained in the usual way to separate the whey. Curing or ripening of the curd is effected under aerobic conditions in a curing room. Such a room should be at 50° to 75°F. and at a relative humidity about 95%. The room should be equipped with perforated shelves on which the cheese to be ripened is placed. Salt in the form of solid of as an aqueous brine is added periodically to the ripening cheese and the cheese is stirred periodically. It is critical to the success of the process that the curing cheese be stirred to prevent matting of the curd and to maintain it in a divided condition.

The pH of the curd varies somewhat during the curing period. It is preferred that the curd be at a pH between 5.7 and 5.9 in the latter stages of the ripening process. A satisfactory cheese will result as long as the pH remains between 5.5 and 6.1. If the pH falls too low, the flavor of the cheese will not be satisfactory; if the pH is too high, a longer curing period will be required. The addition of salt has a tendency to lower the pH of the ripening cheese and some care must be exercised in the manner in which the salt is added.

The fully ripened divided product may be formed into larger lumps by pressure with or without prior granulation. The product may be emulsified and processed to a powder for sale as such or for use in the manufacture of cheese spreads or salad dressings. The product or the powder may be mixed with milk or with skim milk and/or cream in consumer product formulation.

1,000 pounds of pasteurized and homogenized milk containing 3.6% of butterfat is brought to about 85°F. and about 1/2% by weight of starter solution is added to the milk. About one hour later, 35 ml. of chlorophyll decolorizer, 116 g. of active P. roquefortii mold powder and 60 ml. of diluted rennet solution are added to the milk with thorough mixing. When the acidity of the whey reaches 0.11 to 0.12%, the curd is cut into 1/4 to 1 inch cubes and stirred gently with the whey. About 90 minutes after cutting of the curd, it is drained to separate the whey.

The cut and drained curd is placed in a curing room held at about 62°F. and having a relative humidity of 95%. The curd is stirred 2 or 3 times the first 24 hours in the curing room and once or twice daily thereafter until ripening of the cheese is complete. About 2 pounds of salt is added in portions during the 10-day curing period, for example, one-fourth on the 4th day, one-fourth on the 5th day, and one-half on the 6th day. The fully ripened cheese is stored under refrigeration at about 32° to 45°F. to maintain its flavor and prevent spoilage.

A panel of experienced tasters rated the flavor and body of the cheese made by this method in comparison with control cheeses made by conventional methods. Results with various experimental batches showed the blue cheese made by the short cure method was as good as or better than the control blue cheese. When these cheese samples were incorporated into a salad dressing, the taste judges preferred those dressings made from the short cured blue cheese.

SWISS

Rapid Cooling

When Swiss cheese is made by any of the known processes, the body of cheese often has the characteristic known in the trade as "one-sidedness." A cheese which exhibits this characteristic develops eyes or cures more rapidly on one side than on the other side, and as a result one side of the block may have eyes that are either incompletely developed or overdeveloped by the time that the other side of the cheese has obtained proper eye development.

J.B. Stine; U.S. Patent 3,020,163; February 6, 1962; assigned to National Dairy Products Corporation has found that if the cooked curd at the time it is ready for whey separation is rapidly cooled to a temperature materially lower than the final cooking temperature, but which temperature is above that at which the curd absorbs enough water to substantially interfere with the curing of the cheese, the one-sided condition can be minimized or avoided. The Swiss cheese process generally involves a cooking temperature between 120° and 135°F. at the conclusion of the cooking and stirring steps of the process. In accordance with this process, the curd in whey is rapidly cooled from the cooking temperature to a temperature from about 110° to 90°F. With some curd it may be found that cooling to 90°F. will cause the absorption of too much water and that the curing characteristics of the cheese may be somewhat impaired. Any chance of impairment of the curing characteristics may be avoided by cooling to the preferred range of 110° to 100°F. Excellent results may be obtained for a curd if the cooling is carried out to reduce the temperature of the curd and whey mixture to about 105°F.

The rapid cooling is preferably accomplished by indirect heat transfer such as by the use of a cooling jacket around a pipe through which the curd and whey are pumped or the use of coils located within the hollow walls of the make-vat through which a coolant may be circulated. If the curd is cooled in the make-vat, the stirring of the mixture is preferably continued so that the entire volume of curd and whey is subjected to rapid cooling without subjecting any localized portion of the contents of the vat to too low a temperature. The cooling is preferably carried out as rapidly as possible but in any event it is desirable to accomplish the cooling in a period of less than about 15 minutes.

After the cooling step is completed the curd may be molded into the desired shape and cured to produce a curd having little or no tendency to cure into a cheese having a so-called one-sided condition.

PASTA FILATA

Mozzarella Mold

F.L. Langford; U.S. Patent 3,242,571; March 29, 1966; assigned to Western Foods Machinery Company describes a machine for the molding of Mozzarella-type cheeses into various configurations by the use of a vacuum or air pressure differential.

The apparatus has the advantages of rapid molding in a simple series of steps, while water or moisture is simultaneously drained and squeezed from the curd as it is readied for, and during the molding step. The machine may also be easily and rapidly cleaned. It is relatively simple in construction and operation so that it may readily be operated by unskilled labor, and kept in continuous operation with a minimum of "down time" for cleaning and/or servicing. Moreover, a series of molding elements may be simultaneously or sequentially operated to mold cheeses at a rapid rate. Further, provision is made for molding two different types of cheese curd at the same time.

The workings of the machinery are described in conjunction with Figure 4.1a through Figure 4.1o. To operate the apparatus, each of the cutter blades which serves to close a mold inlet (81) is placed in closed position, and the air switches (43) are moved into positions which block off passage of any air to the vacuum accumulator and vent the housings surrounding the actual mold shell units to the atmosphere. One of the switches (34) or (34a) is then thrown to start up a pump (32) or (32a) thereby evacuating substantially all of the air from the vacuum accumulator (22). Cheese curd in moist, hot plastic state may be poured into the two chambers of the hopper (20), preferably before the insertion of the containers (106), to allow surface water to be forced away from the fill neck passage (83), thereby avoiding the collection of water in the container (106).

In the case of each of the box-like molding elements (84), the door (100) is opened and an open top inner container (106) is inserted as far as possible into the rectangular box-like housing (98). Upon such insertion, the wheels (112), (112a) will ride up upon the ramps (115), (115a) and then on to the tracks (116), (116a). In this posture, the upper rim of the inner container (106) will be positioned against the underside of the cutter blade (128), and as this container (106) is pushed further rearward, the upper edge of its back wall (106b) will press the cross block (128b)

FIGURE 4.1: EQUIPMENT FOR MOLDING MOZZARELLA-TYPE CHEESES

a.

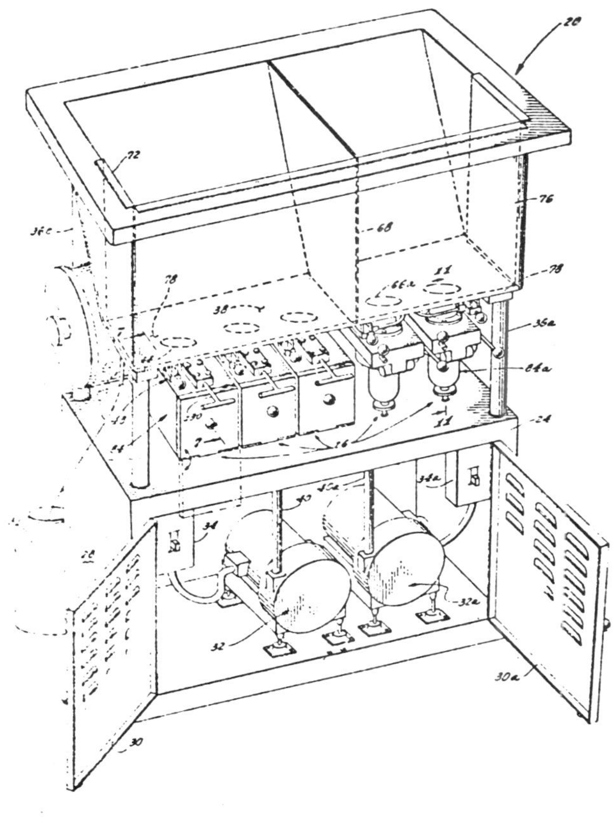

(continued)

FIGURE 4.1: (continued)

b.

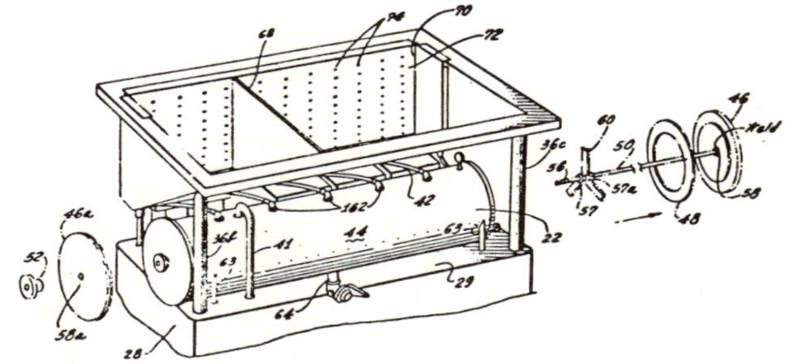

c.

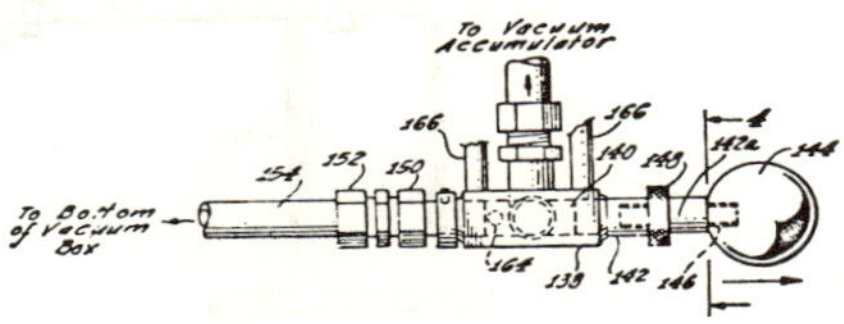

d.

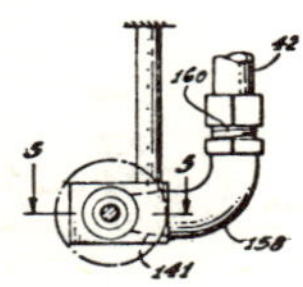

e.

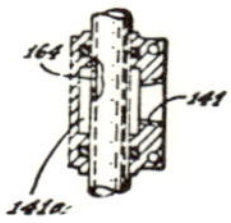

f.

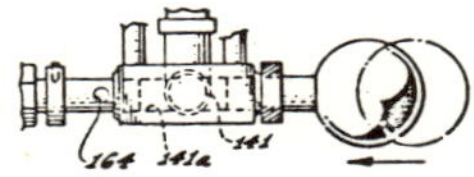

(continued)

FIGURE 4.1: (continued)

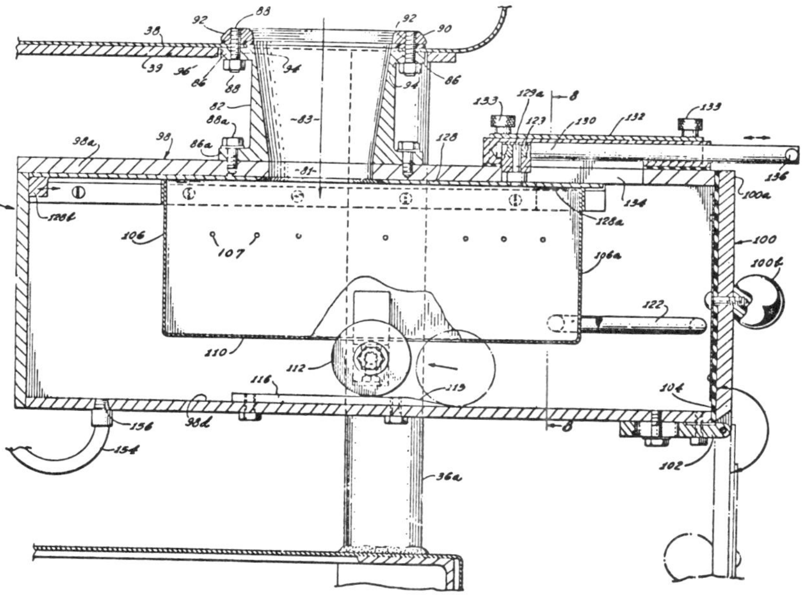

g.

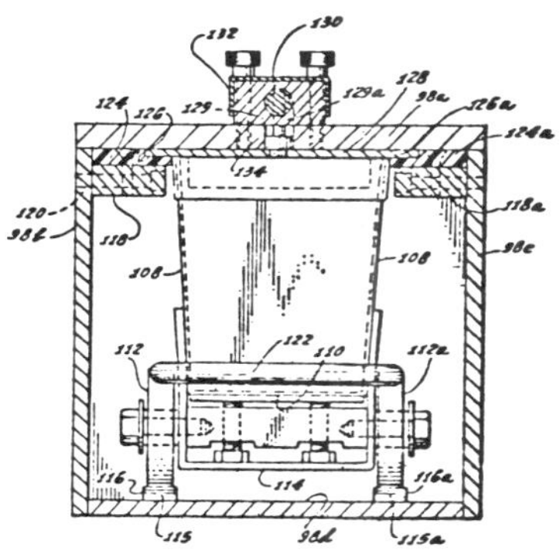

h.

(continued)

FIGURE 4.1: (continued)

i.

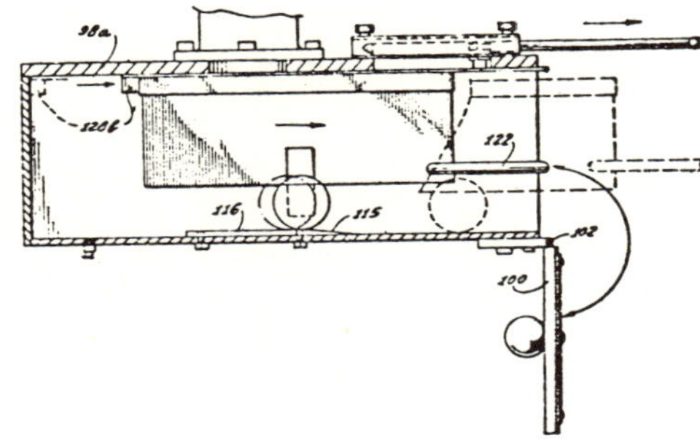

j.

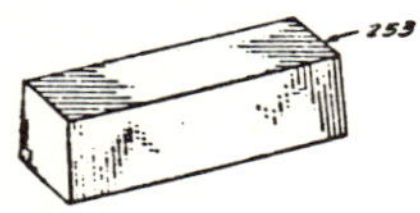

k.

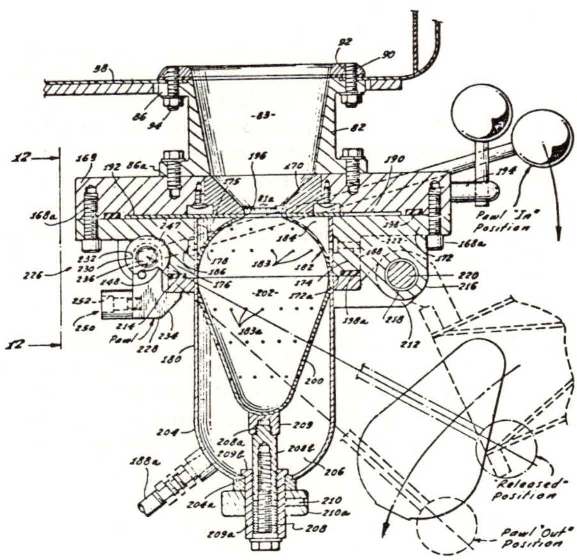

(continued)

FIGURE 4.1: (continued)

l.

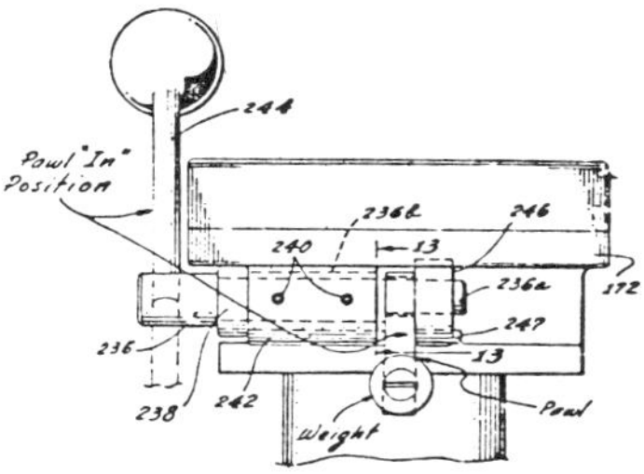

n.

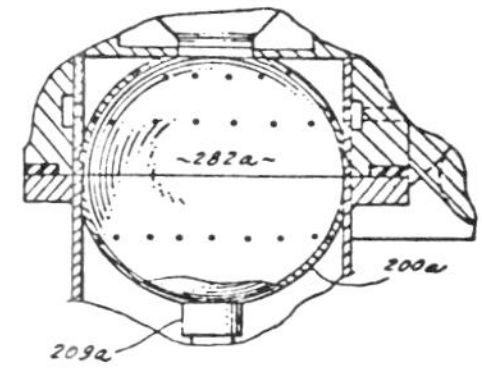

m.

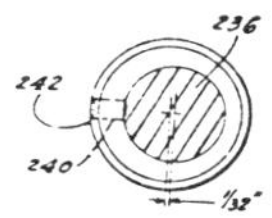

o.

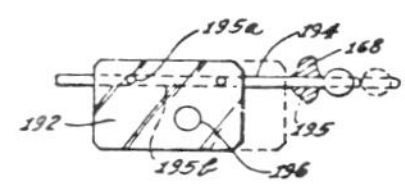

Source: F.L. Langford; U.S. Patent 3,242,571; March 29, 1966

to the left, thereby moving the cutter blade (128) into the open passage position shown in Figure 4.1g. The handle (144) of the air switch (43) for this readied box-like molding element (84) is then pulled forward from the front of the machine, thereby moving the tube (142) into the position shown in Figure 4.1c. In this position, the atmospheric vent hole (164), and therefore the tube (142) and hose (154), are placed in communication with the lateral bore (141), through the annular recess (141a). Since the bore (141) is connected by means of the elbow (158), fitting (160) and the conduit (42), to the vacuum accumulator (22), the air inside of the housing (98) is immediately exhausted to the vacuum accumulator through the flexible air hose (154) and conduit (42).

As the air pressure inside the housing (98) is thereby greatly reduced, atmospheric pressure, acting upon the top surface of the curd in the hopper (20), presses upon the curd to force it through the hopper outlet (66), the passage (83), and the mold inlet (81). The flow of cheese curd will continue until the container (106) is completely filled with curd. It has been found that such filling is accomplished within somewhere between 10 and 15 seconds. The gradual inward taper of the walls defining the fill neck (83), during the filling process, actually effects a pulling and squeezing of the curd, such that the water remaining in the curd is forced back out over the rims (90) and passed into the trough (78) for drainage through the tube (80).

At this point, the air switch handle (144) is shoved inward to effect switching of the tube (142) from communication with the vacuum accumulator (22), to a position in which the vent hole (164) is opened to the atmosphere, as shown in Figure 4.1f, thereby restoring atmospheric pressure to the inside of the housing (98). At about the same time, the rod (130) is pulled forward. This effects a movement of the cutter blade (128), in a righthand direction, thereby closing off the passage (81), as seen in Figure 4.1g. Simultaneously, the transverse block (128b), secured to the cutter blade (128), presses against the upper portion of the lefthand wall (106b) of the container (106), and moves the container to the right until the handle (122) strikes the inside of the access door (100), knocking it open. The operator of the apparatus then simply grasps the handle (122) and pulls out the container (106), turns it over and dumps out the molded brick (253), which has taken the shape shown in Figure 4.1j.

This process may be repeated in rapid sequence by the operator with each of the elements (84), preferably utilizing the waiting time of 10 to 15 seconds for molding in one element, to get up and operate the next molding element or unit.

Operation of the elements (84a) is similar in certain respects to that of the box-like molding elements (84) which have been provided to mold pear-shaped or rounded cheeses. The air switches (43) operate in an identical manner to those involved in the molding through the box-like elements (84). Each of these switches is initially pushed inward to an atmospheric venting position in which the space (206) within the housing (98) is cut off from communication with the vacuum accumulator (22).

The pivotable portion (180) of the molding element (84a) is pushed up to the position shown in Figure 4.1k wherein the annular member (214) of the casting (212) abuts the lower rim (172a) of the stationary mold housing portion (172) and is secured in this position by the L-shaped member (228) by pivoting the rod (244) upwardly about the axis of the eccentrically ended axle (236). This eccentricity of the axle rod, when the axle is thus rotated, effects a pressurized closing of the hook member (228) over the annular rim (214).

When the two mold halves are thus secured together, the rod (194) is moved to shift the cutter blade (192) into open position and the handle of the air switch (144) may be pushed inward, thereby placing the upper housing space (175), and the lower housing space (206), in communication with the vacuum accumulator (22) through the passage (188) and the conduit (188a), respectively. As in the case of the box-like molding element, such communication with the vacuum accumulator (22) results in an immediate exhaustion of all air in the spaces (175) and (206), and through the perforations (183) and (183a), also from the cavity (202) defined by the mold walls (182) and (200).

Also, as in the case of the box-like molding element (84), this exhaustion of air and reduction of air pressure results in cheese curd being forced through the passage (83) and mold inlet (81a), until the pear-shaped cavity (202) is completely filled, whereupon the air switch handle (43) is pulled out to vent the spaces (175) and (206) to the atmosphere; the cutter blade (192) is moved to a closed position; and the rod (244) is pulled down to unlatch the L-shaped member (228). With such unlatching, the movable casting (212) pivots downward and sideward in the manner shown in dotted lines in Figure 4.1k, and the pear-shaped molded cheese may readily be dumped or removed from the mold.

It will be seen that the lower molding shell (200) is secured within the housing portion (204) by the cap (209) and combination of the fitting (208b), threaded into the cap (209), the sleeve (209a), collar (210a) and bolt (208). If it is desired to increase the volume of the mold cavity (202), and therefore the weight of the molded cheese, it is only necessary to unscrew the bolt (208), and then rotate the collar (210a) to draw the fitting (208b) axially outwardly of the mold cavity (202), through the action of the threaded sleeve (209a) which rotates with the collar (210a). Since the threaded end of the fitting (208b) is engaged with cap (209), such withdrawal of the cap (209) forces the shell (200) to slide axially within the housing (204), thereby increasing the volume of the cavity (202).

If it is desired to mold the cheese in a ball, instead of a pear shape, it is simply necessary to withdraw the threaded end of the fitting (208b) from the cap (209) and then remove the lower inner mold shell portion (200) from the lower section (204) of the housing (180). There may then be substituted a hemispheric shell portion (200a) with its own cap (209a) into which the threaded end of the fitting (208b) is then screwed and appropriate adjustment of the fitting (208b), sleeve (209a), collar (210a) and bolt (208) is then made to locate the shell (200a) as shown in Figure 4.1n. It would, of course, be possible to change more radically the capacity of this molding unit by replacing the upper and lower mold housings and their associated parts, and this may be readily accomplished by removing the bolts (168a).

It will thus be seen that the apparatus provides a very effective method for molding Mozzarella-type cheese of different sizes and shapes at a rapid rate. It would, of course, be possible to provide all identical molding elements of either type shown, instead of some of the box-like type (84) and others of the pear or round type (84a).

Molding and Forming Apparatus

J. Stanga; U.S. Patent 3,421,220; January 14, 1969 describes an apparatus and method for molding and forming plastic curd cheese such as provolone into a cheese with uniform texture, a smooth skin, and with the characteristic size and shape in which it is marketed.

The method includes the steps of kneading the cheese, forcing the cheese under pressure through a chamber maintained at a temperature sufficient to melt the fats in the surface of the cheese, extruding the cheese through a constriction into formers which are cooled in a cold water bath to firm the skin and facilitate removal of the cheese from the former. The apparatus includes a hopper for containing a supply of cheese curd which is fed into an auger housing containing a pair of counter-rotating augers which urge the cheese column through a smooth-walled working chamber and into a series of removable cheese formers which are mounted on a turntable for successive registration with the outlet of the working chamber.

As each former is filled, the turntable is rotated, shearing the cheese in the former from the column. The auger housing and working chamber are maintained at an elevated temperature by warm water circulating through a water jacket which surrounds the working chamber and the auger housing.

The method includes the steps of kneading the cheese, forcing a column of cheese through a smooth-walled working chamber maintained at an elevated temperature and extruding the cheese through a constricted opening into formers which are in the desired shape and length for the particular cheese being formed. As each former is filled the cheese contained in it is sheared from the cheese column. The former is then placed in a cold water bath to cool the cheese to allow removal of the cheese from the former. The use of individual formers for each cheese permits removal of the former from the apparatus for cooling. Warm cheese sticks to the walls of the former and is not easily removed without damage to the cheese skin.

The apparatus includes positive pressure counter-rotating augers which knead the cheese while in an auger housing and which force a column of cheese through a working chamber. The working chamber has two bends which offer resistance to the flow of the cheese column through the apparatus resulting in compaction and compression of the cheese and cementing of the curd providing a cheese with uniform texture which is free of cracks or holes. A plurality of formers are removably carried by a turntable which is rotated for successive registration of the formers with the working chamber.

The auger housing and working chamber are maintained at a raised temperature by a water jacket which contains circulating water at a temperature sufficient to melt or soften the cheese fat to provide a smooth nonporous outer film.

Referring to Figures 4.2a and 4.2b, the operation of the apparatus can be described as follows. After the cheese curd is mixed it is placed in hopper (12) in large lumps which may weigh up to 200 pounds. The cheese as it comes from the mixing vat is watery and thin. It will flow by gravity from the hopper into the auger housing (17). The augers (14) and (16) have intermeshing flights and are rotatably driven by a variable speed gear train (19). Auger (14) rotates counterclockwise and auger (16) rotates clockwise with respect to each other. The gear train (19) is provided with a lever (29) for varying the speed of the augers (14) and (16).

The cheese curd (25) enters the auger housing as a string or column (26). The rotating auger flights produce a mixing or kneading action on the cheese column which cements the curd and eliminates mechanical holes or cracks. The auger housing is generally elliptical in cross section with the auger flights substantially filling the cross-sectional area. Cleaning of the auger housing and removal of the augers is facilitated by a plate (27) which is removably attached to the end of the auger housing. The positive pressure of the augers on the cheese column moves the cheese to the end cavity (100) of the auger housing. The auger flights terminate short of the cavity. The cheese packs into and accumulates in the cavity until it is filled. The cheese then passes through outlet (30) into the working chamber (20).

The working chamber is generally tubular and smooth walled and has a first portion (33) beneath the auger housing, a bend (34), a second portion (35) generally at right angles to the first portion, a bend (36) and a third portion (37) extending generally at right angles to the second portion. The cross section of the working chamber is generally cylindrical and slightly larger than the cross section or diameter of the largest cheese that is to be made. The auger housing and working chamber are desirably constructed of stainless steel.

An aperture (38) is provided adjacent bend (36) and covered by a removable plate (39) to afford cleaning of the working chamber. Best results have been obtained with a working chamber of a length of at least 1 1/2 to 2 times the length of the cheese former (32) into which the cheese is ultimately extruded. When a shorter working chamber is used surface cracks in the cheese column are not completely removed. The bends are provided to offer resistance to the flow of the cheese

FIGURE 4.2: MOLDING AND FORMING APPARATUS

a.

b.

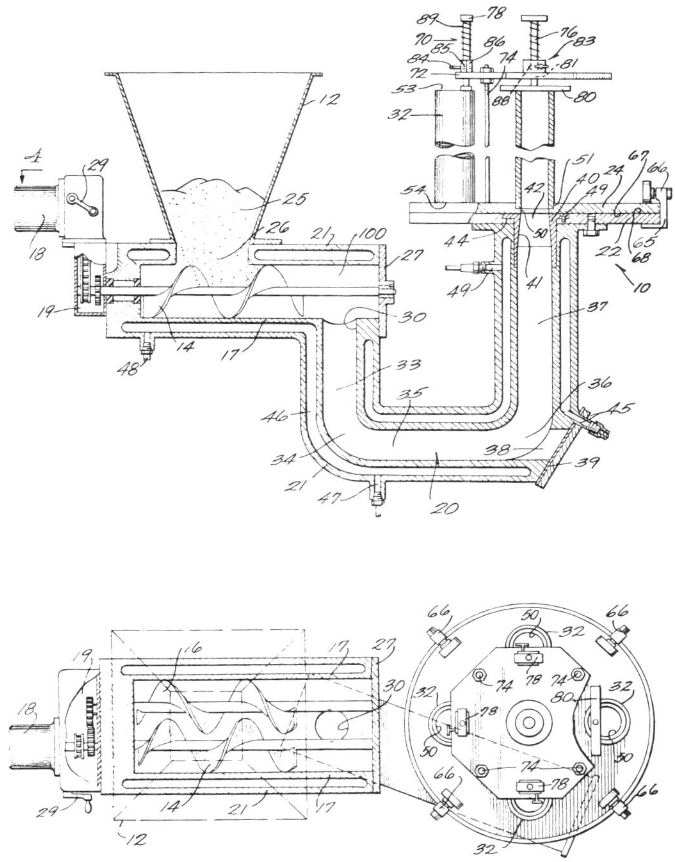

Source: J. Stanga; U.S. Patent 3,421,220; January 14, 1969

column. Auger pressure will supply the necessary force to move the cheese through the bends which compress and compact the cheese and provide a uniform texture with a completely cemented curd.

The upper end of the working chamber is relieved at (37) to receive a removable sleeve (40) which has an inside taper (41) which provides an outlet port (42) for the working chamber. This outlet port is of a smaller diameter than the working chamber and corresponds to the diameter of the cheese formers. The taper provides the final compression on the cheese column as it leaves the working chamber. Mounting of the sleeve to turntable support (22) is facilitated by an annular flange

(43), with a plurality of apertures (35). The flange seats in an annular recess (44) in the turntable support and is secured to it by bolts (49) extending through apertures (57) and threaded into the turntable support. Various shaped sleeves are utilized for different cheese shapes and sizes. If only one cheese size is being made, the apparatus can be used without interchangeable sleeves by providing an appropriately constricted outlet for the working chamber.

The water jacket (21) surrounds the working chamber and the auger housing and comprises a thin passage or duct (46) for circulation of water. The inlet (49) for the water jacket is as shown in Figure 4.2a located at the outlet end of the forming chamber. Three outlets (45), (47) and (48) are provided to drain the water jacket. Inlet (49) is connected to any suitable water source to maintain a constant flow of water around the augers and working chamber. The temperature of the circulating water is desirably controlled at 90° to 100°F. This temperature is sufficient to melt or soften the fats in the column of cheese as it moves through the auger housing and the working chamber. Thus the cheese curd is kept fluid. The flow of cheese along a smooth, warm wall promotes the formation of a smooth, nonporous exterior finish and the removal of any surface cracks or marks in the cheese column caused by the auger flights. Alternate heating means could be employed to maintain the elevated temperature of the working chamber.

The cheese extruded from the working chamber is measured and cut into the desired length by an arrangement which includes a turntable (24) carrying a plurality of cheese formers. The turntable comprises a circular disc or plate of a diameter equal to the diameter of the turntable support and is provided with four apertures (50). A cheese former is removably secured above each aperture (50). The cheese formers consist of thin-walled cylinders with open top (53) and botton (54). The turntable apertures have an annular recess (51) shown in Figures 4.2a and 4.2b to provide a seat for the bottom (54) of a cheese former.

The apparatus can be operated for continuous periods by one person. After a cheese former is filled it is removed from the turntable and placed in a cold water bath. The bath cools the cheese and creates a stiff skin which promotes sliding removal of the cheese from the former. An empty former is placed on the turntable to replace the removed former and the sequence is continued.

Mozzarella Without Cheddaring

L.O. Kielsmeier and J.G. Leprino; U.S. Patent 3,531,297; September 29, 1970; assigned to Leprino Cheese Mfg. Co. describe a preparation of pasta filata cheese, particularly Mozzarella or pizza cheese, in which the cheddaring process is substituted by soaking the curd, after draining the whey, in a water bath to reduce its internal pH and lactose content. The curd is then separated from the water bath and mixed in a confined zone under pressure.

In the process, the mixture of cooked curd and whey can be prepared in the conventional manner. Then, however, instead of the complicated stir-out and dipping procedures to prepare the curd for piling, the whey is simply drained from the curd by pumping the mixture over a separating screen. In other words, it is not even necessary to drain the curd in the vat, but instead the mixture of cooked curd and whey can be pumped directly from the vat to a continuous separating screen such as a flat reciprocating screen or a cylindrical rotating screen. The whey passes through the screen, and the curd flows off the screen efficiently free of whey for further processing according to this method.

In the essential next step in the process, the drained curd is immersed in a water soaking bath, maintained at an elevated temperature, such as 110° to 120°F. In the soaking treatment, lactose is extracted from the curd into the water bath, preferably together with the extraction of lactic acid. The soaking is continued until the internal pH of the curd is reduced to a pH of at least 5.5 but no lower than 5.0. The curd should be protected from contact with highly acid solutions, such as those which have been proposed for treatment of Cheddar curd (see U.S. Patents 2,325,217 and 2,743,186). However, a small amount of acid, such as lactic acid can be in the water of the bath during the soaking of the curd, as will result from the extraction of lactic acid from the curd.

The critical control point is the final internal pH of the curd, although this will have a relation to the external pH of the water bath. At the start of the soaking, the curd will typically have a pH in the range of 6 to 6.5, and as the soaking continues, the internal pH of the curd should be reduced to a pH of at least 5.5 but not lower than 5.0. Correspondingly, then, the external pH (pH of the water bath) should not be so low that it is difficult to terminate the soaking without making the curd unduly acid. In general, the pH of the water bath should also be above 5.0 at the termination of the soaking step. For Mozzarella or pizza cheese, it has been found that a final pH on the completion of the soaking step of substantially 5.4 is best.

During the soaking, the lactose content of the curd is drastically reduced. This occurs by the extraction of lactose into the water, and may also occur where active bacteria are present, as will usually be the case, by the bacterial conversion of lactose into lactic acid. The lactic acid formation can reduce the internal pH of the curd while at the same time being partially extracted into the water bath.

After the lactose has been largely extracted and the critical internal pH of the curd has been achieved, the curd is separated from the water of the bath. The separated curd can then be formed into a pasta filata cheese product by standard procedures, including the conventional mixing or stretching of the curd in the presence of hot water at about 180°F., followed by the steps of molding, brining, and curing, if desired. The resulting product will be of uniformly high quality, and is well adapted for use in making pizza. In particular, the shelf life is greatly extended, and the quality of the cheese is actually improved by storage prior to use. In the baking of pizzas, the cheese has less tendency to char or burn, and in sliced form remains relatively flat on the pizza dough during baking.

Figure 4.3 is a flow sheet showing the distinctive steps of this process which will be further illustrated by the following examples.

Example 1: The milk used was produced in the area around Denver, Colorado. Pasteurized milk having a butterfat content between 1.5 and 2.8% was used in an amount of 10,000 lbs. per batch. The milk had the normal pH for milk, that is, about 6.6 to 6.7.

A 10,000 lb. batch of milk was heated in a steam jacketed vat to a temperature of 89° to 90°F. and a conventional Mozzarella starter added. The temperature was

FIGURE 4.3: FLOWSHEET FOR MANUFACTURE OF PIZZA CHEESE

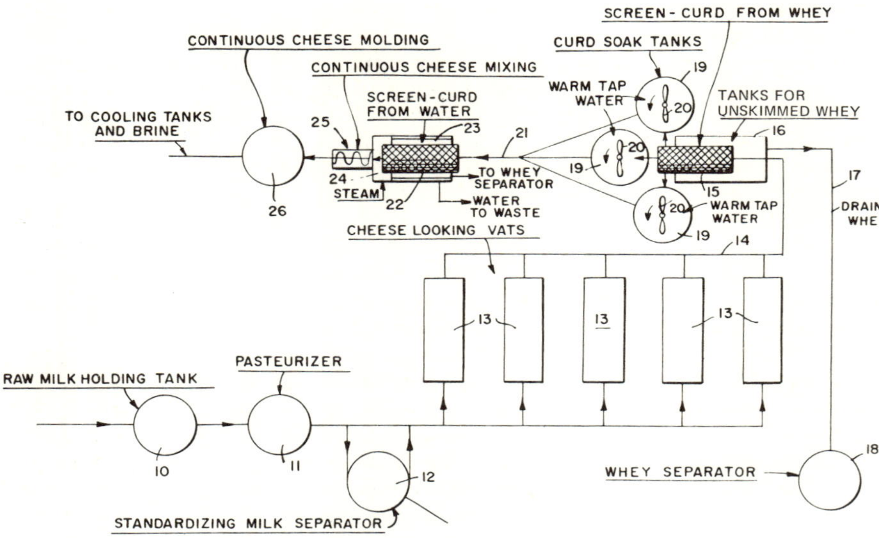

Source: L.O. Kielsmeier and J.G. Leprino; U.S. Patent 3,531,297; September 29, 1970

maintained for about 30 minutes at which time rennet was added in an amount from 3 to 4 oz. per 1,000 lb. The temperature was maintained for about 15 minutes to permit the batch to set. The above operations up to the addition of the rennet were accompanied by stirring. After 15 minutes it was noted that the curd was sufficiently formed, at which time it was cut and cooked for 35 minutes at a temperature of about 110°F.

When the curd had attained a pH of 6.2 separation of the whey from the curd was started. The curd and whey were conveyed through a shaker screen duct connecting the vat and a water bath. The duct was provided with an opening in its bottom near the water bath over which was secured a fine mesh screen so that the water would drain from the curd particles and the curd particles pass on into the water bath.

The water bath had a temperature of 110°F. and a neutral pH. The curd was allowed to remain in the water bath for 45 minutes, at which time it had developed a pH of 5.4. The water had developed a pH of 5.5. At this point the curd was separated from the water, and the further conventional processing steps of forming the cheese were performed on it, including mixing, pressing and setting. If necessary, to halt further acid development, the molds used in subsequent processing steps to mold the cheese may be placed with the cheese in them in cold water at 40° to 74°F.

It was found that the elapsed time for processing the batch had been reduced 30 minutes as compared to that for the conventional batch process originally used, and the pizza cheese product was of superior quality.

Example 2: In this example, 30% FDB low moisture part skim Mozzarella cheese (maximum moisture 52%) is made as follows: Adjust butterfat level of milk to 1.5% in the vat if average test is at 3.5% or over. Add 0.75% S. lactis starter and 1% S. thermophilus and L. bulgaricus grown in pasteurized whey or milk. The high temperature bacteria should show a balance of approximately 3 thermophilus to 1 bulgaricus on microscopic examination. The S. lactis is being used in this case to condition Grade A milk or, in other words, help create conditions in the milk so that our actual working-bacteria can grow under optimum conditions in order to reach maximum activity.

In this case, S. lactis will be killed off in the final high cooking temperature. Using S. lactis in this system may not be necessary under conditions similar to average manufacturing milk. In this case, a certain amount of acid production has taken place by streptococcus types that are regularly found in this milk.

Ripen the milk for approximately 30 minutes at 90°F. and coagulate with 3 oz. 50-50 rennet and pepsin per 1,000 lbs. milk. Make the usual long and cross cuts with 1/4" curd knives. Let curd rest for 5 to 10 minutes. Curd will start to lose moisture rapidly at this point. Start to stir at low speed, wipe sides with squeegee, and bring to 114°F. in 25 minutes. Stir for 5 minutes and draw whey, leaving approximately 40% of initial volume in vat. Break up curd by stirring and proceed to remove both curd and whey from vat.

The pH of curd at this point should be 6.1 to 6.2. Curd and whey mix is pumped through a device to separate curd from whey with the curd being discharged into vertical tanks that have been charged with 118°F. water. A volume of water is used equal to 4 times the volume of curd added. The curd and whey mixture is kept suspended by a paddle arrangement rotated on a shaft that extends vertically to the bottom of the tank. At approximately 2.15 hours after setting the milk with rennet, the pH of the curd should be approximately 5.4, at which time the curd is ready to mix and mold.

The curd and water are spearated by pumping over a device that will discharge water to sewer and curd to continuous mixing machinery. Hence, the curd is brought in contact with water at approximately 140°F. and by the action of an auger is stretched and compressed into a solid homogenous mass, and forced into the proper sized and shaped molds. The cheese is next cooled in flowing tap water sufficiently to retain its shape and dimensions when placed in cold 40°F. saturated salt brine, sodium chloride. After 24 hours, the cheese is put into plastic shrinkable bags. All the air is withdrawn and the end is closed with a clip to exclude entrance of oxygen. The package cheese is passed through hot water to shrink the bag and placed in corrugated boxes for storage and shipment.

Cheese prepared in this way shows uniformly superior melting characteristics. The limited amount of lactose present in the finished cheese had the effect of limiting the reduction in pH and changes in texture. Total moisture content can be kept within legal limits. The cheese shows no tendency to develop the usual sour whey type flavors associated with aging of this type. Shelf life with good handling characteristics is at least double that of the usual Mozzarella or pizza traditional cheese.

Plasticizing Process

N.E. Pontecorvo and W.A. Shaffer; U.S. Patent 3,554,765; January 12, 1971; assigned to American Foods Machinery Corporation describes a method and apparatus for processing raw cheese curd into fine, plasticized, substantially homogenous cheese such as Mozzarella, in which the curd is carried by a stainless steel conveyor belt through an elongated body of water heated by steam or hot water circulated in heat-exchange ducts in the bottom of the tank which holds the water.

The curd, after being softened and moistened by the heated water, being elevated by its conveyor belt up an incline, out of the body of water and then discharged over the end of the tank and downwardly onto a kneading belt which carries it beneath successive kneading rollers and finally delivers the fully processed cheese to a receiving apparatus such as an elevator or conveyor. Successive kneading rolls flatten the cheese into ribbons form. Between successive rolls the cheese ribbon is gathered and folded over upon itself preparatory to being again flattened beneath the next roll. Salt may be applied to the cheese ribbon at a suitable station along the kneading ribbon.

This process is a continuation in part of the method described by N.E. Pontecorvo and W.A. Shaffer; U.S. Patent 3,445,241; May 20, 1969; assigned to American Foods Machinery Corporation and discussed below with reference to the drawings, where Figure 4.4a is a front elevational view of a processing machine exemplifying the process; Figure 4.4b is a longitudinal sectional view of the same; Figure 4.4c is a fragmentary detail longitudinal sectional view thereof; Figure 4.4d is a fragmentary, multilevel plan view of portions of the machine, facing in a direction opposite to that of Figure 4.4a; Figure 4.4e is a schematic diagram of the machine and its water circulating system, viewing the back side of the machine.

Referring now to the drawing in detail, there is shown a processing machine comprising, in general, a frame structure (F) in which is supported a hopper (H) at one end of the machine, to receive raw cheese curd from a suitable elevator (E) (Figure 4.4e); a conveyor system (A) including a horizontal belt to carry the curd through a curd softening heated water bath in a tank (B) extending longitudinally in the upper area of frame (F), and an elevator belt to carry the softened curd (C) out of tank (B), discharging it into a downspout (D) at the other end of the machine, from which the curd (C) is delivered in two laterally separated ribbon-like webs.

A uppporting belt (G) extends full length of the machine in the lower area thereof and onto this the curd webs are directed by downspout (D). A series of kneading rolls (K) are arranged in longitudinally spaced array above supporting belt (G) for rotation on transverse axes. Beneath these the curd webs are carried by belt (G) and kneaded by rolls (K) into webs (C1) (Figure 4.4c) which are of decreasing thickness and increasingly compact, plasticized texture.

The machine also comprises of plurality of infrared lamps (L) mounted in transversely extending, longitudinally spaced array above the spaces between rolls (K), for maintaining the temperature of curd web (C1) in a suitable state of plasticity. A hot water injection system (I) is present above tank (B). There are a plurality of spray heads (S) for delivering fine mist sprays on the upper sides of respective rolls (K), for lubricating them to prevent adhesion of curd webs (C1) to the rolls; doctor blades (Q) for clearing rolls (K) of any adhering curd particles; a plurality of baffles (J) for crowding

FIGURE 4.4: CONTINUOUS PROCESS FOR CONVERTING RAW CHEESE INTO PLASTICIZED CHEESE

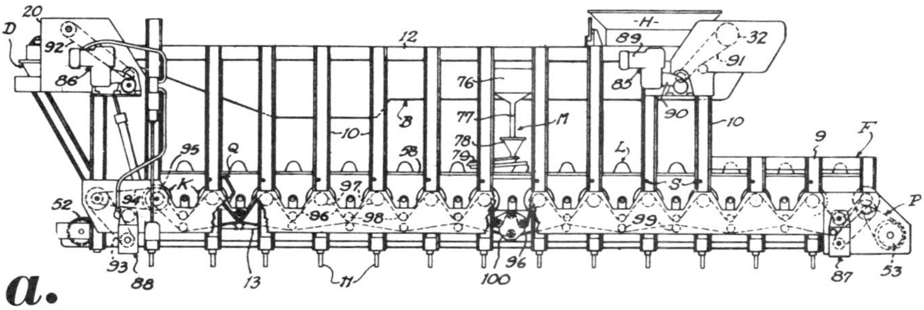

a.

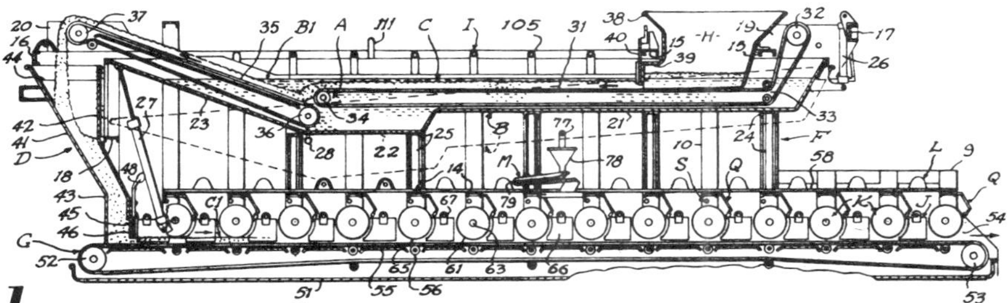

b.

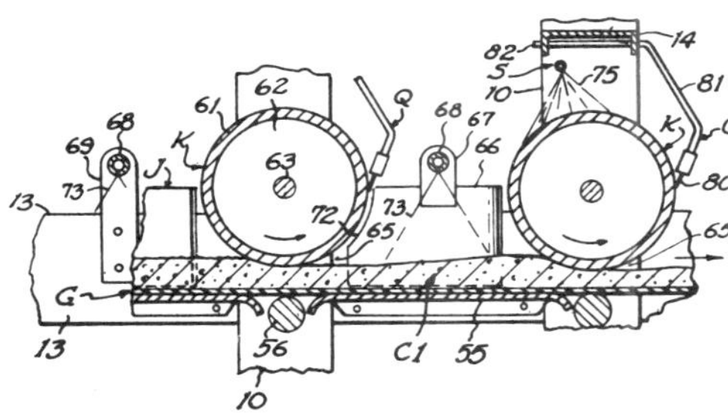

c.

(continued)

FIGURE 4.4: (continued)

d.

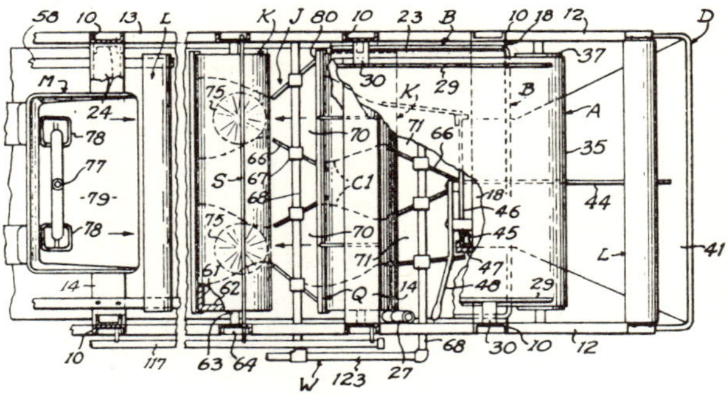

e.

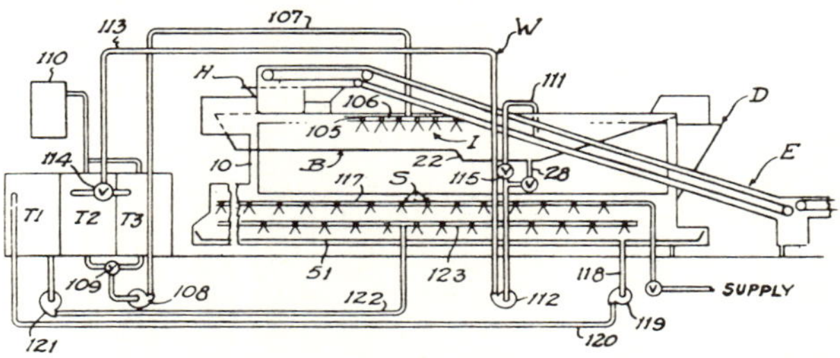

Source: N.E. Pontecorvo and W.A. Shaffer; U.S. Patent 3,445,241; May 20, 1969

the sides of webs (C1) inwardly so as to reduce the width and increase the thickness after each pass under a roll (K) in which they are thinned and widened. There is a removable salting apparatus (M) for sprinkling granulated salt upon the webs (C1); a water circulating system (W) (Figure 4.4e) for recirculation of water to softening tank (B) and spray heads (S); and a drive system (P) for power driving the rolls (K) and belts (A) and (G).

Frame (F) comprises a plurality of longitudinally spaced vertical legs (9)(10) having respective adjustable support feet (11), for levelling; aligned bridge bars (12) and (13) connecting the upper and lower ends of legs (10) along respective upper and lower side margins of the machine; and crossbars including among others, bars (14), (15), (16), (17), and (18), extending transversely between and securing the respective sides of the frame in laterally spaced relation. Frame (F) may also include upward extensions (19) and (20) at its respective ends. Crossbars (14) are disposed above respective rolls (K), their ends being attached to respective legs (9)(10). Tank (B), of stainless steel or equivalent noncorrosive material, includes a horizontal receiving portion (21) extending from beneath receiving hopper (H), an intermediate sump portion (22), and a ramp portion (23) extending from sump (22) to its discharge end which may slightly overhang downspout (D). The tank is supported normally by removing yoke shaped props (24)(25) having legs resting on crossbars (14) and transverse headbeams engaged beneath the tank (B), the upper ends of the legs being hinged to the ends of the headbeams so as to be foldable for removal when the tank is to be cleaned.

During the cleaning periods, involving draining the water from the tank, it is raised and lowered by operation of a pneumatic jack (26) which suspends its receiving end from crossbar (17); and a pair of pneumatic struts (27) which are propped between the two corners of its discharge end and respective pivotal connection to frame (F) in spaces between the ends of the first kneading roll (K) and the adjacent sides of the frame (Figure 4.4d). In addition to removal of props (24)(25), the salting apparatus (M) is removed in preparation for lowering the tank.

In its lowered position, shown in phantom in Figure 4.4b, the tank is tilted, with its discharge end lowered more than its receiving end, and with its receiving portion (21) and its ramp portion (23) both having their bottoms inclined downwardly toward a low area located at the junction between sump (22) and ramp (23). At this junction, a drainpipe (28) is connected to the sump. Cleaning is thus performed by flushing from both ends of the tank to the drain (28).

The pneumatic jack (26) and struts (27) are swingable on their pivotal mountings to accommodate such longitudinal shifting movement of tank (B) as may be required to clear its discharge end from slightly overhanging relation to downspout (D) during the lowering of this end of the tank. A hot water bath (B1) is maintained in tank (B) at a level just above the upper surface of the layer of curd (C) carried by the conveyor (A), fully immersing the curd layer.

Conveyor system (A) comprises a receiving belt (31) operating in a horizontal position in receiving portion (21) of tank (B) for its full length except for its receiving end which extends from a drive roller (32) downwardly into the receiving end of the tank behind the hopper (H). From this downwardly extending portion, the belt extends around idler rollers (33), then forwardly beneath hopper (H) to the drain end of sump (22) and around a return roller (34), with an upper stretch moving toward the sump as indicated by an arrow, and a lower return stretch moving toward the hopper (H).

An elevating belt (35) has an upper conveying stretch extending from an idler roller (36) in the drain area of sump (22) beneath the return roller (34), along an upwardly inclined path of movement indicated by arrow, generally parallel to the ramp (23), and beyond the discharge end of the tank to a drive roller (37) disposed over an idler roller beneath, and back to idler roller (36) in a return stretch. Drive rollers (32) and (37) are mounted in frame extensions (19) and (20) and are driven by drive system (P). Retainer strips (29) (Figure 8.2d) mounted on legs (10) by means of spacer blocks (30), have their lower edges fitted to the upper surfaces of belts (31) and (35) so as to contain the curd on the belts, preventing loss of curd over the side margins of the belts.

Hopper (H) has a throat extending downwardly from a receiving mouth (38) into tank (B) between crossbars (15) on which it is supported, and a delivery nozzle (39) projecting toward the discharge end of conveyor belt (31). The throat and nozzle (39) have a common open bottom fitted to the belt (31) to deliver curd thereon, and nozzle (39) has an open discharge end which can be closed or partially closed by a vertically adjustable gate (40) for feeding forwardly on belt (31) at a controlled depth. Gate (40) is provided with operating mechanism similar to that of downspout delivery gate (45) described later.

Downspout (D) includes a receiving hopper portion (41), disposed beneath the discharge end of conveyor system (A) and hung upon a suitable bracket beam (42) which may extend diagonally between a supporting attachment of its lower end to crossbar (18) and an attachment (not shown) of its upper end to a side of frame (F) near the top thereof. From hopper portion (41), the downspout (D) extends downwardly to a lower portion (43) which has an open lower end disposed above the starting end of supporting belt (G) and closely associated therewith. Lower spout portion (43) provides an agglomerating chamber in which cheese curd (C), fed downwardly from the discharge end of elevator belt (35), settles into a matted body in which the curd particles adhere to one another under a packing action of the descending curd.

A separator panel (44), disposed in a vertical median plane midway between the sides of the downspout (Figure 8.2d) divides the curd into two descending streams, and thus two laterally separated streams (C1) of agglomerated curd are fed from the lower end of the downspout, through a transverse outlet (45) onto the supporting belt (G). Outflow is regulated by a vertically adjustable gate (46) slidably guided in ways (47) at the sides of lower spout (43), and is raised and lowered by an operating lever (48).

Supporting belt (G) extends full length of the machine above a drip pan (51) of corresponding extent, with an upper carrying stretch traveling as indicated by arrows, from an idler roller (52) at its receiving end beneath downspout (D) to a drive roller (53) at a discharge outlet (54) at the opposite end of the machine, from which the ribbons of completely processed cheese may be conveyed by a suitable elevator (not shown) to suitable molding and packaging apparatus.

The carrying stretch of belt (G) is supported by a succession of alternate transversely extending slide plates (55) and rollers (56) which cooperate to support the belt solidly in a substantially horizontal plane, thus providing an unyielding supporting surface upon which the curd ribbons (C1) are kneaded by rollers (K) and baffles (J). The ends of rollers (56) are journaled in frame legs (10) and the ends of slide plates (55) are secured to lower side bridge bars (13). Belt (G) is driven by roller (53) which receives its drive from one of the power units (P).

Kneading rolls (K) (Figure 4.4c) may comprise tubular cylinders (61) of noncorrosive metal (e.g., stainless steel) with external surfaces that are smooth to minimize adhesion of the curd ribbons thereto, with closed ends (62) to exclude the entry of curd particles, and with mounting shafts (63) which have projecting ends journaled in bearings (64) carried by frame legs (10)(11).

For satisfactory operation in feeding the curd ribbons beneath them while compressing the curd, the rolls (K) are of fairly large diameter (e.g., in the range of 10 to 12 inches). Rolls of smaller than 10" diameter are less satisfactory, especially in the initial group of 4 to 6 rolls above the receiving end of belt (G), where maximum compression and reduction in depth of the curd takes place. Smaller rolls tend to captivate (lose their driving grip on the curd ribbons) with the curd piling up ahead of the rolls instead of being drawn beneath them. For adequate stages of kneading operation, a substantial number of rolls (e.g., 12 to 15) spaced apart somewhat less than roll diameter are employed. Rolls (K) are all individually driven by mechanism (P). Rolls (K) are spaced above supporting belt (G) to provide a succession of kneading gaps (65) which vary from maximum depth beneath the first roll adjacent downspout (D), to minimum depth at outlet (54). Substantially all the reduction in depth occurs beneath the first six rolls, the depth remaining substantially constant (or decreasing only slightly) beneath the remaining rolls. The maximum gap depth beneath the first roll is within the range of 2 to 4 inches, a satisfactory depth being 3 to 3 1/2 inches, and minimum gap depth, at outlet (54), is in the range of 3/4 to 1 1/2 inches, a satisfactory depth being 1 inch.

Baffle system (J) (Figure 4.4c and 4.4d) comprises a plurality of baffle blades (66) affixed to respective shanks (67) fixedly mounted on tubes (68) which are supported at their ends in brackets (69) secured to frame side members (13). Blades (66) (Figure 4.4d) with the exception of the first set which is disposed between downspout outlet (45) and the first roller (K), are arranged in sets, each consisting of two pairs of forwardly converging blades defining two laterally spaced passes (70) (Figure 4.4d), in which the respective curd ribbons (C1) are squeezed horizontally to narrowed width.

The first set of blades consists of two pairs, all extending diagonally outwardly from the longitudinal medial axis of the machine (one pair toward one side, the other pair toward the other side), whereby they define two passes (71) which diverge from the downspout outlets (45) so as to spread the curd ribbons (C1) apart as they issue from the downspout. Thus the ribbons commence their travel beneath the rolls (K) along paths located substantially midway between the longitudinal medial axis of the belt (G), and its side edges, the direction of travel being indicated by the arrows.

The blades (66) traverse the spaces between rolls (K) and may have tapered ends (72) projecting beneath rolls (K) and conformed to the peripheral curvature thereof. The baffle mount tubes (68) may in some installations be provided with spray apertures for delivering downward sprays (73) of heated water which may be optionally utilized for maintaining temperature at the correct processing level, in lieu of the use of radiant heaters (L). Also, where acidity may tend to rise above the correct level, sprays (73) may be utilized for leaching excess acid from the curd.

Heater group (L) comprises a plurality of infrared heat units extending transversely, their ends supported upon bars (58) extending along the respective sides of the machine, carried by crossbars (14). The spray heads (S) consist of small tubes extending between respective pairs of frame legs (10) and having two or more extremely fine apertures for delivering atomized sprays (75) of water mist upon the upper surfaces of rolls (K) to lubricate them sufficiently to inhibit adhesion of the curd ribbons to the rolls.

Salting unit (M) comprises a salt supply tank (76) which delivers granulated salt through a branching metering tube (77) and two distributing hoppers (78) onto a vibrator-mounted shaker pan (79) which discharges the salt over its edge as indicated by arrows in Figure 4.4d, into a space between two of the rolls (K) and then onto the curd ribbons (C1). Unit (M) is located at a position sufficiently in advance of discharge outlet so that salt which is sprinkled thereby in a narrow strip or strips across the widths of both curd ribbons (C1), will be thoroughly worked by the kneading operations into the cheese in uniform dispersion therein by the time it reaches the outlet (54).

Doctor blade units (Q) (Figure 4.4d) comprise respective blades (80) carried by the lower ends of arms (81) which extend downwardly from respective shanks (82) mounted in the side webs of respective crossbars (14). The blades (80) have scraping contact with the sides of rolls (K) facing in the direction of travel of curd ribbons (C1), where the roll peripheries are moving upwardly.

Power drive system (P) comprises a number of variable-speed electric drive units (85), (86), (87) and (88), mounted to frame (F) at the four corners of one side thereof. Each unit embodies a variable speed electric motor and a reduction gear [e.g., motor (89) and gear (90) of unit (85)]. From unit (85) a chain drive (91) extends to and drives conveyor drive roller (52). From unit (86), a chain drive (92) drives elevator drive roller (37).

From unit (87), a two-stage, double-reduction chain drive (93)(94) carries drive to a sprocket cluster on the shaft of the first kneading roll (K), from which it is transmitted by a common drive chain (96) to all remaining rolls (K), the chain (96) alternately passing over sprockets (97) on the respective rolls and dipping beneath idler sprockets (98) on lower side frame bars (13), until it passes around the last roll sprocket, from which it returns in a stretch passing alternately over and under idler sprockets (99), all as indicated in phantom by dotted line in Figure 4.4a. Midway between the ends of the machine, the upper and lower stretches of the chain pass through a tension-adjusting unit (100).

The water circulating system (W) and the machine described above, in their relation to one another and to feed elevator (E), are shown schematically in Figure 4.4e. Elevator (D) delivers raw curd from its upper end into receiving hopper (H). Hot water injection system (I) comprises a plurality of discharge heads (105) traversing the machine over tank (B) between the upper ends of frame legs (10) in which they may be mounted. Each head has one end closed and its opposite end coupled to a feeder pipe (106) which is supplied with hot water by a delivery line (107). A pump (108) draws water selectively from one of two alternative hot water tanks (T2) and (T3), under control of a selector valve (109).

The water is heated by heater (110) (e.g., by steam injection from a steam generator). Water is injected into tank (B) through heads (105) and is returned through an overflow line (111) which regulates the level of hot water in tank (B) at an adequate height well above the level of the curd (C) carried by conveyor belt (31). Water is pumped out of the tank by a pump (112) and through a return line (113) and a valve (114) to the tank from which it is being drawn. Overflow pipe (111) and drain (28) are connected selectively to pump (112) by parallel alternative connections through valves (115).

Mist spray heads (S) are supplied with water from an external water supply line, so designated, through a connecting line (117). Drippings from belt (G) are collected in drip pan (51) and discharged through a drain outlet (118) from which it is pumped by a pump (119) through a return line (120) into a tank (11). From tank (T1) this drainage may be directed, when required, through a pump (121) and a recirculating line (122), to a manifold pipe (123) connected to the ends of baffle mount tubes (68) so as to deliver the sprays (73).

In the process, raw cheese curd is fed continuously from elevator (E) into receiving hopper (H), into the hot water bath contained in tank (B). From hopper (H) it is fed at a rate controlled by adjustment of gate (40), onto the stretch of conveyor belt (31) moving from hopper (H) into sump (22). From the end of belt (31), the curd drops onto elevator belt (35) which carries it up the ramp (23) and over the end of the tank (B), dropping it into the hopper portion of downspout (D). The curd is then in a softened, moistened condition as the result of moisture and heat absorption from its immersion in the hot water bath in tank (B). The bath is maintained at a temperature in the range of 119° to 145°F., depending upon curd acidity.

The bath temperature is necessarily higher than the temperature to be imparted to the cheese (e.g. 128°F.) due to the lag in heat transfer [e.g., in a five minute interval between delivery from hopper (H) to discharge into downspout (D)] a satisfactory curd temperature range being 122° to 130°F. (commonly 128°F.). Where the acidity is at a proper level, temperature may range toward the lower limit, but if the acidity level is abnormally high, correction is made by adjusting the temperature to a higher level.

The normal acidity level for best results is substantially in the range of 70 to 90 by sodium hydroxide titration of the whey. If curd has high fat content, it will work best at a low titration level (e.g., near 70). Where it has less fat, a higher titration level is required. Temperature is sensed by a suitable sensor in tank (B) and is controlled by regulating the operation of heater (110). Hot water is continuously circulated from a tank (T2) or (T3) into tank (B) through injection heads (I), and back to the tank (T2) or (T3) through overflow line (111) and return line (113).

The softened, moistened curd particles, still in a loose state when discharged from elevator belt (35), drop through the two throats of downspout (D) and collect in the agglomerating chamber (5) in lower portion (43) thereof, where they settle into two agglomerated masses of adhering particles, resting upon the supporting belt (G). From the agglomerating chamber the two masses of curd are fed forwardly in two streams of depth regulated by adjustment of gate (46). These streams are spread apart by travel through passes (71) between pairs of baffle blades (66) of the first set, and are delivered beneath the first kneading roll (K) in positions substantially centered over the respective halves of belt (G). The streams (C1) are drawn through the gap (65) by the forward travel of belt (G) and the peripheral rotary movement of roll (K) which is synchronized to the belt travel.

As they pass beneath the first roll (K), the streams are compressed, flattened and widened into curd ribbons (C1) as indicated by the dotted lines delineating the margins of the ribbons (C1) in Figure 4.4d. The flattened, widened ribbons are gathered by the next set of baffle blades (66) and are narrowed and thickened by such gathering action, the narrowing being indicated in Figure 4.4d and the thickening in Figure 4.4e. From the second set of baffles, the ribbons (C1) are fed beneath the second kneading roll (K), and again are flattened, further thinned and widened by the rolling action between the roll and the synchronously moving belt.

In passing beneath the first group of (e.g., 6) rolls (K), the ribbons are reduced to successively lesser average thickness until the voids between curd particles are largely squeezed out and the curd material commences to assume a compacted, homogenous consistency. Without substantial further reduction in the thickness of the ribbons as they pass under the rolls, they are alternately subjected to the flattening and widening action of the rolls (vertical kneading) and the thickening and narrowing or gathering (horizontal kneading) action of the baffles (J). Such alternate vertical and horizontal kneading reduces the curd material to the completely plasticized, homogenous consistency which is characteristic of Mozzarella cheese, and the product which is delivered from outlet (54) is of Mozzarella quality.

Excess water carried by the loose curd material as it leaves the hot water bath is squeezed out by rolls (K) and drains over the edges of belt (G) into drip pan (51). Heaters (L) function to restore heat loss, maintaining the curd ribbons (C1) at the temperature at which the curd leaves the tank (B). Water mist sprays (S) maintain wet surfaces on rolls (K), preventing adhesion of ribbons (C1) to the rolls.

Homogeneous Cheese

N.E. Pontecorvo; U.S. Patent 3,580,161; May 25, 1971 describes an apparatus for attaining a more homogeneous cheese while processing cheese curd. This patent is a continuation of U.S. Patent 3,476,036. The general object of this process is to overcome problems that have been encountered with the earlier processes and apparatuses. Among these are: the difficulty of attaining uniform distribution of heat in the vat process, and the tendency to overheat portions of a cheese batch, with consequent loss of fat content, while other portions of the batch remain underheated and inadequately softened; the difficulty of maintaining proper operation of the kneading belt in the belt and roll-type of apparatus; and the lack of adequate kneading action in the auger and tube type of apparatus.

In achieving these objectives, the apparatus provides a method in which raw curd in small pieces is fed from a hopper into a large heating tub rotating slowly beneath the hopper, and is laid onto the bottom of the tub in a layer about 3 inches in depth, which is softened by the heating and moistening action of a body of water maintained in the tub at a temperature of approximately 160°F. When adequately softened the curd is removed from the tub by an elevating conveyor and transferred to a kneading unit which in one form of this process may be a turntable below the tub, on which it is kneaded by radially disposed rolls.

After being flattened beneath several rolls, the ribbon of curd on the turntable is folded inwardly upon itself by curd-folding plates, and it is then carried beneath another stage of rollers where it is flattened to a thinner ribbon than before. After traversing a nearly 360° path, it is deflected outwardly by an exit guide shoe and discharged over the periphery of the turntable into a receiver.

In the operation of the apparatus of Figure 4.5a through 4.5d, the apparatus provides a continuous cheese-processing operation in which raw cheese curd in small pieces is fed into hopper (B) so as to maintain therein a body of curd resting on the tub-bottom (10) within the confines of the hopper bottom opening (31). From this hopper-confined body of curd, a stream of curd having a depth regulated by the height-adjustment of gate (33), is drawn by rotating tub (A).

The tub is rotated slowly, at a speed such as to provide a satisfactory time-period for softening the curd in approximately one revolution of the tub. As the curd stream is carried by the tub in a circular path, it forms an annular ring of curd on the tub bottom, commencing at the front of hopper (B) and extending around to the conveyor (C) which is located adjacent the back of hopper (B). As this curd ring reached conveyor (C), it is moved radially outwardly to the lateral wall (12), elevated along the wall (12), and discharged over the rim of the tub, by the scraping of conveyor paddles (38) along the surfaces of tub bottom (10) and lateral wall (12). The diagonal

FIGURE 4.5: HOMOGENEOUS CHEESE

a.

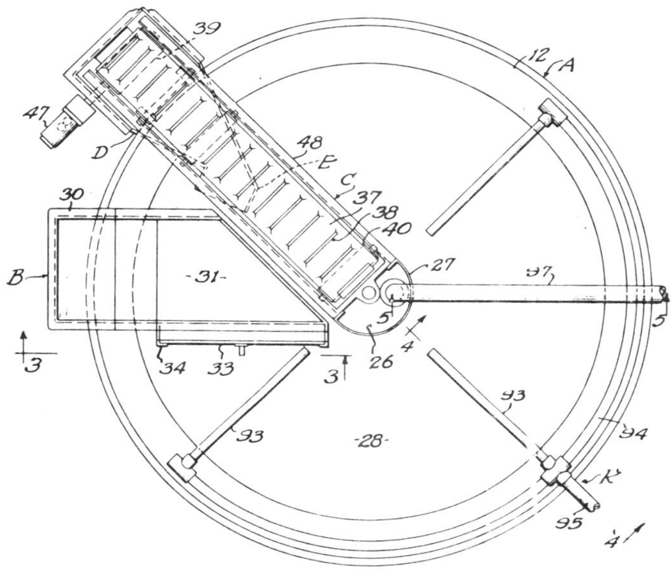

b.

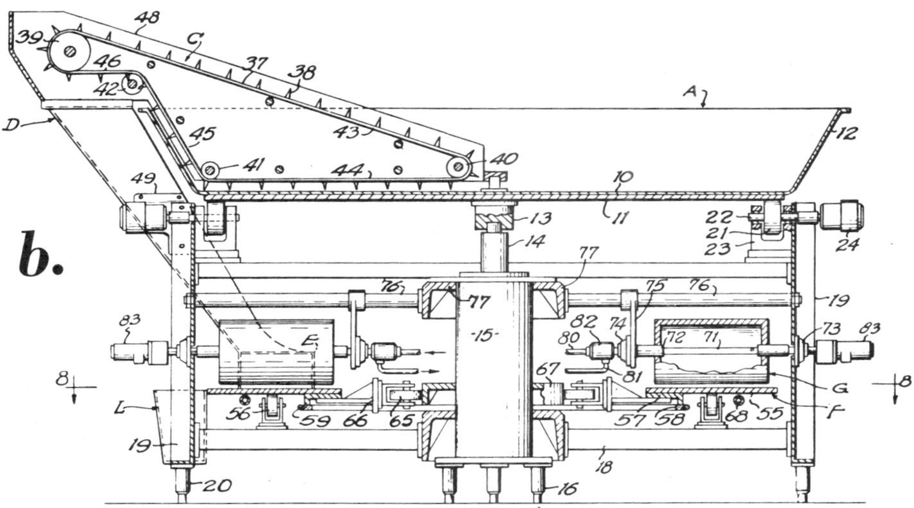

(continued)

FIGURE 4.5: (continuted)

c.

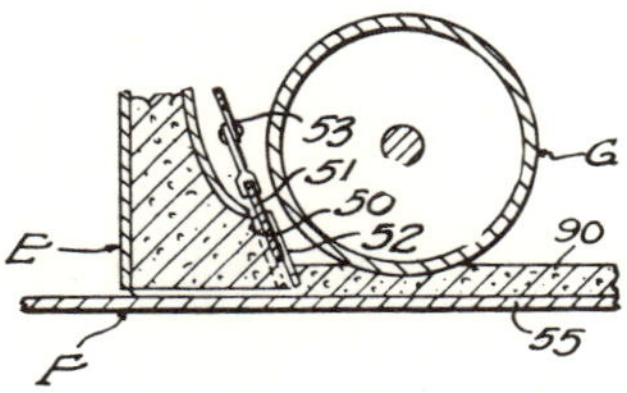

d.

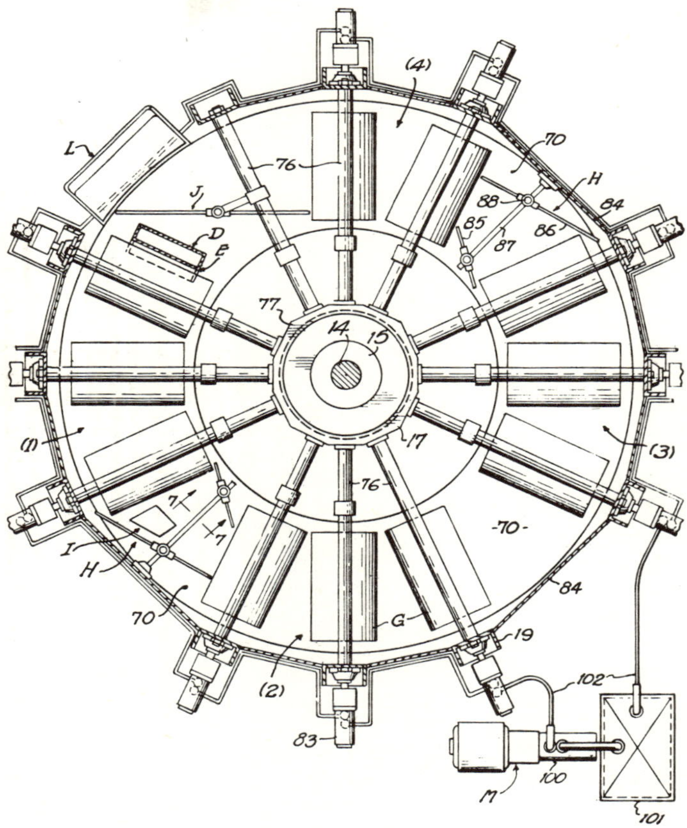

Source: N.E. Pontecorvo; U.S. Patent 3,580,161; May 25, 1971

adjacent side of hopper (B) functions as a barrier to largely arrest the rotary travel of the ring of curd past the conveyor (C). This barrier can be extended to the lateral wall (12) if desired. It will now be apparent that the curd remains in the tub for somewhat less than full revolution of the tub and is discharged from the tub just short of the the position of entry.

As the curd is discharged over the rim of the tub, it will drop into the chute (D) which will guide it into metering hopper (E), the open bottom of which is immediately adjacent the mixing disc (F). From this hopper the softened curd is metered onto the revolving heated disc (F) through the discharge opening (50), beneath metering gate (51) (Figure 4.5c).

The stream of curd carried away from the hopper (E) will immediately pass under the first mixing drum (G), which will compress the loose, softened curd into a ribbon. This ribbon will pass successively beneath the remaining drums of the group, being progressively flattened thereby, and will then be carried between a pair of the folding blades (H) (Figure 4.5d) which will fold over the margins of the ribbon so as to narrow and thicken it. The ribbon will then pass beneath the drums of the second group, again being progressively flattened, and upon emerging therefrom, will again be folded back upon itself by a second pair of folding blades (H).

These stages of alternate flattening and folding will be repeated until the ribbon approached the back of metering hopper (E) after nearly a complete revolution of mixing disc (F). The ribbon then will be deflected off the disc (F) by the diagonally positioned exit guide shoe (J), into the curd receiver (L).

Kneading Apparatus

N.E. Pontecorvo; U.S. Patent 3,580,162; May 25, 1971 describes a cheese kneading apparatus. The general object of the apparatus is to provide a kneading apparatus embodying a relatively simple arrangement of moving parts, of easily cleaned and sterilized materials, eliminating the use of belts, small diameter rolls and numerous bearings, and which is consequently reliable in operation, requiring a minimum of servicing and repair problems, and easy to maintain in a clean and sanitary condition.

Specific objects are (1) to avoid curd contamination; (2) to attain better sanitary conditions with less effort and time expenditure; (3) to avoid breakdowns and frequent repairs; (4) to minimize servicing requirements; (5) and to decrease operational expense generally. This apparatus is related to cheese kneading techniques previously described by N.E. Pontecorvo and W.A. Shaffer; U.S. Patent 3,403,030; September 24, 1968.

In the present mechanism a plurality of substantially coaxial kneading rings, encased in separate annular casings, spaced axially along the length of a common cylindrical mandrel which extends through the rings in eccentric relation thereto so as to define a plurality of annular kneading chambers having restricted passes at the lower side of the mandrel and wider areas on the upper side of the mandrel, are rotated by a drive which also rotates the mandrel so that the opposed cylindrical surfaces of the chambers move in the same direction and at the same circumferential speed to convey premoistened and softened cheese curd around the mandrel in a succession of circumferential movements through the passes to effect successive compressions of the curd, followed by release of pressure in expansion areas of the chambers. From

FIGURE 4.6: KNEADING APPARATUS

a.

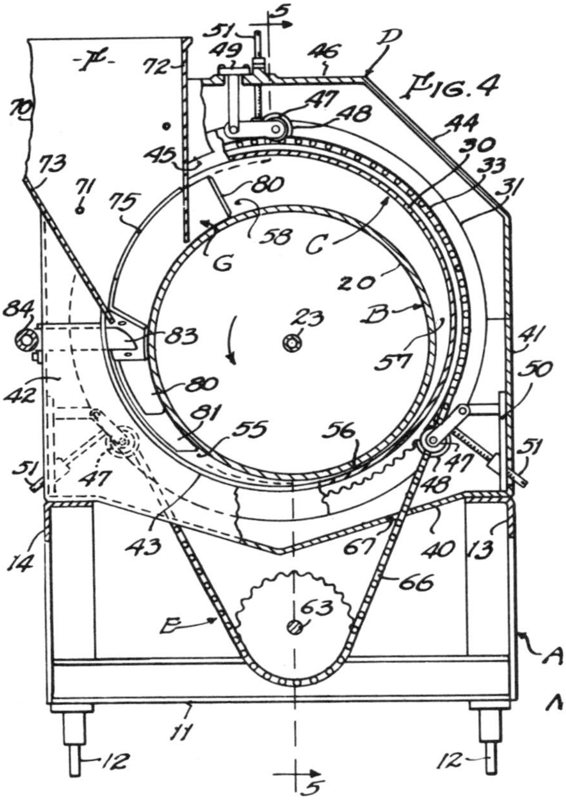

b.

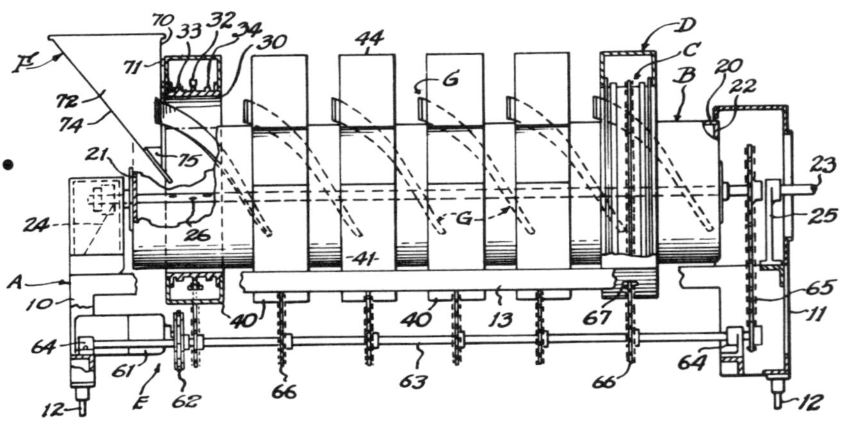

(continued)

FIGURE 4.6: (continued)

c.

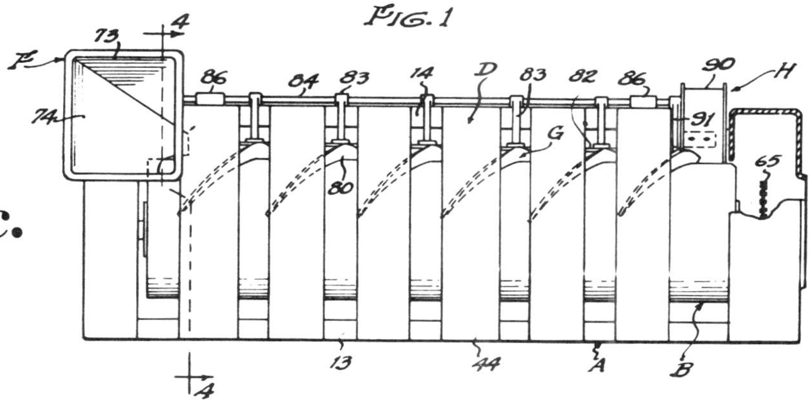

Source: N.E. Pontecorvo; U.S. Patent 3,580,162; May 25, 1971

each expansion area the curd is transferred axially into a contracting feed throat defined between converging surfaces of a succeeding chamber, the axial movements being effected by the movements of the curd stream against deflector blades extending helically from each chamber to the next.

As viewed from the front end of the machine (Figure 4.6a) the mandrel (B) and kneading rings (C) are rotated in a counterclockwise direction as indicated by arrows. Hence, the raw curd delivered into the feeding area (55) of the first kneading chamber will be drawn through the kneading pass (56) where it will be compressed between the closely spaced surfaces of mandrel (B) and ring (C). Thence the curd will be carried into the widening area (57) where it will expand upon release of the compression.

From the expansion area (57) the curd will be carried by the mandrel into the transfer area (58) at the top of the machine. At this time the curd will be moving circumferentially in the form of an annular ribbon in the vertical plane of the first kneading ring (C). In the area (58) it will be moved against the helical deflector blade (80) which extends through that area, and since the arcuate inner edge of the blade is in scraping contact with the mandrel surface, the blade will deflect the curd ribbon into a helical path which will lead it across the gap between the first two kneading rings and into the feeding area (55) of the second kneading chamber, where it will resume an annular path in the plane of the second kneading ring after being released past the end of the deflector blade (80) which has guided it across the gap between the kneading rings.

The compression and expansion steps will be repeated in the second and succeeding kneading chambers until the ribbon processed cheese is delivered from the last blade (80) into the delivery chute (H) (Figure 4.6c). The continuing rotation of the mandrel (B) will feed the now homogenous ribbon of cheese downwardly through chute (H) into a suitable receptacle or conveyor for transferring it to another processing apparatus (e.g., molding machine for shaping it into blocks or patties).

Preferably the series of kneading rings (C) are adjusted to successively closer spacing from mandrel (B) in the succession from the front (receiving) end to discharge end of the machine, so that as air voids are successively squeezed out of the curd body, it may be compressed to increasingly thinner ribbon in successively narrower passes (56).

A major operational advantage provided by this process is in the elimination of conveyor belts which, when of rubber or plastic material are virtually impossible to maintain in a sterile, sanitary condition and which, constructed of metal, are difficult to maintain in satisfactory operating condition and require a high power expenditure for driving them as well as frequent servicing. Rubber belts, if subjected to temperatures required for full sterilization, will be rapidly deteriorated and rendered useless. The parts of the machine of this process can be made largely of stainless steel or equivalent corrosion and contamination resistant and easily sterilized material.

Molding Under Vacuum

In earlier work done B.B. Elder and N.E. Pontecorvo; U.S. Patent 3,041,153; June 26, 1962; assigned to Cheese Molding Machine Company devised a device for carrying out a method of molding Mozzarella cheese under vacuum. In the process air is withdrawn from the cheese during the molding process to achieve the desired consistency normally obtained through hand kneading. Various other objects and advantages of the process will become apparent in the following drawings.

Figure 4.7a is an elevational view, partially in cross section, of the apparatus of the process in the loading position wherein the lower portion of the mold is coupled to the filling mechanism; Figure 4.7b is a cross-sectional view taken on line (2 – 2) of Figure 4.7a; Figure 4.7c is a cross-sectional view taken on line (3 – 3) of Figure 4.7a; Figure 4.7d is a partial cross-sectional view of the apparatus in the molding position with the two sections of the mold coupled together and Figure 4.7e is a partial cross-sectional view on line (5 – 5) of Figure 4.7d.

Referring now to the drawings in detail, the authors have shown a cheese processing apparatus comprising generally a loading unit (A) supported by a suitable support (14) in a fixed position at loading station (so designated by legend in Figure 4.7a), a lower mold unit (B) mounted for movements back and forth between that station and a molding station (so designated by legend in Figure 4.7d) and an upper mold unit (C) supported in a fixed position at the molding station. Figures 4.7a to 4.7c show the apparatus in its loading position with the loading mechanism (A) coupled to the lower mold assembly (B). Figures 4.7d and 4.7e show the lower mold assembly at the molding station, coupled to the upper mold unit (C).

The Apparatus: The loading mechanism (A) comprises principally a cheese curd hopper (10) coupled to a dehydrating jacket (12) which is separated from a concentric mold mounting ring (14) by a cutoff gate or cutter (16). Hopper (10) includes a rim (18), and a downwardly tapering body (20), terminating at its lower end in a discharge throat (22) defined by a tubular neck (24), in which a stopper valve (26), for regulating the flow of the curd from the hopper, is slidably mounted between guideplate (27) and guideplate (29) joined along their lateral edges at (31). A continuation of throat (22) is provided by the inner wall of dehydrating jacket (12) which consists of concentric inner sleeve (32) and an outer sleeve (34). Sleeves

FIGURE 4.7: MOLDING OF CHEESE UNDER VACUUM

a.

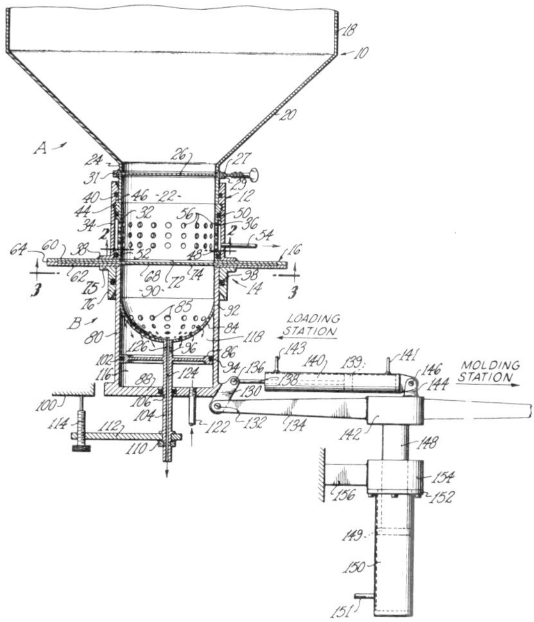

b.

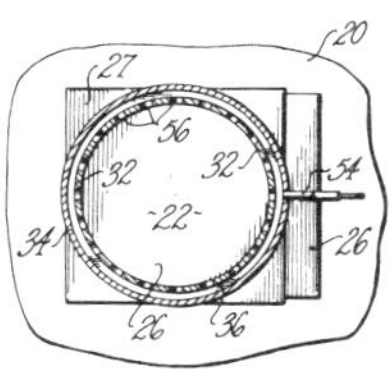

c.

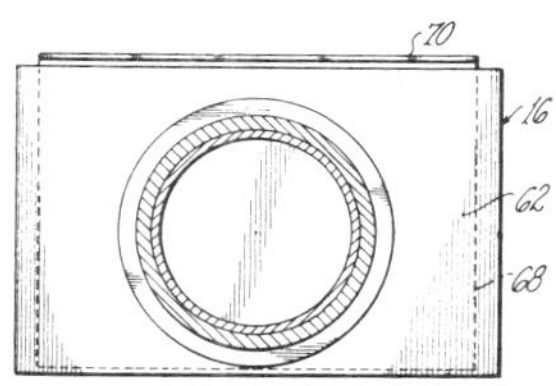

(continued)

FIGURE 4.7: (continued)

d.

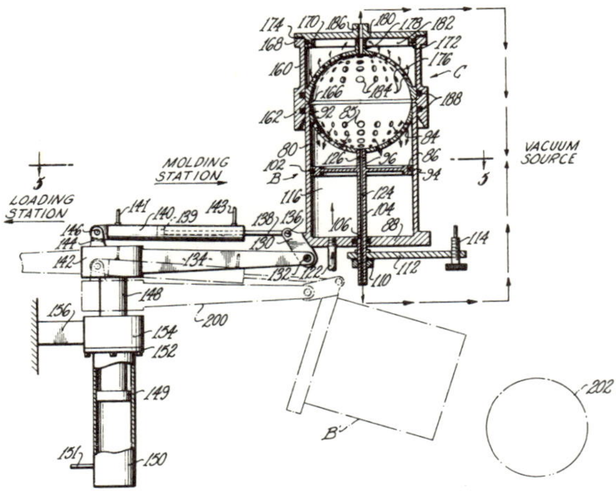

e.

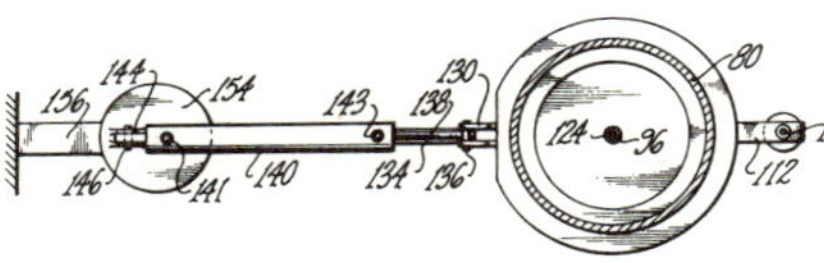

Source: B.B. Elder and N.E. Pontecorvo; U.S. Patent 3,041,153; June 26, 1962

(32)(34) are spaced to define an annular cylindrical vacuum chamber (36). In more detail, outer sleeve (34) has at its lower end an outwardly extending flange (38) for mounting jacket (12) to gate (16), and at its upper end an inwardly offset collar (40) which receives neck (24). Threaded into sleeve (34) is a flange (44) outwardly extending from the upper end of sleeve (32).

The upper end of flange (44) abuts a shoulder (46) on the underside of flange (40) and also abuts the lower end of neck (24), thereby serving as a support for hopper (10). Sleeve (32) has at its lower end an outwardly extending radial flange (48) which engages the top of cutter (16) and the inner wall of sleeve (34). Vacuum chamber (36) is defined by the inner wall of sleeve (34) and the circumferential groove in the outer wall of sleeve (32) bordered by flanges (44) and (48). O-rings (50) and (52) serve to prevent any leakage between the inner wall of sleeve (34) and flanges (44) and (48), the latter having annular grooves in which the O-rings are retained.

Through a fluid line (54), vacuum is applied to chamber (36) which in turn receives moisture drawn from cheese in throat (22), through a plurality of circumferentially spaced apertures (56) in sleeve (32) which communicate vacuum chamber (36) with throat (22). Seated against the underside of flanges (48) and (38), and welded to the latter, is the upper of two rectangular plates (60) and (62) which are joined along their lateral edges at (64) to provide a slideway for a corresponding rectangular cutter blade (68) of cutter (16).

Blade (68) has a turned-up edge at (70) (Figure 4.7c) which limits the travel of the blade on the cutoff stroke and facilitates its manual operation. Plates (60) and (62) have concentric openings (72) and (74), respectively, which are of equal diameter and form a continuation of throat (22). Mounted to the bottomside of lower plate (62) is the outwardly extending flange (75) of mold mounting ring (14), having a bearing collar (76) depending from flange (75).

Lower mold assembly (B) is comprised of a cylindrical housing (80) the inner wall of which serves as a cylinder for an integrally connected cheese mold (84) and piston (86). Mold (84) is perforated by a plurality of apertures (85). Housing (80), closed off at its lower end by a head (88), is slidably received at its open upper end by the cylindrical inner wall of mounting ring (14), and its upper end abuts the underside of plate (62), thereby supporting the entire loading mechanism (A).

A mold (84) is hemispherical in shape and at its equator it has a cylindrical rim wall (92) slidably bearing against the inner wall of housing (80). A loading chamber (90) is defined by the inner surface of mold (84) and that portion of the wall of housing (80) extending from its level of contact with the mold to the cutoff blade (68). Threaded into the bottom of mold (84) is a tubular stem (96), coaxial with the major axis of housing (80) and integrally connected to the web of piston (86).

In the rim of piston (86) is an annular groove (94) accommodating an O-ring (102), in slidable contact with the inner wall of housing (80). Forming a continuation of stem (96) below piston (86) is a tubular stem (104) integrally connected to the underside of piston (86) and extending through an opening (106) in head (88). Secured by a nut (110) against the underside of an integral flange on stem (104), is a horizontally extending arm (112) which has threaded into it a stop screw (114) for limiting upward travel of mold (84) and piston (86), by contact with a fixed support (100). Nut (110) is threaded onto the lower extremity of stem (104).

In addition to loading chamber (90), housing (80) has contained within it a pressure chamber (116), defined by the lower side of piston (86), the inner wall of housing (80) and the upperside of head (88), and a vacuum chamber (118) is defined by the upperside of piston (86), inner wall of housing and the lower side of mold (84). For upward movement of piston (86) and consequently mold (84), pressure is applied in chamber (116) through a fluid (e.g., air) pressure line (122) coupled to head (88). By means of a vacuum applied to chamber (118) through a passage (124) which extends from the extremity of stem (104) through said stem and terminates at ports (126) in stem (96), moisture from cheese in mold (84) is withdrawn through the mold apertures (85) into vacuum chamber (118), then through ports (126) and passage (124). Housing (80) is sealed to bearing collar (76) by an O-ring (98) retained in an internal annular groove in collar (76) and sealed against the outer surface of housing (80).

Welded to the periphery of flange (88) of housing (80) is a short tilting lever (130) by means of which housing (80) is pivoted at its lower end on a pivot (132) mounted in the left end of a horizontally extending support arm (134). Lever (130) is connected at its upper end by means of pivot (136) to one end of a piston rod (138) of a fluid actuator (140) having a piston (139) on the other end of rod (138). Integral with and supporting the arm (134) is a mounting head (142) which, by means of a bracket (144) welded to its upper side, pivotally supports actuator (140) at (146). For support of the lower mold assembly (B) through tilting lever (130) and arm (134), and for vertical and horizontal movement of these elements, the upper end of a piston rod (148) connevted to a piston (149) in a cylinder (150) is fixed to the underside of mounting head (142). Cylinder (150) has for its rigid support a flange (152) bolted to the underside of a mounting ring (154) which is in turn supported by an integral wall bracket arm (156).

Referring now to Figures 4.7d and 4.7e, lower mold assembly (B), with its tilting and supporting mechanism [after rotation of piston rod (148) through 180° about its axis] is positioned at the molding station, coupled to upper mold assembly (C). Assembly (C) includes a housing (160) which has as its lower end a thickened cylindrical collar (162) receiving the upper end of housing (80) of lower mold assembly (B).

Collar (162) has an inwardly extending annular shoulder (166) with a diameter equal to that of the inner wall of housing (80). Shoulder (166) engages the upper end of housing (80), thereby serving as a positioning stop for housing (80). Upper mold assembly (C) is supported by any suitable support (not shown) in a fixed position at the molding station. At the upper end of housing (160) is a rim (168) which supports a cap (170) having a downwardly extending flange (172), fitted within rim (168) and seating against a rubber gasket (174) to seal off the upper end of housing (160). Mounted within housing (160) is a hemispherical upper mold section (176) which is seated on shoulder (166) at its lower end, thereby forming a complete spherical mold in conjunction with mold (84) of lower mold assembly (B) when the latter has been moved upward from its original position in housing (80) to engage beneath the shoulder (166).

Integral with and extending upwardly from the top of upper mold section (176) is a hollow stem (178), the upper end of which is sealed to the underside of cap (174) and engages it to maintain mold (176) in position. The vertical passage in stem (178) communicates, at its lower end, through lateral ports (180), with a vacuum chamber (182) defined by housing (160) and upper mold section (176). Through apertures (184) in upper mold section (176), vacuum chamber (182) communicates with the interior of the mold for withdrawal of moisture from and attenuation of cheese within the mold.

Vacuum chamber (182) is evacuated by means of an outlet (186) in cap (170), communicating with the open upper end of stem (178). Vacuum (subatmospheric pressure) may be applied by a suitable vacuum pump, designated Vacuum Source in the drawings. Suitable suction lines may be utilized to connect the outlets (186) and (124) to the vacuum source as indicated by the lines of arrows in Figure 4.7d. Alternatively, separate vacuum sources for each of these outlets could be utilized. O-rings (188) in collar (162), retained in internal annular grooves therein, serve to seal the mold sections (176) and (84) to the housings (80) and (160), respectively.

The Method: In referring to the operation of the aforescribed apparatus, first in the loading station position (Figure 4.7a), mixed Mozzarella cheese curd is placed in hopper (10), butterfly valve (26) and cutter blades (68) being in open positions so as to allow the curd to flow through throat (22) of hopper (10) into dehydrating jacket (12) to which a vacuum is constantly applied through line (54) for withdrawal of moisture from the cheese through holes (56) in sleeves (32), and into loading chamber (90) of lower mold assembly (B). At the same time atmospheric air pressure is applied to the bottom of piston (86) through pressure line (122) to maintain lower section mold (84) through pressure line (122) to maintain lower section mold (84) within housing (80) at a position determined by the setting of stop screw (114) which abuts stationary support (100).

Upon complete filling of, lower mold section (84) and loading chamber (90), cutter blade (68) is actuated to the closed position, thereby cutting off a measured amount of cheese in the lower mold assembly and preventing the flow of any more curd from hopper (10). Simultaneously, a vacuum is applied to lower vacuum chamber (118) through ports (126) and passage (124) for continual withdrawl of moisture through holes (85) in mold (84), and with resultant attenuation of the curd.

In transferring from the loading station to the molding station, lower mold assembly (B) is lowered so as to be completely disengaged from the coupling collar (76). This is accomplished by release of pressure, applied through line (151) and cylinder (150), to the bottom of a piston (149) on the lower end of piston rod (148). After dropping to the position indicated by broken line (200) (Figure 4.7d) lower mold assembly (B) is transferred to the molding station by rotating piston rod (148) about its axis, through 180°, vacuum being continually applied to the cheese in lower mold section (84), throughout the station-changing operation. At the molding station, lower mold housing (B) is again elevated to engage the upper end of housing (80) against the underside of annular shoulder (166) in stationary hopper mold assembly (C).

Upon thus coupling the upper and lower mold assemblies (A) and (B), vacuum is applied to vacuum chamber (182) of the upper assembly through ports (180) and outlet (186), thereby evacuating air from the cheese in the mold through apertures (184), and causing mold section (84) to move to its upward limit position as shown in Figure 4.7d, by the action of atmospheric pressure through line (122) on the underside of piston (86). The withdrawal of moisture and the attenuation of the cheese in the mold is uniformly effected throughout the body of cheese by continuing the application of vacuum to the lower vacuum chamber (118) simultaneously with the application of vacuum to the upper vacuum chamber (182).

The two mold sections (84) and (176) are held in coupled position for a period of time to allow for the shaping of the cheese into the desired ball form. After the molding period is completed, the vacuum in chamber (182) of the upper mold assembly (C) is released and the lower mold (84) is withdrawn into cylinder (80) by the vacuum still being applied to it in chamber (118).

Upon complete withdrawal of the mold and piston (86) within cylinder (80), the vacuum applied to chamber (118) is stopped, lower mold assembly (B) is again lowered by releasing of pressure under piston (149) in cylinder (150) through line (151), to the position indicated by dotted lines (200), and pressure is applied to a piston in cylinder (140) through line (141), thereby actuating piston rod (138) so as to pivot tilting arm (130) about pin (132), and subsequently place lower mold assembly (B) in the position indicated by broken line (B'). At this point, pressure is applied through line (122) to the underside of piston (86) in cylinder (80) whereupon it and mold (84) are moved towards the open end of cylinder (80), thereby ejecting the cheese in the form of a ball as indicated at (202). By applying pressure through line (143) to the other side of the piston is cylinder (140), lower mold assembly (B) is subsequently righted and transferred to the loading station whereupon the whole process is repeated again.

Process Without Working Step

Pasta filata type cheeses are made by R.R. Mauk; U.S. Patent 3,117,008; Jan. 7, 1964; assigned to National Dairy Products Corporation without substantial handworking. In this method a low acid curd (less than 0.08% acidity in the whey) is provided containing biologically active acid-producing agents. The low acid curd is subjected to pressing under vacuum conditions to provide desired body and texture.

The curd is then cured at about 45°F. for about four weeks to permit the biologically active acid-producing agents to break down the curd to provide desired body and texture. While this method is good, it would be desirable to provide a method of making pasta filata type cheese not only without handworking and but also without a lengthy curing period.

It is with this object in mind that R.R. Mauk; U.S. Patent 3,692,540; September 19, 1972; assigned to Kraftco Corporation describes a process in which pasta filata cheese is prepared by forming pasta filata curd in whey, holding the curd in the whey until the acidity of the whey is from 0.25 to 0.30%, draining the whey from the curd, washing the curd, salting the curd, pressing the curd, dividing the pressed curd into pieces, heating the curd pieces to 130° to 160°F. while at a pH of 5.0 to 5.4 in the absence of working while immersed in water to form a soft plastic curd mass and packaging the plastic curd mass.

In the process, a fluid media is provided containing nonfat milk solids and fat. Depending upon the type of cheese to be made, fat is present at a level of from 1 to 4% by weight of the fluid media and nonfat solids are present at about 8 to 11%.

Acidity is developed in the fluid media by the addition of a lactic acid producing culture, such as *Streptococcus lactis* and ripening the media. Ripening is continued until the acidity of the fluid media increased by at least about 0.01% above that of the initial acidity of the fluid media. The fluid media usually has an initial acidity of 0.14 to 0.18%. The lactic acid culture will normally contribute an immediate increase in the acidity of the fluid media of about 0.01%.

Therefore, after ripening, the total acidity of the fluid media is usually from 0.16 to 0.20%. The fluid media is usually ripened in about 30 minutes to one hour when performed at 90°F. After ripening, the fluid media is set with rennet to provide a coagulum. The rennet is added at about three fluid ounces of commercial single strength rennet per 1,000 lbs. of milk.

A firm coagulum is usually obtained about 30 minutes after the rennet is added. The coagulum is then cut to provide curd and whey. The acidity of the whey after cutting about 0.10 to 0.12% and the pH of the curd is about 6.3 to 6.5. After cutting, the coagulum may be stirred for a brief time to initiate whey expulsion from the curd. The curd is then heated in the whey gradually over a period of about 30 minutes to about 95°F. to 120°F. The curd is stirred in the whey until the acidity of the whey is at least about 0.25% equivalent lactic acid to effect acid conditioning of the curd.

The curd and whey are then transferred to a drain table. Acid conditioning of the curd is then effected by holding the curd immersed in the whey until the acidity of the whey has increased to at least about 0.25%. Acid conditioning of the curd may be effected up to a whey acidity of about 0.30%. Acid conditioning of the curd while immersed in the whey to within the indicated range of whey acidities permits subsequent acidity development of the curd while on the drain table to be more easily controlled. Also, it is believed that during acid conditioning of the curd while immersed in the whey the protein of the curd is rearranged and calcium ion is freed from the protein. It is believed that the freeding of calcium ion and its subsequent removal by curd washing is at least partially responsible for providing a pasta filata curd that may be directly heated and packaged.

The acid conditioning step may be effected after a portion of the whey has been removed. This permits recovery of a portion of the whey as low acid whey suitable for processing into an edible grade whey. Removal of a portion of the whey also permits the acid conditioning step to be effected in less time. Up to about 85% by weight of the whey may be removed prior to effecting acid conditioning of the curd in the whey. If a portion of the whey is removed prior to acid conditioning it should be removed before a whey acidity of about 0.18 is developed.

The curd and whey are then transferred to a drain table and the whey is drained from the curd to provide moist curd with some retained whey. The drained curd is held on the drain table until the acidity of the curd reaches a pH of about 5.1 to 5.3. The pH of the whey retained by the curd at this time is about 5.05 to 5.6 and the percent acidity of the retained whey is about 0.35 to 0.50.

The curd is then washed with water at about 45° to 60°F. The wash water is preferably added at a level of about 15 to 30% of the whey that was drained. The absorption of moisture is desirable to leach from the curd free lactose and free calcium ions. However, it is not desirable to have too much moisture absorbed by the curd. Washing the curd with water within this range provides the desired level of moisture absorption in the curd.

The curd is retained in the wash water about 15 minutes until the curd temperature is reduced below about 75°F. Cooling of the curd shrinks the curd and expels whey and free lactose.

The curd acidity after the wash step is such that the pH of the curd during the subsequent heating step is from 5.1 to 5.3. The wash water is then drained and the curd is salted by direct addition of salt to the curd. The salt is added at a level sufficient to provide about 1.7 to 2.6% salt in the final cheese product. Usually about 2.5 lbs. of salt per 1,000 lbs. of milk is sufficient to provide this level of salt.

The curd is then transferred to cheese hoops or drums and is covered and pressed for about one hour or longer. The curd may be stored overnight or may be cured for longer periods of up to several months.

After this point the curd is ready for heating to convert the curd to the smooth, close-knit condition associated with pasta filata cheese. The curd is removed from the hoops and is broken into small pieces. The pieces are then heated by direct steam injection. The curd is heated to about 140° to 155°F. Lower temperatures within the above range are sufficient to establish the curd as a soft, plastic mass. Higher temperatures within the range and strenuous agitation may produce a rough textured product having a broken emulsion. If the acidity of the curd during heating is below about pH

5.0 the heated curd will have an undesirable texture and lack stringiness. If the acidity of the curd during heating is above about pH 5.4 the heated curd becomes rubbery and the emulsion may be broken.

The curd should not be maintained at the elevated temperature of heating longer than about 30 minutes. The curd is preferably maintained at the temperature of heating for as short a time as possible, preferably less than about 5 minutes. After the temperature of the curd is reduced below heating temperatures, however, further cooling of the curd may be effected over a sufficiently long period of time to pasteurize the curd. Pasteurization provides pasta filata cheese with extended shelf life. However, the curd may be cooled immediately after packaging if desired.

The plastic curd, which is now in a flowable condition, is filled into containers by extrusion. A sausage stuffer or piston type filler may be used to package the curd in shrinkable, plastic film tube packages which are sealed at points along the length of the tube.

Example: One thousand gallons of skimmed milk is standardized to provide a level of 2.6% milk fat by the addition of cream. *Streptococcus thermophilus*, at a level of 0.5%, and *Streptococcus lactis*, at a level of 2.0%, cultures are then added. The milk is warmed to 90°F. and is permitted to ripen for one hour.

At this time, the titratable acidity of the milk is 0.18% equivalent lactic acid. The titratable acidity of the milk before the starter cultures are added is 0.15% equivalent lactic acid. Three fluid ounces of commercial single strength rennet is then added for each thousand pounds of milk. After about 25 minutes the milk is set to provide a coagulum, and the coagulum is cut to provide curd and whey. The curd is stirred in the whey for 15 minutes and is then heated in the whey to a temperature of 98°F.

The curd is maintained in the whey while stirring is effected until the titratable acidity of the whey is 0.16%. The curd and whey are then pumped to a drain table and the whey is one inch above the surface of the curd. The curd is held in the remaining portion of whey until the titratable acidity is 0.27% equivalent lactic acid. The free whey is then drained from the curd. The curd, containing residual whey in the curd particles, is then held on the drain table until the pH of the curd is 5.3.

Water at 60°F. is then added to the curd. The water is added until the level of liquid in the cheese vat is about 20% of the level of liquid which was originally present in the vat before partial draining of the whey. The curd is held in the water until the curd temperature is 72°F. and the wash water is then drained. Salt is then added to the curd at a level of 2.5 pounds of salt for each 1,000 pounds of milk which is used. The curd is stirred to disperse the salt and is then transferred to hoops which are lined with cloths. The curd is then pressed for a period of one hour at which time the pH of the curd is 5.2.

The curd is then removed from the hoops and is broken into pieces and is transferred to a process cheese cooker. The curd is then heated to a temperature of 153°F. by injecting steam directly into the curd. At this temperature a soft plastic curd mass is formed.

The plastic curd mass is then transferred to the hopper of a sausage stuffer. The curd is forced from the stuffer into a Saran plastic tube. The tube is sealed at intervals to provide individual packages of cheese curd. The packages are then dipped into water maintained at 200°F. to shrink the film about the curd. The curd packages are then immediately removed from the hot water bath and are placed into a cold water bath, at 45°F. for 30 minutes. The curd is then ready for use identical to Mozzarella curd produced by a conventional procedure.

The pasta filata cheese has a smooth, striated, slightly loose texture with a medium firm body. The flavor is bland, slightly acid and slightly salty. When the pasta filata cheese is shredded, spread on the surface of a pizza and baked, the cheese is stringy, has good spreadability and has a slightly oily sheen. The taste, appearance, color and baking properties are comparable to or better than pasta filata cheese made by conventional hand working methods. The stability of the pasta filata cheese produced with this method with respect to shelf life and body is superior to pasta filata cheese produced with conventional procedures.

Uniform Curd Ricotta Cheese

Ricotta cheese, a product of new milk added to the whey residue from the processing of cheese of other types, is customarily processed in vats of large diameter and volume, by steeping the mixture of whey and milk at an elevated temperature (cooking) then charging the mixture with acid (e.g., vinegar) to curdle the mixture into a suspension of finely divided curd particles in a suspending body of whey, and allowing the mixture to stand while the curd particles rise to the surface, where the curd is skimmed off by hand ladling into a suitable receptacle.

The Ricotta curd is very delicate and will readily disintegrate and remix with the residual whey body if subjected to shock from rough handling during ladling. Also, in hand-skimming from a large vat, it is difficult to capture the remaining vestiges of curd after the bulk of the floating curd body has been ladled off. As a result of these factors, considerable proportion of the curd is lost from each batch. A related problem arises from the time consumed in the hand-skimming operation.

During the skimming operation, the curd remaining in the vat is subjected to the continued cooking action of the underlying body of whey (the temperature of which remains elevated subject only to a slow gradual heat loss) and due to the continued cooking in the presence of the acid, the character and texture of this remaining curd is gradually altered, so that the total content of skimmed-off curd is a mixture of curd portions of varying texture and quality. Thus the aggregate quality of the batch of recovered Ricotta cheese is dependent upon the time consumption and nature of handling of the curd by the workman during the skimming operation. This is not conducive to accurate control of uniformity of product.

This process by N.E. Pontecorvo; U.S. Patent 3,732,110; May 8, 1973; assigned to Pontecorvo R & D Company avoids these difficulties by stimultaneously processing a plurality of small volumes of whey-milk mix, collectively equal to a normal batch, in separate processing cells in the form of vertically standing tubes of small diameter, in a series of steps comprising, in general, preparing the mix by adding milk to whey (e.g., in a proportion of 1 part milk to 2 parts whey); steeping it at elevated temperature (cooking); injecting the mix into the battery of cells; acidifying the mix to develop a colloidal suspension of curd in a residual whey body; chilling the suspension to arrest the cooking action; allowing the suspension to stand at reduced temperature sufficiently long for most of the curd to rise to the surface; draining the residual whey downwardly through respective strainer baskets (in which the curd is collected and retained) at a rate sufficiently slow to allow the fine curd particles to settle in the baskets without being pulled through them; and finally elevating the baskets and emptying their contents into a receptable or onto receiving surfaces. All of these steps are performed simultaneously in the several cells so as to develop a uniform aggregate product.

The bodies of collected curd are sufficiently small in volume to avoid any weight-induced deterioration of the curd. The sudden cooling (a new step) provides efficient control over curd development, by arresting the cooking operation to enable a processor to attain a uniform product from a selected schedule of processing factors.

Arrest of cooking also makes it possible to utilize time control in the slow draining of the whey from the curd. By utilizing small diameter cells in a fairly large number, it becomes possible to effect charging, acidification, chilling, and draining much more rapidly than would be possible in a single vat having a volume equivalent to the aggregate volume of the plurality of small cells. The injection of mix and the application of steam or hot water for heating, and of cold water for chilling, is effected through common manifolds communicating with all of the cells. The baskets are manipulated in unison, preferably being tied together for that purpose.

The process thus provides a Ricotta cheese with a maximum recovery of curd from the processed mixture, with maximum predictable uniformity of the resultant product, with minimum dependency on the skill of the operator, and with maximum rapidity.

EQUIPMENT

APPARATUS TO CONDUCT ENTIRE PROCESS

Mechanization of Cheddaring Steps

Arthur M. Sadler; U.S. Patent 2,942,983; June 28, 1960; assigned to the U.S. Secretary of Agriculture discuss a mechanical procedure for packing, matting and draining excess whey from the curd which in this method, reduces the variables encountered from one batch of cheese to another, thereby promoting uniform quality of product.

In this method, pasteurized milk is inoculated with a starter culture of lactic acid-producing microorganisms, the milk is coagulated with rennet and subsequently cut into cubes giving a mixture of curd and whey which is heated and stirred. The mixture is transferred from the vat into a cloth lined, perforated curd retention and matting device, preferably constructed of stainless steel or other inert material, where the curd, submerged in whey, is collected in this container until the desired amount, part or all of the curd in the vat, has accumulated in the container, excess whey being withdrawn from the container during this procedure.

Pumping is halted, the cloth liner of the curd retention and matting device is folded over the curd and whey, a perforated stainless steel follower is introduced into the container and a pressure of about 30 pounds per square foot is applied, usually for a period of two hours. Free whey is allowed to drain during the pressing. After pressing, the matted block of curd is removed from the curd retention and matting device and cut into slices and milled in a Cheddar cheese mill. The milled curd is then salted, hooped and pressed into forms.

The process provides many advantages in regard to savings in labor and equipment. For example, the cheese vats are available for reuse as soon as the pumping of curd and whey has been completed. In addition, the procedure yields a better more uniform quality product.

Example 1: Whole milk containing 3.9% butterfat was pasteurized and cooled.

This milk was pumped into a regular 4,000 pound steam jacketed cheese vat. In this example 2,550 pounds of milk was used. The temperature of the milk was adjusted to 88°F. Seven-sixteenths of 1% of S. lactis was added as a starter and the milk was allowed to ripen for 1 hr. After ripening, 306 cc of rennet was added. When the milk was properly coagulated, in this case 30 minutes, it was cut into cubes with regular 1/4 inch curd knives. The cut curd was agitated slowly to prevent breaking and clumping of the soft particles.

About 15 minutes after cutting, the curd and whey was heated to 100°F. in 30 min. time. Stirring of the curd and whey continued for an additional 60 minutes. After heating and stirring, the mixture of curd and whey was transferred by pumping into the large sized cheesecloth lined curd retention and matting device, sufficient whey being first transferred to the curd retention and matting device in order that the curd particles in the curd and whey which followed would at all times during pumping be kept completely submerged in whey. After pumping was completed the cheesecloth was folded over the top of the curd and whey in the curd retention and matting device. A perforated metal follower and press weights of about 30 lbs. per square foot were applied and the free whey was allowed to drain.

After about 2 hours of pressing, the curd block was removed from the curd retention and matting device, cut into slices suitable for milling (3/4 of an inch thick), and milled with a conventional Cheddar cheese mill. The milled curd was stirred intermittently for 10 to 15 minutes and a total of 7 lbs. and 7 oz. of cheese salt was added in three applications.

After the salt was adsorbed, in this case 15 minutes, the milled salted curd was hooped in conventional cheese hoops and placed in a horizontal cheese press. After pressing for about 30 minutes the cheese was dressed and further pressed for 16 to 20 hours. The cheese after pressing may be wrapped or surface dried for paraffining and placed at 50°F. for curing. At the time of wrapping the cheese contained the following.

	Moisture, percent	Butterfat, percent	Salt, percent
20 lb. block	35.38	35.75	1.61
40 lb. block	35.12	36.00	1.56

Example 2: The process of Example 1 was repeated using the same amount of milk, 2,550 lbs. At the time of wrapping the analyses below were obtained. Comparison of the analyses shows that the process results in a product of uniform composition from one batch to another.

	Moisture, percent	Butterfat, percent	Salt, percent
20 lb. block	34.93	35.5	1.64
40 lb. block	35.26	36.0	1.54

Whey Expulsion and Curd Shredder

W.W. McCadam; U.S. Patent 3,078,169; February 19, 1963; assigned to Ched-O-Matic Corporation provides a method and apparatus for making cheese that makes the final quality of the product less dependent upon the skill of the operator. In one phase, the tedious and inherently unsanitary curd block turning process is replaced by a whey removal and curd operation that minimizes loss of heat and contact of the curd with the atmosphere and provides excellent control over the acid development and expulsion of moisture. In another phase, all of the operations following the pressing of the curd are made continuous and substantially automatic, the curd being continuously and automatically shredded, washed and uniformly salted prior to hooping.

The apparatus generally comprises a whey expulsion press, a conveyor for continuously receiving blocks of curd from the press and transporting them to a shredder, means for washing and salting the shredded curd and a screw conveyor mechanism for removing the curd for hooping. Figures 5.1a through 5.1d illustrate the apparatus.

The whey expulsion press comprises a rectangular vessel (10) provided with a vertically moveable platen (11) whose movement within the vessel is controlled by hydraulic cylinders (12). The platen is provided with a suitable filling pipe (13) for receiving curd and whey from suitable means such as a rotatable foraminous drum (14) for preliminarily removing part of the whey by tumbling the suspension of curds in whey. As shown, part of the area of the platen is perforated at (16). The other nonperforated area (17) of the platen is separated from the perforated area by a screen (19).

The bottom of the vessel comprises a fixed plate (20) having perforations (21). Under plate (20) there is a vertically moveable plate (22) having upstanding plugs (23) which, when plate (22) is in its normally elevated position, fit within the perforations and form a substantially unbroken plane surface with the top of the fixed plate. Plate (22) is positioned by guide pins (24) and actuated for vertical movement at a preselected time in the operation of the press by means indicated at (26). Below the bottom assembly (20) and (22), there is provided a suitable drainage board (27) for collecting fluid leaving the vessel by way of the perforations.

The lower portion of the vessel is provided with a parting knife and partition (29) moveable horizontally in a slide or track (30) and actuated by a hydraulic cylinder (31) or equivalent means. Immediately below the parting knife and in a cutout portion of the side wall there is provided an extrusion piston (32) flush with the side wall and actuated by the hydraulic cylinder (34). At the opposite side in the lower portion of the vessel there is a vertical cutoff knife (36) actuated by hydraulic cylinders (37) for cutting a block of curd into a mill slab (39), shown in Figure 5.1a as resting on a conveyor (40).

A preferred curd treating apparatus is shown in Figure 5.1c where a slab conveyor (44b) is provided for transporting the slab blocks (39) to a shredder (42), from which shredded or milled curd is deposited directly in a hopper (50) for a screw conveyor (51), the curd being washed and salted by spraying with one or more brine sprays (52).

The underside of the conveyor is provided with a groove (54) for draining off excess moisture as the curd is advanced through the conveyor. In operation, a suspension of curd and whey having a total acid content (the total titratable acidity of the whey expressed as percentage of lactic acid) between 0.16 and 0.25% is introduced into the vessel from a cheese vat. With this process it is desirable to modify the cooking method to the extent of cooking at slightly lower temperatures and thus achieving

FIGURE 5.1: WHEY EXPULSION AND CURD SHREDDER

a.

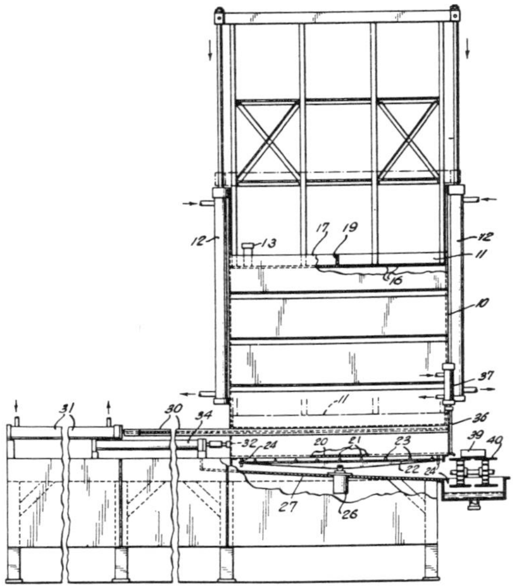

Whey Explusion Press with Means for Cutting Curd Block into Mill Block

b.

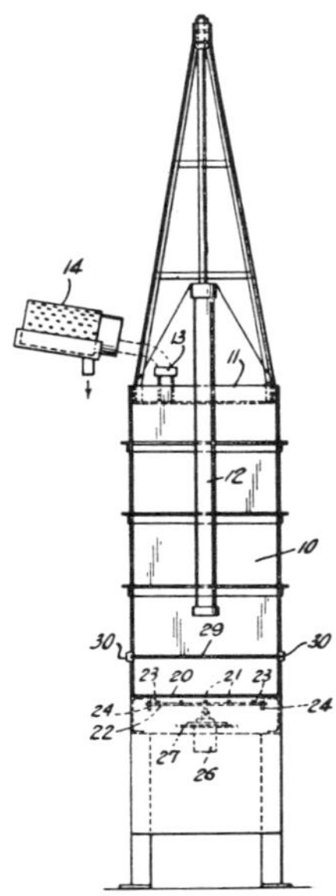

End View of (a) with Whey Drainage Means (continued)

FIGURE 5.1: (continued)

c.

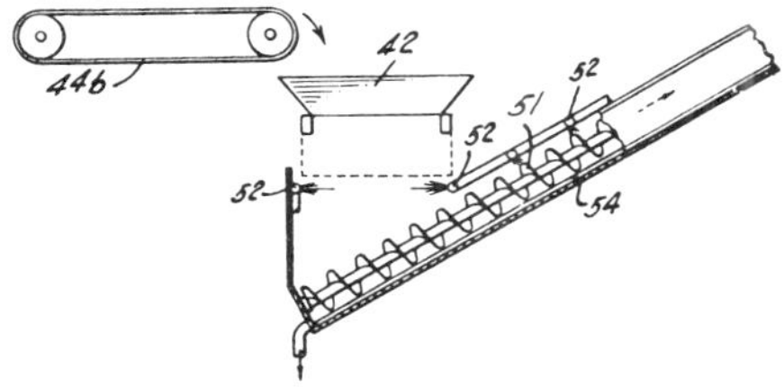

Conveyor for Continuous Processing

d.

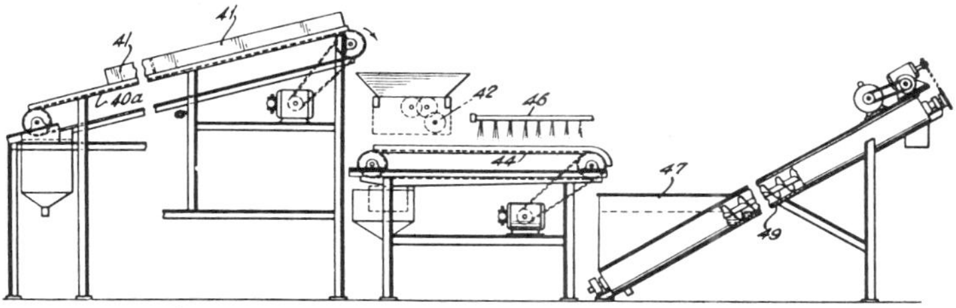

Continuous Slab Processing Apparatus

Source: W W. McCadam; U.S. Patent 3,078,169; February 19, 1963

better texture than is obtainable at slightly higher temperatures. Also, it is desirable to drain off as much of the whey as possible, usually about 50%, in the cheese vat and in a rotating foraminous drum before introducing the suspension into the vessel, care being taken however not to drain off so much whey in the vat that the suspension becomes unpumpable or otherwise resistant to transfer from the vat to the drum.

When the drained contents of a cheese vat have been passed through the drum and emptied into the vessel, the platen is moved downward by actuation of the cylinders to drain off any free whey at the top of the suspension through the perforations. The free whey passes to the nonperforated portion by way of the filter screen which filters out particles of curd that may have passed through the perforations From the nonperforated portion of the platen the free whey is siphoned out. An additional quantity of free whey is removed from the bottom of the curd mass by lowering the moveable plate sufficiently to allow free whey to drain through the perforations for collection in the drain.

When the suspension of curds and whey has been introduced into the whey expulsion press, periodic or continuous acidity determinations are carried out to follow the rate of acid development and correlate it with the moisture content. It is the general objective in the press to develop an acid content between 0.6 and 0.75%, more or less simultaneously with a reduction of the moisture content in the curd between 35 and 40%. If the desired acid content is achieved before the moisture content is reduced to the desired range, the final cheese product tends to be sour. If, on the other hand, the desired moisture level is reached before the acid value reaches the desired range, the development of acid is curtailed and perhaps stopped entirely so that the desired acid content may never be achieved.

Application of relatively high pressure before the acid content has increased to about 0.4% is generally undesirable because in that stage the curd particles are fragile, they release moisture only with difficulty and a reduced yield, because of the escape of casein with the whey, would ensue. Inasmuch as the bacteria convert the sugar of the whey in the curd into acid and the sugar content is related to the whey content, adjustment of acid development and hence final acidity, and correlation thereof with the moisture content is obtained by controlling the sugar content through control of the rate of whey expulsion.

It is generally necessary to increase the pressure on the curd from an initial value of approximately 0.45 psig to final values as high as 1.75 psig, depending upon the rate and manner in which acid development takes place. Thus, for example, a suspension having an initial acid value when introduced into the press of 0.18% was subjected to an initial pressure of 0.48 psig, the pressure was increased an average of 0.086 psig every 10 minutes for 80 minutes at the end of which time the acid level was 0.54% and the pressure was 1.19 psig. Thereafter, both the rate of acid development and the rate of increase in pressure were reduced until at the end of 170 minutes the acid value was 0.75%, the moisture content was between 35 and 40% and the pressure was approximately 1.56 psig.

In another run, the acid content rose rapidly from an initial value of 0.25% to a value of 0.6% in 80 minutes, the pressure in the meantime being gradually increased from 0.57 to 1.30 psig. Thereafter, the acid developed still more rapidly to a value of 0.85% after a total of 100 minutes, the pressure in the meantime being increased just as rapidly to 1.75 psig.

These two runs represent extremes, it generally being preferable for optimum quality of the final cheese to start with a suspension having an acid value higher than 0.16 and lower than 0.25 and to so adjust the increase in pressure as to attain an acid value between about 0.6 and 0.75% and a moisture content of 35 to 40% in approximately 130 minutes.

It is to be understood, of course, that if the milk used initially has a fat content higher than the normal 3.5 ± 0.5% of cheese milk, generally higher pressures must be employed to compensate for the slower release of moisture in order to obtain the desired final 35 to 40% moisture content unless compensation has already been effected through the use of somewhat higher cooking temperatures in the cheese vat.

When the whey expulsion operation has been completed and the desired acid value and moisture content have been achieved, the curd remaining in the press is in a sufficiently coherent state to form a block having a toughness and general characteristics similar to those exhibited by blocks of curd that are obtained as the end of a conventional cheddaring operation. With the plate returned to its higher position, the hydraulic cylinder is actuated to advance the parting knife and partition across the bottom of the vessel as indicated in Figure 5.1a.

When the knife has completed its motion across the vessel, the vertical knife is elevated to its upper retracted position and the plunger is advanced several inches. The vertical knife is then lowered by the action of the hydraulic cylinders to cut a block of pressed curd in the bottom of the vessel into a mill slab which is deposited, as shown in Figure 5.1a, upon the conveyor (40). The knife is retracted to its uppermost position and the operation is repeated, the plunger and the knife being operated alternately until the plunger reaches the right side of the vessel.

The plunger is then retracted to its leftmost position and the parting knife and partition are likewise retracted so as to allow the curd block in the vessel to move down into contact with the bottom member. The sequence of operations is then repeated beginning with the movement of the parting knife and partition and proceeding with the alternate movements of the plunger and the vertical knife until the entire curd block in the press has been extruded, cut and deposited upon the conveyor. The mill blocks of curd deposited on the conveyor may be any desired size.

In Figure 5.1d, the blocks were cut to have dimensions, when lying flat on the conveyor, of 24" x 6" x 2". Upon being deposited on the conveyor the blocks passed under the hood in which they were maintained at a temperature of 98° to 102°F. as in the curd press and were subjected to pressure by rollers or equivalent means in order to increase the coherence of the curd particles for one another. Four rollers weighted successively to exert pressures of 1.27, 1.91, 3.05 and 2.55 psig on the mill blocks were employed and this resulted in flattening out the blocks until they had dimensions of approximately 30" x 10-12" x 1".

The pressed mill blocks were fed continuously into a shredder or milling machine (42) and the shredded curd particles were deposited on a second conveyor (44). The shredded curd particles were then washed with water while on the conveyor. While rinsing with water is not critical or necessary to the method, it is preferred for optimum cohesion of the curd particles and texture of the final cheese, removal of butterfat from the surface of the shredded curd particles, and most uniform absorption of salt in the next step.

The conveyor deposited the shredded and washed curd in a salt brine solution having a salt concentration of about 5% and maintained at about 100°F. for about 1 minute in the tank (47). Generally, the brine solution should have a salt content in the vicinity of 5%. Its temperature should be about 90° to 110°F., 90°F. being minimum for further promoting cohesion of the curd particles and a practical rate of penetration of the salt into the interior of the curd. The residence time for the curd particles in the brine solution should be about 1 minute when the temperature is between 90° and 110°F. and the salt concentration is about 5%.

For higher salt concentrations the residence time should be correspondingly short as the object of the treatment with salt brine is to achieve a salt concentration in the curd of 1.7 to 1.8%. The screw conveyor (49) continuously removes the salted curd from the brine tank for hooping and curing operations.

In the preferred method the mill blocks of curd are, as illustrated in Figure 5.1c, moved by the conveyor (40b) to the shredder. The shredded or milled curd is dropped into the hopper (50) of a screw conveyor (51), being simultaneously washed and salted by sprays of saturated salt solution, preferably maintained at 90° to 110°F., from brine sprays (52). This is preferred because it lends itself to maximum control over the amount of salt taken up by the curd and avoids problems of varying residence times of curd in a brine bath. If sampling indicates that more salting is required, it is a simple matter to bring one or more additional brine sprays into operation or to slow down the rate of advancement of curd.

One of the most outstanding advantages of this method and apparatus is that while the curd and whey are in the whey expulsion press, it is possible for the operator to follow the development of acid in the curd as reflected by the change in acid content of the whey and thus to regulate the increase in pressure on the mass so that the removal of moisture will properly keep pace with the development of acid. The finished cheese is characterized by uniformly excellent flavor, a superior firm and waxy texture and an improved translucent appearance. This is believed to be due in part to the fact that it is unnecessary to cook the milk at a temperature higher than about 100°F.

Another reason for the increased uniformity of good texture and flavor is that in the whey expulsion press the regulation of pressure is correlated with the development of acid and the removal of moisture on the basis of recorded observations rather than simply on visual inspection and judgment of a given operator. Another important advantage is derived from the uniform rinsing and salting of the shredded curd particles.

H. Leber; U.S. Patent 3,078,170; February 19, 1963; assigned to Ched-O-Matic Corporation cites the further advantage of this method of cheese-making in that it lends itself not only to the making of Cheddar cheeses, wash curd cheese and Colby cheese, but also to the semisoft and soft cheeses, such as Limburger, Munster, Camembert, Brie and cottage.

For the stirred curd cheese the compression chamber may be avoided as there is little need of curd fusion, however, most of the free whey should be drained rapidly from the curds as by the rotating drum and the washing and salting carried out in accordance with the preceding discussion. For the other cheeses fusion of the drained curd will be carried out in the compression chamber although with pressures appropriate for the particular cheese curd. With all of these cheeses a superior product will be obtained when most (90%) but preferably substantially all (98%) of the free whey is rapidly removed from the curd when the whey has developed an acid content appropriate for the particular type of cheese and at which fusion of the curd readily occurs.

A further improvement in this cheese manufacturing method and apparatus is proposed by M.W. Pinckney; U.S. Patent 3,140,185; July 7, 1964; assigned to The De Laval Separator Company. In the process the separated curd is delivered to a confined space where additional whey then is withdrawn from the curd mass at a plurality of spaced locations throughout the area of the confined space to form a block of dry curd. After the curd mass has been drained for a sufficient period of time to indicate that further whey removal cannot be obtained, the curd mass is mechanically cheddared.

The curd mass is mechanically or power cheddared by segregating a portion of the curd mass from the remainder of the mass and applying an internal force on the segregated portion to cause it to flow through a restricted path into the remainder of the mass. This operation is repeated until the entire mass of curd has been cheddared at least once and until the curd has matted into a block and has a chicken breasted texture which permits continuous strings to be pulled from the remainder.

This mechanical cheddaring performs the function of pressing the curd to squeeze out free whey and to physically displace all portions of the curd mass so that any pockets of whey will be moved into engagement with a side of the confined space where the whey may be drained.

After the mechanical cheddaring of the curd mass has been completed, it is left in a quiescent state at a temperature suitable for acid development for a period of time required to produce the desired acid content. The mechanical cheddaring provides for dry acid development in the curd mass as distinguished from wet acid development where free whey is present. This reduction in moisture and the elimination of pockets of whey produces a superior cheese product both in texture and flavor. After the acid development in the cheese has been completed, it is removed from the confined space and salted. Preferably, salting is performed after the cheese mass has been shredded by either subjecting it to dry salt or warm salt brine.

The draining is performed by perforated drain tubes located at spaced points and extending throughout the height of the curd mass. The drain tubes act as sieves or strainers to permit the flow of whey but prevent the flow of curd into the tubes. Whey separated from the curd then flows through the perforations into the tubes which, in turn, are connected to a vacuum source to withdraw the whey. The curd mass is in a rather fluid state and is adapted to flow from one location to another in the confined space and, due to the low pressure, causes free whey to flow toward the vacuum source.

It has been found that any pockets of free whey in the curd mass located within 6" of the periphery of a perforated drain tube will migrate to and flow into the drain tube. It is therefore desirable to locate the drain tubes at a distance no greater than 12" from periphery to periphery. The draining step can usually be completed in a period of about 20 minutes but should be continued until it is evident that the curd mass is relatively free of whey. The perforated drain tubes are then withdrawn from the curd.

About 20 minutes after the perforated drain tubes have been withdrawn and when the curds have started to fuse and the dry acid has started to increase the mechanical or power cheddaring step is initiated. This is accomplished by segregating a layer of curd from the remainder of the mass as, for example, by passing a horizontal blade through the mass until its forward end is about two-thirds the distance from the rear to front wall so as to leave an opening one-third the distance between the front wall and forward edge of the blade.

Pressure is applied to the segregated layer by means of a plunger through the opening and into the remainder of the curd mass. This produces a displacement of successive layers of the curd mass toward the bottom and front walls of the confined space to separate any additional pockets of whey from the curd which can then drain from these areas.

Successive segregating and displacing steps are performed by withdrawing the blade and plunger to permit the curd to fall into the open space and then operating the blade and plunger successively. This mechanical working of the curd mass, in addition to expelling additional whey, further augments fusion and matting of the curd into a uniform cohesive texture characteristic of a gocd cheese. The cheddaring operation is repeated until the entire mass has been mechanically worked or cheddared.

This may require the cheddaring steps to be repeated 6 or 8 times or until the operator is satisfied with the texture of the curd. The mechanical cheddaring should be continued until a chicken breast texture is acquired. The cheddaring operation requires a relatively short period of time, for example, 20 minutes.

Upon completion of the mechanical or power cheddaring of the curd mass within the confined space, the mass is allowed to stand quiescent for from 30 to 60 minutes while the dry acid develops by biochemical action. Dry acid development is required if a cheese having a flavor and aroma of the desired quality is to be produced. After treatment of the curd mass has been completed it is removed from the confined space and salted usually after milling and hooped for storage.

Use of Gravitational Force

The method and apparatus described by J. Czulak; U.S. Patent 3,167,862; Feb. 2, 1965; assigned to Commonwealth Scientific and Industrial Research Organization, Australia provides a process for the manufacture of Cheddar or like cheese, in which granular curd is converted into a fused, partly fibrous mass by gravitational force, and in which the uniformity is enhanced by inverting the mass at some stage of the conversion to allow the gravitational force to act on the mass in the opposite direction. Thus the usual practice in the manufacture of Cheddar, Cheshire or like cheese to work the curd mechanically or manually is not necessary in this described technique which may then simplify and cheapen production.

The process consists of feeding cheese curd in whey in a traversing stream against a perforated wall to effect a substantial separation of the whey from the curd, stirring the curd material to effect further drainage, forming the drained material into a series of slabs each of which is small enough to allow efficient drainage of the whey therefrom, moving the slabs along a predetermined path to allow partial consolidation by gravitational force, curving the path through an angle of approximately 180° to cause inversion of the slabs, supporting the slabs on all sides during inversion and further moving the inverted slabs along a predetermined path to further consolidate the curd. The slabs while travelling along the last mentioned path are supported on an upwardly convex surface to cause the curd to spread and flow laterally and thereby develop a fibrous structure.

Figures 5.2a through 5.2d more specifically demonstrate the procedure. Material in the form of small pellets of curd in whey is fed into vat (1) (Figure 5.2a) This material is then continuously stirred while liquid is allowed to drain through the perforated inner walls (9) and the perforated plates (16) to the side troughs (18) Since the nozzles (7) are mounted at an angle to deliver the curd and whey slurry against the perforated walls, a high degree of separation of the whey from the curd takes place along the top portions of the perforated walls.

This ensures that the curd needs less mechanical stirring than would otherwise be required to free the whey from the curd and also provides a self-cleaning action on the perforations in the walls. While the curd is thus being dried or firmed, its temperature is controlled by the circulation of warm water through the hollow retractable floor (10). When the initial drying or firming is complete, the retractable floor is moved by means of its rack and pinion drive. The drained curd is thus discharged in a body from the vat into slat conveyor (19) which consists of an endless chain of compartments made up of L-shaped slats (21) (Figure 5.2b).

FIGURE 5.2: USE OF GRAVITATIONAL FORCE

a.

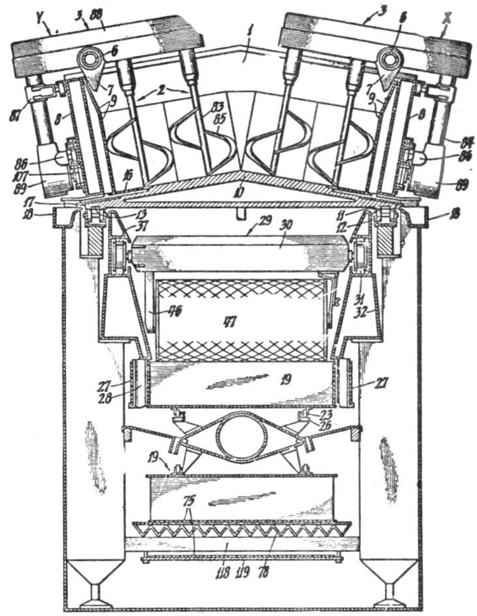

b.

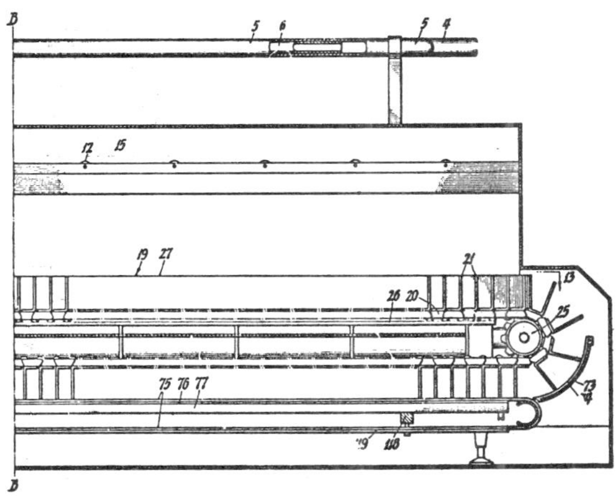

(continued)

FIGURE 5.2: (continued)

c.

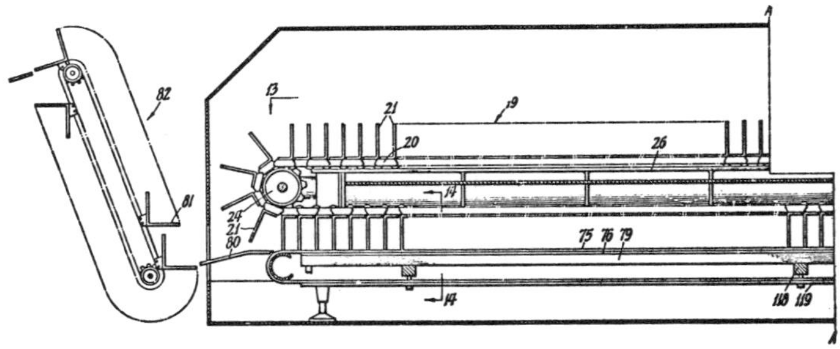

d.

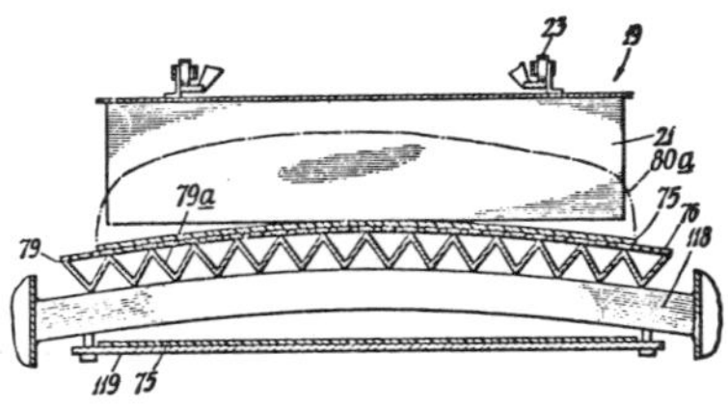

Source: J. Czulak; U.S. Patent 3,167,862; February 2, 1965

After the drained curd is discharged from the vat into those compartments directly under the vat, the top surface of the curd is leveled off by means of a leveling device. As the compartmented and leveled curd is thus being carried by the upper span of the conveyor, the curd is consolidated under its own weight and drains further. To ensure efficient drainage, the conveyor slopes gently downwards towards its left hand end. In this way drainage can take place through the spaces between successive slats (21).

As each curd-carrying compartment passes downwards at the right hand end of the conveyor, the curd is prevented from falling out of the compartment by means of a curved stationary retaining plate (73), which is provided with a plastic inner covering (74). In this way, the curd in each compartment is brought into an inverted position. After a compartment has passed beyond the plate, the curd rests on the top span of an endless plastic belt (75) which is slidably mounted on stationary curved plates (76).

The upper span of the plastic belt is slidably supported on water tanks (77) and (79). Warm water is passed through these tanks to control the temperature of the curd while it passes along the lower span of the conveyor. The upper surfaces of the water tanks are perforated to provide a lubricating water film between the plastic belt and the upper surfaces of the water tanks. In this way, the frictional forces tending to prevent movement of the plastic belt are reduced to such an extent that the frictional grip of the curd on the belt is sufficient to move the belt with the curd without slippage between them.

The upper surfaces of the water tanks become increasingly convex in cross-section in the direction of travel of the lower span of the conveyor. Thus tank (77) is flat at its right hand end while at section (14—14) (Figure 5.2c) the tank (79) has the cross-section shown in Figure 5.2d. Because of this convexity, the tanks are divided into separate longitudinal water compartments as shown at (79a) in Figure 5.2d to maintain a constant water head across their width. In this way a uniform lubricating film of water is provided between each tank and the top span of the plastic belt. The water tanks are supported by rigid beams (118) which also carry plates (119) upon which slides the lower span of the plastic belt.

When a block of consolidated curd reaches plate (80) at the end of the lower span, it is pushed onto the sloping portion of the plate by the following slat (21). The block then slides onto one of the plates (81) by which it is raised and finally tipped into a tearing or shredding device or into a mill and is then treated as may be required to produce Cheddar, Cheshire or similar cheese.

While traveling along the upper span of the conveyor, the curd in the various compartments and particularly the curd in the lower ends of the compartments, becomes converted into a fused, partly fibrous mass under the action of its own weight. While traveling along the lower span of the conveyor in the inverted position, the process of conversion is continued, particularly in the portion of the curd which occupied the upper ends of the compartments while traveling along the upper span. In this way, the degree of conversion of the curd to a fused, partly fibrous mass becomes substantially uniform.

It is also to be noted that, while the curd filled compartments traveling along the upper span of the conveyor are closed on four sides, the compartments in the inverted position are not provided with side walls. Thus, the blocks of partly consolidated curd are free to spread and flow laterally during the movement of the blocks along the lower span of the conveyor.

The upper surfaces of the water tanks become increasingly convex in cross-section in the direction of travel of the lower span of the conveyor. This means that the effective floors of the inverted compartments take on an increasingly convex configuration and this ensures stretching of the curd blocks due to spread and flow through the open sides of the compartments. Thus the shape of a curd block at the stage of the process represented by line (14—14) in Figure 5.2c may be represented by broken line (80) in Figure 5.2d. For reasons well-known in the art, this stretching is necessary for producing the desirable fibrous structure in the curd.

Uniform Moisture Content

G.E. Krueger, H.F. Mohr and O.C. Huber, Jr.; U.S. Patent 3,355,805; Dec. 5, 1967; assigned to National Dairy Products Corporation describes apparatus used to produce large blocks of hard natural cheese such as Cheddar cheese. The apparatus not only aids in producing uniform moisture content throughout the block but also facilitates the maintenance of hygienic conditions.

The apparatus shown in Figure 5.3a, 5.3b and 5.3c uses a plurality of rectangular containers (10) adapted to receive a large quantity of curd and some whey for pressing and further draining prior to curing. Each container is adapted to be positioned in one of a plurality of pressing frames or holders (14) which supports the container and

FIGURE 5.3: UNIFORM MOISTURE CONTENT

a.

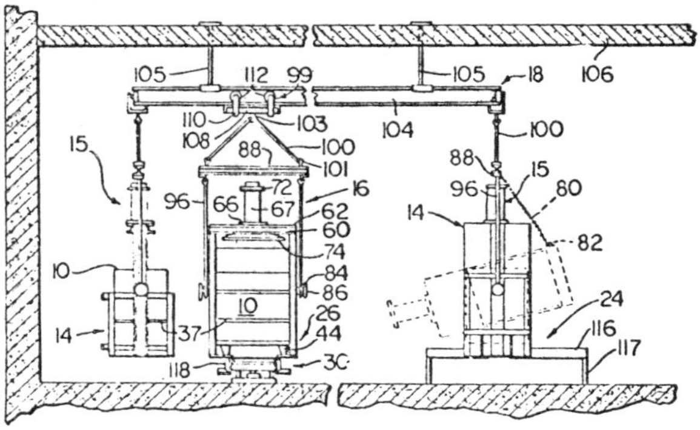

b.

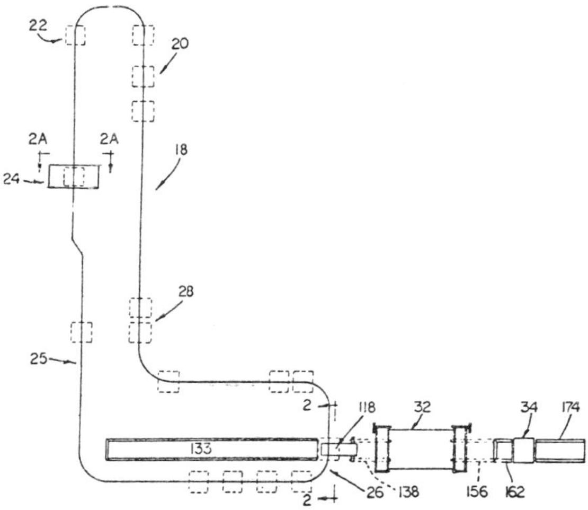

c.

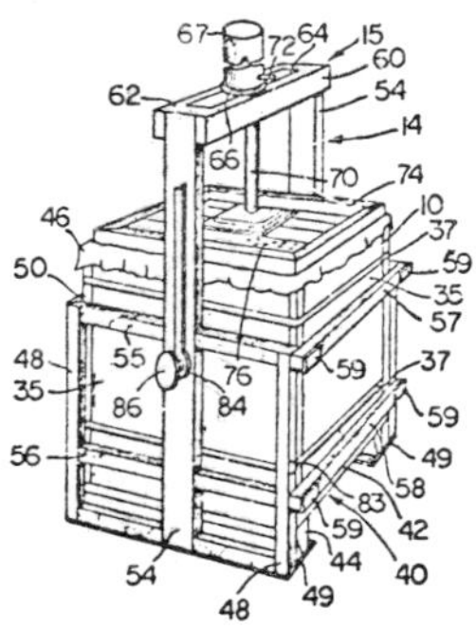

Source: G.E. Krueger, H.F. Mohr and O.C. Huber, Jr.; U.S. Patent 3,355,805; December 5, 1967

includes an air powered press (15) of the reciprocating type for applying force to the contents of the container. Each of the pressing frames is pivotally mounted in a suspension frame (16) so that the container it holds can be rotated about a horizontal axis into a plurality of orientations.

The suspension frame in turn is suspended from and movable along an overhead conveyor (18) comprising an endless closed loop. In Figure 5.3b the overhead conveyor loop extends from a loading station (20) where the containers, filled with curd and containing some whey, are placed in the pressing frames through successive pressing stations (22), (24) and (25), where the press is connected to an air pressure source and to an unloading station (26) where the containers are removed from the pressing frame. The overhead conveyor loop further extends from the unloading station through a frame storage and cleaning area (28) and returns to the loading station.

At the unloading station a second conveyor means (30) is provided for removing the containers from the pressing frames, moving the containers into and out of a vacuum chamber (32) where gas is extracted from the cheese and to a curing area. A turnover means (34) is also provided for inverting the containers after they are removed from the vacuum chamber (32) in order to facilitate closing the containers prior to curing.

An important feature of the apparatus is that it is almost entirely suspended from the ceiling or supported by simple means above the floor of the plant. By this expedient sanitizing of the plant area and prevention of infection of the bacterial starter culture by phage is greatly facilitated. Furthermore, the use of heavy complex machinery such as fork lift trucks to move the containers into and out of presses and to tilt and invert them is eliminated. In addition, more whey may be recovered and utilized for various by-products, than is the case with prior apparatus.

As previously noted, in operation of this apparatus, the frame is tilted into various orientations. The frame is maintained in these positions by a stabilizing chain (80) attached at one end to the suspension frame (16) and having a hook (82) on its other ends. This hook engages the frame by means of a U-shaped bolt (83) which is attached at its ends to one side of the frame adjacent the lower rail (58) and which has its opening projecting beyond the outer face of the rail (58). This arrangement provides an eye which the hook on the stabilizing chain can engage.

As previously mentioned the pressing frame is pivotally mounted in the suspension frame and for this purpose, horizontally aligned trunnions (84) are provided on the side posts (54) and extend outwardly from the pressing frame between the upper and lower side beams (55) and (56), and slightly above the center of the gravity of the pressing frame with a filled container positioned therein. The trunnions (84) are provided with disks or keepers (86) on their outer ends in order to maintain the desired horizontal alignment of the pressing frame in the suspension frame.

The suspension frame includes a horizontal carrier beam (88) which, as illustrated, is an I-beam approximately as long as the distance between the keepers (86) of the pressing frame. Hanger arms (96) are pivotally suspended from opposite ends of the carrier beam and, at their lower ends rotatably receive the trunnions. The hanger arms and carrier beam thus pivotally support the pressing frame so that it can rotate about a horizontal axis which is defined by the trunnions and is parallel to the carrier beam.

The hanger arms are long enough in relation to the pressing frame so that the top of the cylinder housing does not touch the carrier beam as it rotates about the axis defined by the trunnions. Since the trunnions are positioned close to the center of gravity of the pressing frame and filled container, it is relatively easy for a single workman to tilt the frame manually. The stabilizing chain is attached at one end to the carrier beam and is of such length that, when the hook on the end of the chain engages the U-bolt, the pressing frame will be retained in an inverted position with its lower end higher than the cylinder. In this instance, the angle of tilt with respect to the horizontal is small. For different angles, different lengths of chain are used.

The conveyor extends from the loading station through successive pressing stations to the unloading station and through the frame storage and cleaning area back to the loading station. The containers in this instance are filled after the primary draining has taken place with a mixture of curd and the remaining whey from vats before being placed in the pressing frames. The conveyor loop, however, may extend from the loading station to the area of the vats so that the containers are loaded after they have been placed in the frames.

The conveyor loop has the general shape of two L's with their legs parallel to one another and interconnected by cross-rails. The loading station is located at the end of the longer leg of the small L and the three pressing stations are located on the parallel longer leg of the other L.

After loading the pressing frames with the filled containers they are moved via the adjacent cross-rail to the adjacent leg of the other L and to the first pressing station. There a plurality of air hoses are connected to a source of air pressure and are adapted to be connected to the cylinder housings (67) on the pressing frames via coupling means (72).

Desirably, each pressing frame is tilted slightly, on the order of a few degrees from the vertical, to facilitate the drainage of whey from the container and the frame through the perforations in the supporting plate of the pallet. The frame is maintained in this position by inserting the hook of one of the stabilizing chains having a suitable length through the U-bolt. In a specific example, where the container has a height of 4 ft. and sides of 3 ft. in length, a force of 1,000 lbs. is applied to the pressure plate for 1/2 hour.

After pressing and draining at the first pressing station, each pressing cylinder is disconnected from the air pressure source and each pressing frame is moved further along the longer leg of the longer L of the overhead conveyor to the second pressing station where it is positioned directly over one of a plurality of drain pans (116). The drain pans are provided with legs (117) to space them from the floor to facilitate cleaning and disinfecting the area adjacent the drain pans.

Each pressing frame is then tilted more than 90° from the vertical as shown in Figure 5.3a in dashed lines so that the whey will drain through the apertures in the pressure plate at the top of each container rather than through the perforations in the pallet. The pressing frame is maintained in this position by hooking a stabilizing chain having a suitable length through the U-bolt. A plurality of air hoses connected to a source of air pressure also are provided at the second pressing station and are coupled to the presses through the coupling means.

In a specific example, utilizing the 22" x 28" x 28" containers mentioned, approximately 2,000 lbs. force is applied for approximately 10 minutes at this station. Whey runs out the top of the container and is caught by the drain pans from which it may be removed for further utilization.

After pressing at the second pressing station the air hose is disconnected and the stabilizing chain is unhooked from the U-bolt. The pressing frame is returned to a generally upright position and is moved along the monorail to the third pressing station, which, as shown, is on the same leg of the conveyor loop as the first and second pressing stations.

Other air hoses are provided at the third pressing station and are connected to the presses at that station. Each pressing frame is tilted slightly at the third station in the same manner as it is at the first pressing station to facilitate drainage of whey. At the third station, in a specific example, utilizing the containers previously described, a force of approximately 2,000 lbs. is applied for about 3 hours and 15 minutes.

After this final pressing, the hoses are disconnected from the presses and the stabilizing chain is unhooked from the U-bolt returning the frame to an upright position. The pressing frame then is moved along the track to the unloading station, which, as illustrated, is on one of the two cross-rails joining the two L's of the overhead conveyor.

The containers are removed from the pressing frames at the unloading station. The empty frames are then moved onto the adjacent leg of the conveyor where facilities are provided for them to be cleaned and stored in the storage and cleaning area prior to being returned to the loading station.

Steam Release Within Cheese Mass

B.E. Budahn; U.S. Patent 3,449,127; June 10, 1969; assigned to Industrial Plant Service & Mfg. Co. and U.S. Patent 3,154,002; October 27, 1964 describes an apparatus for continuously manufacturing cheese by milling curd slabs into small chunks, applying a proportionate amount of salt and tumbling the salted curd through an open ended drum into cheese hoops, all in a continuous operation. Steam is released within the interior of the mass of curd as it is tumbled to produce a higher grade cheese. The apparatus for making cheese in a continuous operation from slabs of cheese curd comprises:

(a) a power driven milling device for continuously milling slabs of cheese curd;

(b) a continuously rotatable open ended tumbling drum mounted for rotation and rotating about an axis inclined off horizontal which has a receiving end and a discharge end;

(c) means for transferring the milled curd from the milling device into the receiving end of the tumbling drum;

(d) salt dispensing means for adding a predetermined proportionate amount of salt to the curd as it is transferred to the tumbling drum;

(e) live steam dispensing means discharging live steam under pressure into the interior of the tumbling drum as it rotates, and

(f) the discharge of the steam dispensing means is elevated relative to the bottom of the drum and is disposed laterally of the longitudinal axis at the side of the drum towards which the curd is carried by the drum as it rotates and is disposed in the path of the curd as it is tumbled so that the curd covers it continuously as it is tumbled.

The difference in construction between the two apparatuses lies in the arrangement and positioning of the discharge of the steam discharge means and its relation to the direction of the tumbling drum.

The description of the continuous manufacture of cheese using the apparatus designed is covered by B.E. Budahn; U.S. Patent 3,156,567; November 10, 1964. The process of making cheese from slabs of cheese curd in a continuous operation from milling to hooping with a substantial saving in butterfat content of the cheese, a substantial reduction in dusting losses and a substantial reduction in time consumed for the operation entails:

(a) milling slabs of the curd;

(b) spreading a predetermined proportionate amount of salt upon the milled curd;

(c) tumbling the curd immediately as milled and salted for such time and under such conditions as required to coat the salt evenly on the curd and keep potential dusting losses in the so tumbled curd along with the full butterfat content; and

(d) then discharging the tumbled curd and salt into containers for subsequent pressing of the curd into cheese.

Subatmospheric Pressure for Air Extraction

G.M. Robertson and G.K Charles; U.S. Patent 3,468,026; September 23, 1969; assigned to Cow & Gate Limited, England fully describe an apparatus they have devised for a method of continuously making cheese in which a stream of crumbled curd is fed into a chamber to form a pillar or rod of curd which is passed through the chamber, the curd being compressed in the pillar or rod, and blocks of cheese are severed from the pillar or rod as it emerges from the chamber.

The chamber is preferably maintained at a subatmospheric pressure to extract air from the crumbled curd fed into the chamber. The curd may be compressed either by the weight of superimposed curd in a pillar of the curd, or by forcing the curd into a duct leading to an extrusion nozzle which is blanked off during the compression of the curd in the duct.

The absolute pressure in the chamber would depend on the variety of cheese being made and the closeness of the body of the cheese. By the use of a suitable low pressure within the chamber, almost all the air can be removed from the curd before it is compressed into the pillar or rod, so that the blocks of cheese will be free of air pockets.

The blocks of cheese manufactured by this method are in a condition suitable to be wrapped in impervious sheet material and matured under pressure. The object of the process is to provide a method of continuously making cheese from crumbled cheese curd.

Apparatus for carrying out the method can be described by using Figure 5.4. In an operational cycle of the apparatus, a tray is placed on the platform (33), the access door (39) is closed, the guillotine blade (30) is moved to its closed position, closing off the lower end of chamber (12) and air is extracted from the two chambers (12) and (13). Cheese curd, which has previously been crumbled in a mill and mixed with salt, is then fed into the hopper (18) and the slide valves (21) and (22) in the air lock are automatically opened and closed alternately to allow the curd to pass from the hopper into the upper chamber and fall into the column (14).

FIGURE 5.4: SUBATMOSPHERIC PRESSURE FOR AIR EXTRACTION

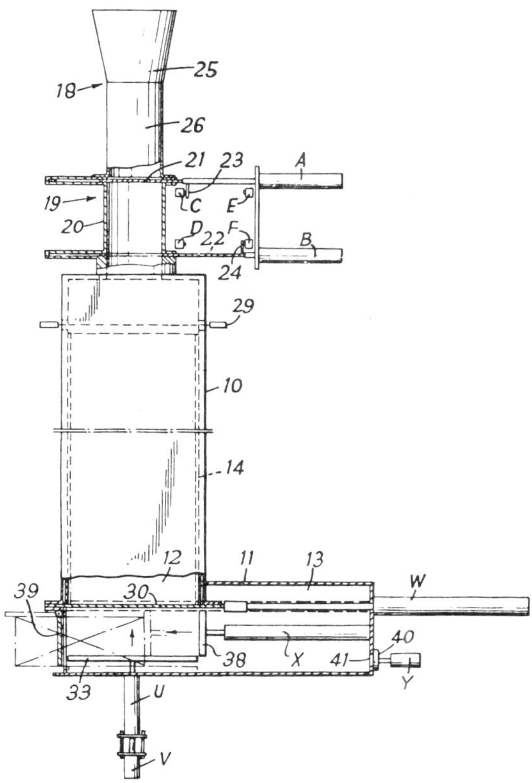

Source: G.M. Robertson and G.K. Charles; U.S. Patent 3,468,026; September 23, 1969

Some of the moisture in the curd vaporizes as soon as it enters the upper chamber due to the subatmospheric pressure therein. The cheese curd accumulates on the blade (30) at the bottom of the column and builds up into a pillar within the column. The weight of curd in the column compresses the curd at the lower end and forces whey out of the curd, the whey passing through the slits in the walls of the column into the space between the column (14) and casing (10). The whey then drains into the manifolds (31) from which it is continuously extracted by a pump.

When the column is almost filled with cheese curd, the platform is raised to a position in which the tray is immediately below the blade and the blade is then withdrawn so that the pillar of cheese within the column slides down inside the column and on to the tray on the platform. Platform (33) is then lowered by pistons within the cylinders (U) and (V).

The blade is then driven into its closed position, thereby severing a block of cheese from the lower end of the cheese pillar. Although the cheese in the block has been compressed by the weight of the superimposed curd in the column, the cheese block is compressed further by forcing the platform upwards and compressing the block against the underside of the blade (30). The cheese can thus be compressed to any desired degree by use of a suitable size of cylinder and operating pressure for the air.

After compressing the cheese block for a predetermined time (for example, 30 to 60 seconds), platform (33) is further lowered to provide a clearance distance and thereby space the top of the block from the guillotine blade. The valve (40) is then withdrawn by cylinder (Y) to admit air into the lower chamber through opening (41) and the tray carrying the block of cheese is removed through the access door (39) by ejector (38) driven by cylinder (X).

The ejector is then retracted, a fresh tray placed on the platform, the door closed and the cycle repeated. During subsequent cycles the vacuum in chamber (12) remains substantially constant and only the vacuum in chamber (13) need be released at the end of each cycle to remove the compressed block of cheese. The temperature of the cheese curd in the column may be raised by injecting hot air into the space between the inner and outer walls of the double walled casing, or by heating the casing by means of steam coils surrounding the casing.

Linear Coagulation and Pressing

A process for making telemea cheese described by V. Nikolic; U.S. Patent 3,518,094; June 30, 1970; assigned to Ministerul Industriei Alimentare, Romania is one in which milk is curdled and coagulated in a continuous manner concurrently with pressing of the curd and portioning of the cheese along a linear coagulation and pressing path such that the curd retains its original stratification.

Milk coagulation, cutting, pressing and portioning the curd and telemea cheese are carried out in a continuous flow by laminar and rectilinear displacement of the product, by using a unit consisting of a feeding means whereby the coagulation solution is injected into milk, the mixture is homogenized and then introduced into a coagulation body in the form of an inclined tube in which the coagulation of milk is carried out. The tube is connected to the treatment body in which are successively carried out the first cutting, the first pressing, the second cutting, the final pressing, the portioning and cheese-cooling stages. The displacement is effected by an inclined conveying belt and the pressing by means of other inclined pressing belts is gradually reduced during the displacement with respect to the conveying belt.

Milk is prepared for coagulation by pasteurization, addition of correction agents (calcium chloride, starting culture and possibly also alkaline nitrates), the acidity is brought (in terms of density degrees in the Thömer system) to 20° to 22°T. by means of diluted lactic or hydrochloric acids and the temperature is established to 28.5° to 30.5°C. The coagulation solution is simultaneously prepared by dissolving rennet in water.

When milk and the coagulation solution are prepared, the pumps (1) and (2) (Figure 5.5) are operated. Milk enters the lower part of the coagulation body (4) by flowing upwards in the form of a laminar, linear and uniform flow over a period of 30 to 35 minutes. When it arrives in front of the knives (7), the curd advances and is cut vertically by the blades of the knife-disks and crosswise by the wires, which join the borders of the outside disks within 10 cm. The cut curd is taken up by the conveyor belt (10) and carried under the treatment devices.

The belt (8) exerts a light gradual pressure and the whey is separated and eliminated in the upper part of curd, by draining it through the orifices (17) and (18) The fused curd is cut again by the set of knife disks (9) which are closer together than the former set. The eliminated whey is drained by decantation through the lateral orifice (19). The curd is then pressed by the belt (11) and the whey separated is eliminated through the orifice (20). A strip of curd is obtained whose width depends on the transversal sizes of the unit and which must be a multiple of the block sizes. The strip of cheese is transversely cut by the guillotine cutter which is hydraulically driven by circulating water. The disk-knives set (14) cuts longitudinally the transversal pieces of cheese by obtaining blocks of telemea cheese.

FIGURE 5.5: LINEAR COAGULATION AND PRESSING

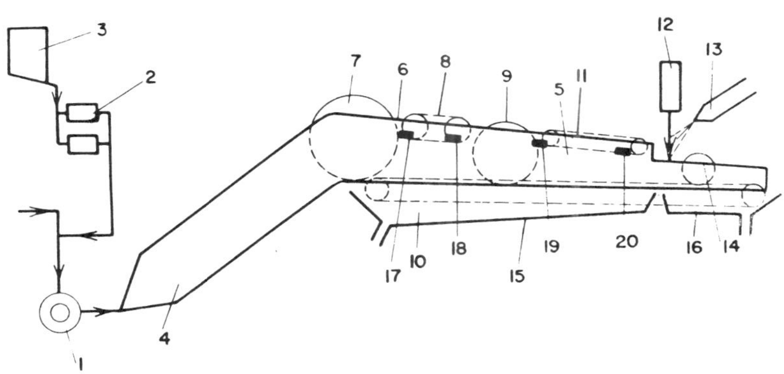

Source: V. Nikolic; U.S. Patent 3,518,094; June 30, 1970

The whey drained is collected by the trough (15) and the cooling water after spraying the blocks obtained by cutting is collected in the trough (16). The last portion of curd is drawn out from the coagulation body by pumping a solution of 10% salt, sodium chloride, through the milk flow line.

Reduction of Butterfat Loss

B.E. Budahn; U.S. Patent 3,522,059; July 28, 1970; assigned to Industrial Plant Service & Mfg. Co. describes the process by which the loss of butterfat normally occurring in cheese manufacture is minimized by treating the cheese milk under conditions which do not fracture the butterfat particles. The treatment is accomplished by separating the major portion of the butterfat from whole milk, concentrating the separated milk under flashing conditions to remove 5 to 10% of the water and reconstituting the milk by adding the separated butterfat in nonflashed condition to the concentrated separated milk.

An apparatus is designed and fully described for properly conducting this process. In operation, the whole milk enters the system through a conduit and is preheated to approximately 90° to 94°F. in a preheater to facilitate the separation of the milk into cream by separators. Since the vapors (principally water vapor) taken from the concentrator by the preheater are hot and primarily capable of ready condensation, these vapors are condensed and the water resulting from the condensed water vapor collects in the bottom of the preheater and is drawn outwardly by a pump. Since the vapors are being constantly condensed, additional vapors are constantly being drawn into the interior of the preheater because the condensation of the vapors will automatically create a negative pressure resulting directly from the condensation of the vapor.

From the separators, the skim milk passes through the regenerative section of the pasteurizer and has its temperature elevated to approximately 140° to 145°F. It is then carried by a conduit upwardly and projected into the interior of the concentrator in which it is directed circumferentially around the interior of the concentrator while under 24 to 25 inches of vacuum.

Since the temperature of the skim milk is relatively high for pairing with a negative pressure of 24 to 25 inches of vacuum, a condition substantially in excess of the flash point of the milk exists and a substantial portion of the moisture is withdrawn from the skim milk as a result of flashing as it falls to the lower portion of the concentrator. In other words, substantial flashing action takes place within the concentrator to remove a substantial portion of the water content from the skim milk.

The temperature of the skim milk is maintained at the relatively low temperature of 140° to 145°F. to limit the extent of the protein breakdown which tends to produce crumbly cheese if the flashing action is excessive. By the time the milk has reached the lower portion of the concentrator it is at approximately 132°F. Since the pump is set to be unable to handle the entire flow of the partially concentrated skim milk as it reaches the lower portion of the concentrator, it collects in the lower portion of the concentrator and is recirculated through a steam chest.

The partially concentrated skim milk is recirculated through the steam chest in that it passes into the header of the chest and into the tubes of the chest that are heated by the hot gases passing into the steam chest from a conduit.

These gases are at a temperature of approximately 190° to 200°F. and since the interior of the steam chest is also subjected to approximately 24 to 25 inches of vacuum, additional water vapor and milk solids are sprayed upwardly as a result of the flashing action and are carried through yet another conduit and reintroduced into the upper portions of the concentrator and are directed circumferentially along its inner wall.

Thus additional water vapor is removed from the skim milk as it is carried upwardly through the dome and conduit of the concentrator. In this manner, 5 to 10% of the water content of the skim milk can be removed without any of the serious consequences of flashing. Actually, it is preferred to remove approximately 8% of the water content as a combined result of the initial flashing and the subsequent flashing which takes place in connection with the recirculation of the partially concentrated skim milk.

It should be noted that in the arrangement described, the pasteurizing action takes place subsequent to the concentration of the milk whereas it is conventional to pasteurize the milk as it enters such systems prior to any concentration of the milk. Thus this particular arrangement has the added advantage that the pasteurization takes place immediately prior to the conveyance of the reconstituted milk to the cheese-making apparatus and therefore avoids the possible contamination of the milk by the concentrating apparatus which could otherwise take place.

The separation of the cream from the skim milk prior to the concentration of the milk enables the skim milk to be flashed at relatively high temperatures (considering the use of 24 to 25 inches of vacuum) to concentrate the skim milk at a more rapid rate than would otherwise be advisable if the butterfat remained in the milk.

This is true because the high degree of flashing which takes place, if the butterfat remains in the milk, will cause the butterfat globules to fracture so minutely as to make it impossible to collect all of them from the whey by separation after the cheese has been made from the milk. By avoiding flashing of the butterfat, the butterfat globules remain in sufficiently sized particles that when cheese is made from the reconstituted milk, a substantial loss of butterfat via the separated whey is avoided.

Precompression Under High Pressure

A method of making Cheddar or similar cheese by a continuous process including the steps of cheddaring, flavoring and homogenizing by precompressing the usual milled and salted curds under relatively high pressure for a short period of time, spraying them with hot water and subjecting them to a final compression at high pressure for a short period of time, before the process of curing is described by H. Koopmans; U.S. Patent 3,615,587; October 26, 1971.

The method provides a considerable reduction of the process time and a mechanization and automation of the process. It is characterized in that the thermal process is carried out on a plurality of endless conveyor belts, which also perform the turnover of the curds while the sliced curds are introduced in measured quantities by weight into metal vessels closed by a metal follower. The curds are enveloped in a towel in the vessel and covered by another towel. The vessel is then conveyed on a conveyor belt to a press where the material is subjected to a high pressure for a short time and subsequently sprayed on all sides with warm water, after which it is again subjected to high pressure for a short time and conducted away. Perforated metal vessels and perforated metal followers are preferred. The endless conveyor belts are arranged in a space in which the humidity and temperature are maintained at a desired level. The pH value of the whey which flows out of the curds on the endless belts is continuously measured. The sliced curds are accurately weighted when conveyed from the so-called brine mill to the perforated vessels.

The precompression period is preferably 3 to 10 minutes, the pressure being 3,000 to 12,000 kg./20 kg. of cheese. Rinsing is carried out on all sides for 2 to 8 minutes with water of 40° to 70°C. After-compression takes 3 to 20 minutes at a pressure of 3,000 to 12,000 kg./20 kg. of cheese.

The method is also suitable for compressing the cheese in vacuo. In this case the perforated vessel is surrounded by an outer vessel, the clearance between them accommodating water inlets and outlets and a suction valve for the air. In this case the inner vessel is preferably not perforated in part of the upper side, for example, over a length of 10 cm. from the top.

When compressing in vacuo, the rinsing tunnel is replaced by a connection with a hot water main for a short time which may be 2 to 5 minutes. The temperature of the water may lie between 40° to 70°C. After 3 minutes the water outlet may be opened so that the rinsing space can be emptied and after-compression can be carried out. During compression the clearance between the perforated metal vessel and the outer vessel is caused to communicate with an air pump, which may be arranged on the compression table. The towels which favor drainage have a particularly useful effect in compressing in vacuo.

Continuous Advancement of Inoculated Condensed Milk

The apparatus described by P. Stenne; U.S. Patent 3,616,536; November 2, 1971; assigned to Paracurd SA, Switzerland was designed to carry out the method of making cheeses devised by P. Stenne; U.S. Patent 3,528,821; September 15, 1970; assigned to Paracurd SA, Switzerland.

The process first concentrates normal (whole) milk, which may have been filtered, standardized, homogenized and pasteurized, until the percentage of solids is from 2 to 4 times that prevailing in normal milk. This concentration is effectuated in an evaporator of a conventional type, preferably a film evaporator, which operates under subatmospheric pressure, using a thin sheet of liquid and a temperature below that at which the proteins in the milk would denature (about 65°C.).

After the concentrating step, the concentrated milk is immediately cooled in a heat exchanger to a temperature less than 10° to 12°C. The concentrated milk is then stored at this temperature in insulated vats having double walls, and about 2% of lactic ferments is introduced to create lactic flora for the protection of the mixture.

After having been stored for several hours, the temperature of the concentrated milk is increased to 30°C. and a supplemental dose of 5% of the same lactic ferments is added to the milk. At this temperature the presence of the lactic ferments brings about a progressive transformation of the lactoses of the milk into lactic acid. The pH of the mixture then becomes acid and when this pH reaches a value between 4.6 and 6.5 the concentrated milk is cooled to a temperature below 10°C.

The desired pH may in certain cases be obtained directly, or may be adjusted in the process just described, by adding to the milk a quantity of lactic ferments such that the desired pH is obtained immediately.

A quantity of rennet corresponding to that normally used for obtaining the type of cheese in question is then added to the concentrated and acidified cold milk. The dose of rennet added to the concentrated milk is the same in absolute value as would have been added to the normal milk. Dependent upon the time of contact between the concentrated milk and the rennet, the enzymatic action of the rennet will have a greater or smaller effect, and after the concentrated milk which has been so treated has been rapidly reheated, an instantaneous or controlled rapid coagulation is obtained. This coagulating step is followed by a draining step, the length and nature of which depends on the type of cheese being made.

In particular, when soft cheeses are to be made, the cold concentrated milk is kept in contact with the rennet for about 5 to 12 minutes. It is then heated rapidly by mixing it with hot water and coagulation results in a short time, between 40 and 120 seconds, the mixture being kept still at the moment of coagulation.

The dilution effectuated during this reheating step lasts for such a short time before the collection or pressing of the curd that the phosphocaseinate particles do not have time enough to return to the state in which they existed in the normal starting milk, and consequently all the advantages resulting from use of the concentrated mixture are retained. When the curd has attained a certain firmness it is sliced to permit a suitable syneresis. This is followed by spontaneous slow draining and the curd is then separated into conventional cheese molds.

When making a semisoft cheese, the cold concentrated milk is left in contact with the rennet for a little longer than in the previous case, from 30 to 60 minutes, for example. The concentrated acidified cold milk is then reheated by introducing it into a stream of nonturbulent hot water. The concentrated milk coagulates almost instantaneously without any dilution. The curd grains are then washed and agglomerated by mechanical means, thus accelerating the drainage.

When making a cured or hard cheese, of the Cheddar type, for example, the cold concentrated milk, acidified to a suitable pH, is kept in contact with the rennet for about 30 to 60 minutes. It is then introduced into a nonturbulent stream of hot water, at an appropriate temperature, the rate of flow of which is from 1 to 3 times that of the concentrated milk, depending upon the type of cheese to be made. Coagulation is then immediate.

The grains of the curd are mechanically agglomerated and the grain is separated from the whey. The whey, after heating in a heat exchanger to an appropriate temperature, is reintroduced into the curd to produce a curing of the curd, and in the case of Cheddar, an appropriate scalding which is a first step in the cheddarization operation.

While the milk is being ripened by the lactic ferments, the required heating and cooling steps to which it must be subjected may be carried out by circulating the concentrated milk in a suitable heat exchanger, heated by hot water or steam, or cooled by a refrigerating fluid. These different operative steps may be programmed or controlled automatically and the ferments may be added manually or by automatic dosing means, with the operator controlling the acidity and temperature of the concentrated milk. In like manner, the rennet may be added manually or automatically, either by leading the acidified concentrated cold milk through batteries of vats having gauged volumes, in which the rennet is added to and mixed with the concentrated milk, and in which the concentrated curdled milk remains for a predetermined time, or by continuously injecting the rennet into a tube through which the concentrated milk is pumped. This tube is so dimensioned that the milk remains in it for from 5 to 60 minutes, dependent on the type of cheese being made.

The apparatus provides a means for coagulating a concentrated, acidified cold milk, to which rennet has been added, by rapidly adding hot water, then separating the curds from the whey and finally washing the grains of curd either with hot or cold water, or with whey, or with a mixture of water and whey, which may be hot or chilled or at room temperature and finally collecting the curd for subsequent molding in conventional cheese molds.

Figure 5.6a is a perspective view showing an installation for manufacturing either soft or pressed cheese and Figure 5.6b shows an apparatus for manufacturing a pressed or hard cheese, both of which can be operated continuously and automatically.

FIGURE 5.6: CONTINUOUS CHEESE PROCESS FROM CONCENTRATED MILK

a.

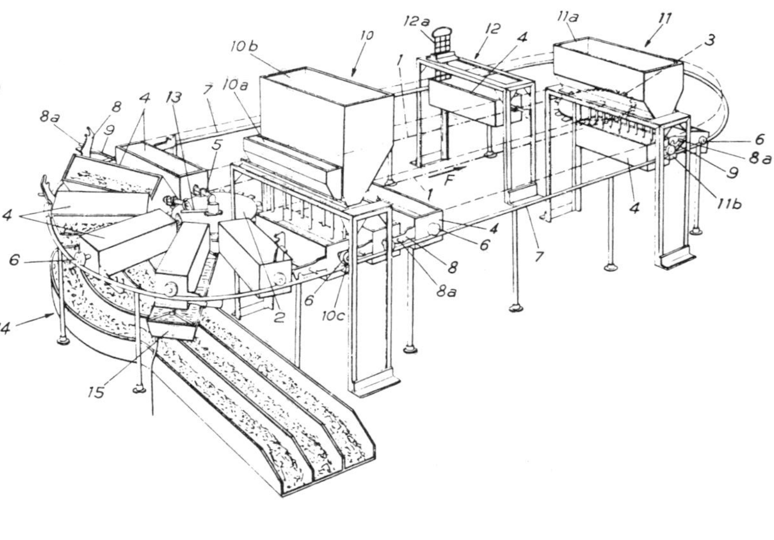

(continued)

FIGURE 5.6: (continued)

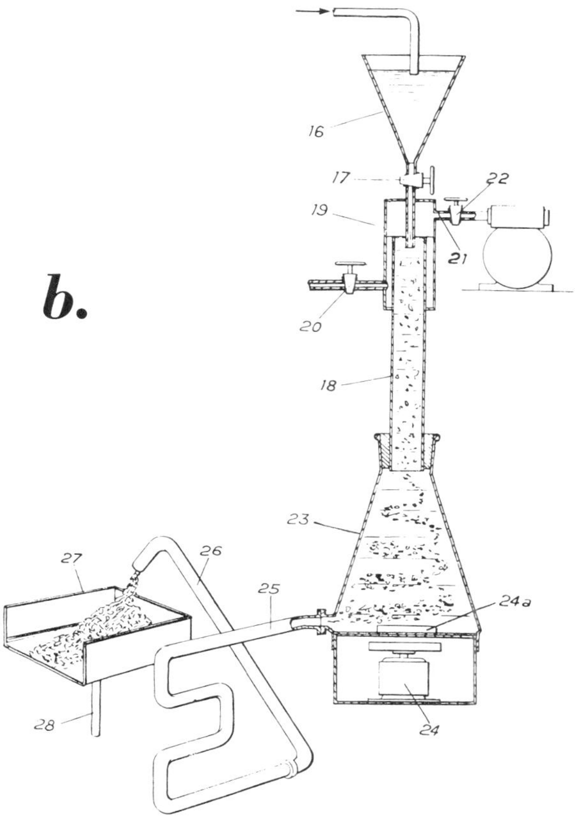

Source: P. Stenne; U.S. Patent 3,616,536; November 2, 1971

Perforated Conveyor

The method and apparatus described by B.E. Budahn; U.S. Patent 3,636,630; January 25, 1972 is designed for use in the continuous, automated manufacture of cheese, commencing with the curd slurry and terminating upon hooping of the matted and salted curd preparatory to pressing. It provides for the manufacture of cheese in a continuous manner from the curd and whey stage to the final hooping stage, which eliminates about 75% of the labor involved in the manufacturing process and also provides better control over the moisture content, giving a more uniform end product.

The apparatus shown in Figure 5.7 provides an inclined, continuous stainless steel perforated conveyor to separate the whey from the curd and deposits the curd upon an imperforate stainless steel belt matting conveyor.

FIGURE 5.7: CONTINUOUS MANUFACTURE OF CHEESE

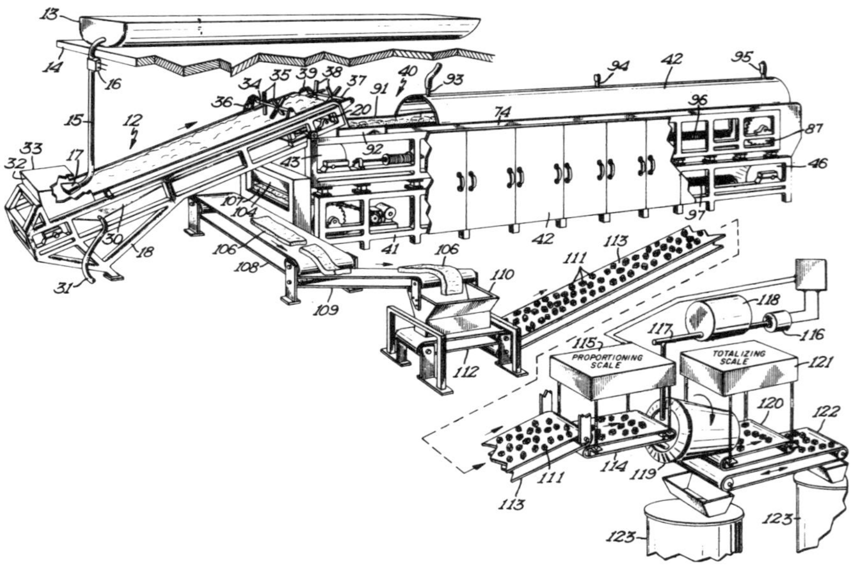

Source: B.E. Budahn; U.S. Patent 3,636,630; January 25, 1972

The curd is forced initially into narrower transverse dimensions and then permitted to expand laterally to facilitate the escape of whey while moving on the matting conveyor. When the matting process is essentially half completed, the curd mat is transferred to a second and similar belt conveyor running in the opposite direction and in so doing, is inverted. Live steam is applied to the matting curd as it moves along the matting conveyors within their housings.

An automated guillotine knife cuts the curd mat vertically into inch slabs at the end of the matting conveyor and another conveyor moves the slabs into a milling machine which cuts the curd into 1 inch cubes from which it is conveyed into an inclined, open ended tumbling drum along with a proportionate amount of salt. As the salted curd leaves the drum, it is deposited in hoops for subsequent pressing and aging.

Multipurpose Ripening Tank

A process described by E. Syrjanen and O. Reijonen; U.S. Patent 3,707,769; January 2, 1973; assigned to MKT-tehtaat Oy, Finland eliminates the manual work in the manufacture of Cheddar cheese. This device is characterized in that the conventional ripening tank has been provided with a perforated bottom movable from the initial to the ultimate end of the tank, with a press plate located above and with a cutter for cutting the cheese mattress which rests upon the bottom. The device comprises a conveyor extending from the ultimate to the initial end of the tank for reintroduction of the minced and salted cheese into the same tank for final pressing of the cheese in the tank. According to the process the cheese goes into the tank twice. The first time the tank operates as ripening tank, and the second time it serves as press for the final pressing of the cheese. In both instances the cheese is in the form of a large mattress, which is an essential feature of the process. After the cheese mattress has been finally pressed, it is cut with a cutter into cheeses of ultimate size, which are taken into storage to ripen. No separate cheese molds or presses are thus required.

Figure 5.8 is a longitudinal section of this apparatus in which Cheddar-type cheeses are manufactured as follows. Into the tank (1) from a cheese kettle, a cheese mass is introduced, which consists of the cheese curd and of whey. At this stage the end wall (5) of the tank is closed, or in its lower position. The whey flows out through the wire (8) and the perforated surfaces of end walls (4) and (5) and possibly also of the side walls (2) and (3), that is through the sieves, and the cheese curd forms, on the bottom of the tank, a cheese mattress (26) which may have a height of as much as 40 to 50 cm.

After the major part of the whey has run off, the cheese mattress is prepressed with the press plate (11) to enhance the removal of whey, gases and air and to condense the cheese mattress so that cheddarizing may be begun rapidly and effectively.

After the desired degree of cheddarization has been achieved, the power cylinders (30) and (31) are used to lift the end of the tank up, and the wire is set in motion so that the cheese mattress begins to move towards the left. From the mattress, pieces (27) are cut with the cutter (33), which have a length equal to the width of the tank. The thickness of the pieces is determined by the ratio of the speeds of the wire and cutter. The pieces fall on the conveyor (14) and come into the gap between the rolls (16) and (18). The rolls squeeze the pieces so that further whey, gases and air are removed from them. At the same time the disk-like cutters (17) and (19) on the rolls cut the pieces up into strips.

FIGURE 5.8: MULTIPURPOSE RIPENING TANK

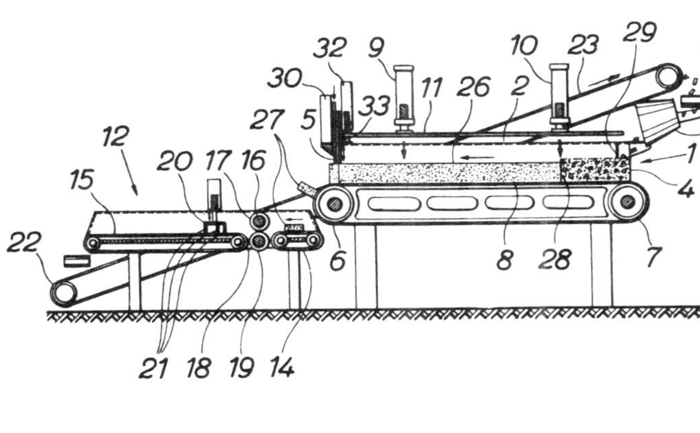

Source: E. Syrjanen and O. Reijonen; U.S. Patent 3,707,769; January 2, 1973

The strips come under the cutting elements (21) of the vertically movable cutter (20) and are cut across into approximately finger-sized small pieces. Since the distances between the cutting elements are different, the length of the small pieces varies. Such pieces stick together better than pieces of equal size when they are later pressed together to form cheeses.

The small pieces go from above the mincing device (12) to the transverse conveyor (22), from which they are further transferred to the oblique conveyor (23), which returns the minced pieces to the initial end of the tank. The small pieces fall from the conveyor onto a transverse conveyor and from this further into the salting device (25). After passage through the salting device, the small pieces fall once more into the initial end of the tank and again form a cheese mattress on the bottom of the tank.

In order that the mattress might become uniform, a stop plate (29) is placed at the initial end of the tank, which is adjustable in its height and which may be entirely detached from the tank. The boundary surface between unminced and minced cheese mattresses in the tank is indicated by number (28). When the unminced cheese mattress has entirely left the tank and returned into it in minced form, the stop plate is removed and the cheese mattress is finally pressed with the aid of the press plate (11). The mattress is left under pressure for the desired length of time.

Upon completed pressing, the wire (8) is once more set in motion and the cheese mattress is cut with the cutter (33) into cheese blocks of desired size, which are received onto an appropriate conveyor and transferred into storage to ripen.

Lateral Spreading Action

R.T. Clark; U.S. Patent 3,716,376; February 13, 1973; assigned to Alfa-Laval AB, Sweden describes a process in which curds substantially freed from whey are fed into a vertical tower and extruded under pressure through a plurality of extrusion apertures in the base of the tower, the curds being subjected to a lateral spreading action in their downward travel through the tower.

The lateral spreading action contributes significantly to the very good grain structure and chicken breast texture obtained. The lateral spreading can be achieved by the use of a tower which increases in cross-section in a direction from the top to the bottom of the tower. The arrangement of the extrusion apertures in the base of the tower also benefits the process in that it assists in the complete dispersal of pockets of whey which may be trapped in the passage of curds down through the tower.

The operation may be carried out in batches or on a continuous basis. In batch operation, the tower is filled with curd particles and, after a certain interval, subjected to air pressure applied to a head space at the top of the tower to extrude the curd in strings which are severed into short lengths by a cutting means below the base of the tower.

If the process is continuous, the curd is propelled through the column under mechanical pressure exerted by an archimedian screw or a reciprocating piston. In either case, the pieces of curd leaving the tower may be salted directly and passed to a filling station for hooping and subsequent pressing. A preferred form of cheese-making apparatus is described in the patent.

CURD MANUFACTURE

Curd Fusion

J. Czulak; U.S. Patent 3,060,575; October 30, 1962; assigned to Commonwealth Scientific and Industrial Research Organization, Australia describes a cheese-curd fusing machine that permits the fusion of cheese curd in an almost continuous process.

Referring to Figure 5.9a, a drainage vat (1) is provided with a pair of opposite false perforated walls (2) and with double walled bottom doors (3) hinged to the outer walls (4). The drainage vat is equipped with traveling curd stirrers (5) and traveling delivery pipes (6). These are carried on a carriage (7) provided with rollers (8) which travel on flanges (9) on the outer walls of the drainage vat.

The function of the drainage vat is the separation of the curd and whey in it and control of the moisture content of the curd by further stirring of the curd particles after the bulk of the whey has been drained off. Curd and whey delivered by the delivery pipes are sprayed against the false walls and a large proportion of the whey passes through them into the space between the false walls and the outer walls. This whey is drained off in any suitable manner.

FIGURE 5.9: CURD FUSION

a.

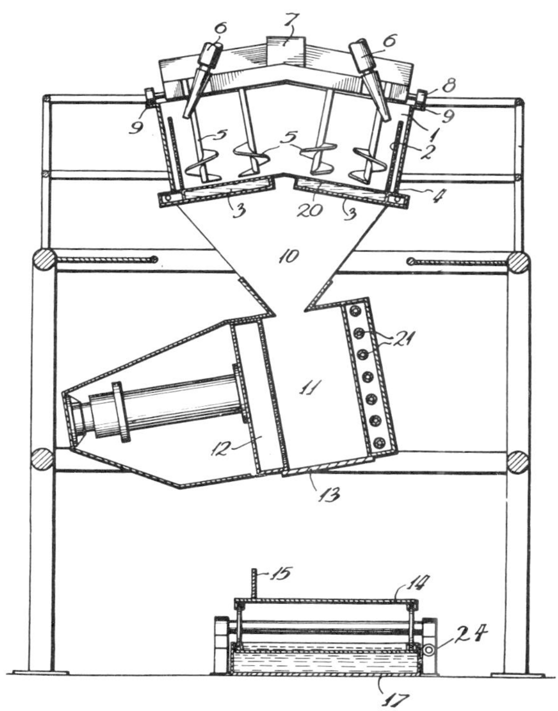

(continued)

FIGURE 5.9: (continued)

b.

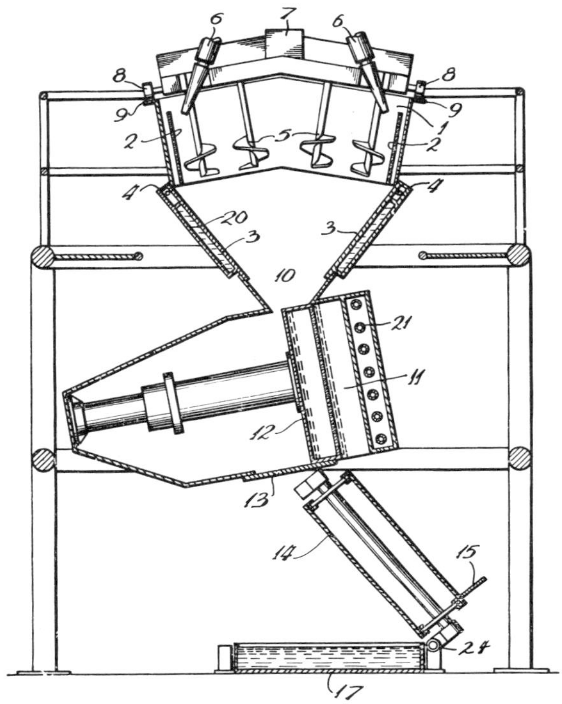

Source: J. Czulak; U.S. Patent 3,060,575; October 30, 1962

As shown in Figure 5.9b, the hinged bottom doors are adapted to be swung down, at the appropriate stage of the drainage, to form portions of a hopper (10) which is positioned beneath the drainage vat, and this hopper delivers the curd into a compression-and-flow chamber (11), one side of which comprises a perforated hydraulic piston (12) and the bottom of which is a slidable plate (13).

The piston then moves forward to the position shown in dotted lines in Figure 5.9b and the curd is thus compressed into a fibrous mass by compression of the particles together and by deformation of the fusing mass by causing it to flow upwards. On completion of the compression and deformation the curd is discharged onto a slat conveyor (14) positioned adjacent to the compression-and-flow chamber. This is achieved by movement of the slidable plate to the open position and by slight withdrawal of the hydraulic piston to the position shown in full lines in Figure 5.9b.

The slat conveyor is pivoted about one longitudinal edge, as at (24), and when curd is to be discharged from the chamber the conveyor is pivoted to a position as shown in Figure 5.9b and curd is permitted to slide out of the chamber onto the surface of the conveyor. To guide the curd as it passes from the chamber to the conveyor a movable plate (15) is provided on the conveyor and before the curd is passed to the conveyor this plate is in the position shown in Figure 5.9a and as the curd passes on to the conveyor the plate is moved to the position shown in Figure 5.9b. When it has received the curd the conveyor is returned to the horizontal position and is moved forward step by step. A guillotine at the end of the conveyor cuts the curd into strips. A whey vat (17) is located beneath the conveyor and excess whey drains from the curd into the vat.

The apparatus described, thus permits the fusion of cheese curd in an almost continuous process. The amount of curd discharged from the vat to the chamber is regulated so that the chamber is not filled. Thus when the perforated ram moves forward it not only compresses the curd but also causes it to flow upwards thus leading to more intimate fusion.

Electrical Voltage to Coagulate Curd

Through the use of the method described by H.G. Foster, Jr. and E.H. Cornwell; U.S. Patent 3,079,263; February 26, 1963; assigned to Swift & Company it is possible to add a food acid directly to a milk-proteolytic enzyme mixture and obtain a curd having good strength and particle size and eminently suitable in the manufacture of a good flavor and texture cheese. The time involved is negligible.

Use of the equipment shown in Figure 5.10a illustrates the process. Storage tank (10) contains the milk to be used in the formation of the curd. The milk is withdrawn from the tank through conduit (11) by a variable speed, positive action pump (12) which forces the milk first through cold milk separator and clarifier (13), where the butterfat content of the milk is adjusted to that desired, and then into heating plate (14), where the milk is warmed to the desired operating temperature (between 85° and 180°F.). Tank (18) contains rennet solution (10 lbs. rennet extract to 100 lbs. of water). Rennet is metered into the conduit containing the heated milk. Tank (21) contains a starter culture which is metered into line (11). A culture may be added at any point in the line, but preferably should be added before the lactic acid enters the milk.

A solution of lactic acid (10 parts acid to 100 parts of water) is placed in corrosion-resistant tank (26) and is metered into mixing zone (30) where the lactic acid is thoroughly and rapidly mixed with the milk-rennet mixture. If it is desired to add calcium chloride to strengthen the curd it can be conveniently added along with the acid.

As shown in Figure 5.10b, the milk-rennet mixture enters section (32) of the mixing zone (30) which comprises T (31). Concentrically positioned within the stem portion of the T is capillary tube (28) which carries the acid into the mixing zone. The tube terminates in cap (33) which has an orifice (34) by which the acid is dispersed in a fine stream such that the particles of acid are intimately mixed with the milk-rennet mixture.

After mixing in the T, the mixture quickly exits into holding tube (35) in the form of a mixture of curd and whey. Any mixing device capable of intimately and thoroughly mixing the acid and the milk mixture is suitable. The apparatus should be of a type that will not be clogged by the formation of the curd. From the holding tube the curd and whey are forced into centrifuge (36) where the whey is separated from the curd.

FIGURE 5.10: ELECTRICAL VOLTAGE TO COAGULATE CURD

a.

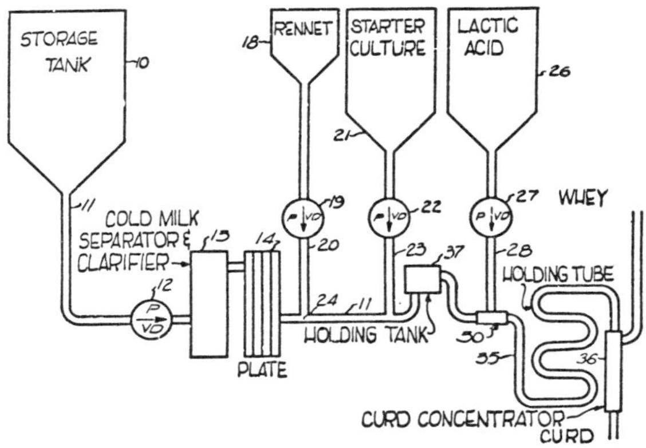

b.

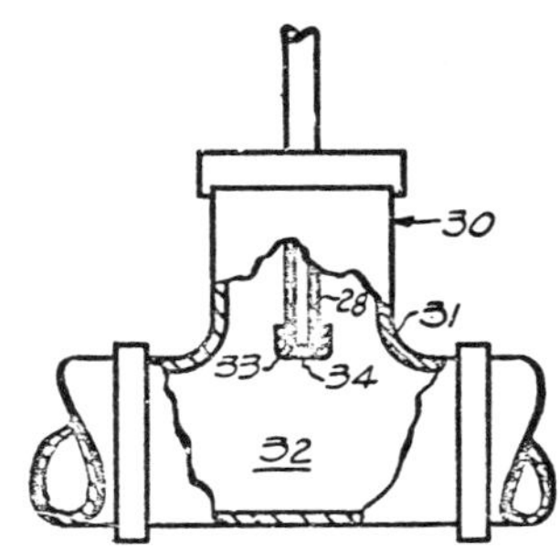

Source: H.G. Foster, Jr. and E.H. Cornwell; U.S. Patent 3,079,263; February 26, 1963

Between the point of entry of the rennet into the line at (24) and the mixing zone (30) there is a holding tank (37) which may be replaced with piping. The only requirement is that the equipment used be such that at the velocity at which the milk mixture is being pumped through the line, the rennet will be maintained in contact before the addition of the acid for between 5 seconds and 5 minutes, with the optimum being about 55 seconds.

Example: Thirty gallons whole milk was run through the continuous system at a rate of 400 lbs./hour. The milk was heated to about 108°F. and a 400 cc pepsin solution diluted with water to 4,000 cc was metered into the milk at a rate of about 7,500 cc pepsin solution per hour. After a milk-pepsin contact time of about 5 seconds, 90 cc of 80% lactic acid diluted to 900 cc was metered into the system. A coagulum formed. The pH was approximately 4.3. Upon separation from the whey the curd formed was of good texture for cheese manufacture. With reference to the amount of acid required for good curd formation, it was found that when only 25% of the normal amount of acid solution (the normal amount being 4.6% acid solution by weight of the milk) was used no curd was formed with whole milk. The pH was 5.8. When 50% of the normal amount of acid solution was added, a curd was formed and the resultant pH was 5.2 in the holding tube and, on separation, the pH of the whey was 5.5 and of the curd was 5.4. Thus it would seem that, while 4.6% acid solution by weight of the milk is optimum, it is possible to form a curd with as little as 2.3% acid solution.

More than 4.6% could be used as far as the formation of curd is concerned. However, taste considerations dictate that the pH not be lower than about 4.0. When using skim milk to prepare the curd, curd is obtained when the pH is as high as 6.3. This pH was obtained using a 2.5% by weight of the milk of a 50% dilution of the 1 to 10 lactic acid solution.

G.H. Richardson and E.H. Cornwell; U.S. Patent 3,394,011; July 23, 1968; assigned to Swift & Co. describes a further improvement in this rapid and continuous method for the manufacture of cheese and designs an apparatus to be used in the process which involves the exposure of a curd-whey mixture to a heating cycle in which the curd and whey are subjected to simultaneous and uniform heat.

The curd-whey mixture is passed through a heating zone, where the mixture is heated by electrical heating. It is thus possible, by applying a moderate electrical voltage across the curd-whey mixture to coagulate the curd and produce a tighter curd rapidly. The resistance presented by the curd as it passes the electrical source is substantial and the curd contracts resulting in the expulsion of the liquid from the mass of the curd.

This curd shrinking and curd-whey separating method is of particular utility in the continuous production of curd since it is possible to provide, in combination, apparatus adapted to permit introduction of cheese-forming ingredients in one end and initiate a continuous flow of ingredients through the apparatus, resulting in good quality, firm curd product at the downstream end of the apparatus.

Thus, cheese-forming ingredients are continuously fed into the apparatus and high quality curd substantially free of whey is continuously produced in a short period of time. The apparatus is capable of handling about 10,000 pounds of milk per hour input, with the production of curd at the rate of about 1,000 pounds per hour.

In operation, milk is warmed to 105° to 110°F. and 0.25 and 1.5% by weight of the milk of an aqueous 10% rennet solution is introduced into the mixing section of the tank. The mixture is vigorously agitated in the mixing zone and the flow through the tank toward the electrical heating system is quiescent. Flow of the mixture at this point is largely a result of the pressure exerted by the incoming ingredients. A light coagulum starts to form almost immediately and by the time the milk, enzyme and acid mixture comes within the area of the sweep of the paddle conveyor, curd is forming in good quantity.

Passage of the solution through the tank can be adjusted to the requirements of the particular milk, enzyme and acid being employed. A flow rate of about 5 to 10 ft. per minute has been found very satisfactory. As the curd-forming mixture reaches the area of a paddle conveyor, the paddles or sweeps gather the forming curd and each sweep or paddle moves the curd at a uniform rate through the tank.

The width of the paddles is such that there is approximately 1/16 inch clearance from the walls of the tank. As the paddles drag or convey the forming curd through the tank, the curd is carried through the path of electrical current provided in the sides of the tank. The effect of the heating cycle is to cause the curd to shrink and expel entrapped water. The curd is then uniformly firm and contains no apparent soft spots. The temperature of the curd is raised to 110° to 115°F. during the passage through the electrical heating zone and this temperature rise seems to be well suited to excellent curd-whey separation.

No overcooking or undercooking is observed since the curd is moved at a uniform rate into intersection with and through the heating zone. Curd temperature can in some instances be increased to as much as 120°F. but the curd must not be exposed to temperatures much above 120°F. unless the holding time at that temperature is very short. In those cases where it is desired that the conventional lactic acid bacteria not be killed as when a culture is to be allowed to grow after matting of the curd attainment temperature of the curd should not be allowed to exceed 104°F.

The sweeps deposit slabs or sheets of firm curd on the takeoff conveyor and this conveyor transports the slabs beneath a bank of heat lamps to a second mesh conveyor also exposed to a bank of heat lamps. The slabs are inverted in the transfer from the first to the second takeoff conveyor so that the final product is a uniformly firm curd substantially free of whey.

It is possible to equip the tank or trough with more than one set of electrodes to provide more than one heating zone. Further, since the voltage is variable, it is possible to adjust the amount of voltage passing through the curd carried through the electrical heating area, thus adjusting the temperature to which the curd is heated to any desired degree. Similarly, the curd drying section of the apparatus can be varied and a drying tunnel can be used.

The ingredients and order of addition of ingredients can also be varied. While in the preferred form, acidified milk is pumped into the feed conduit and enzyme is metered into the milk through a T connector, it is also possible to meter acid into sweet milk before or after the rennet or, in some instances, acid can be eliminated entirely. Another embodiment involves the use of acid alone without using enzyme. It is possible to introduce a cold acidified milk mixture into the apparatus and heat to coagulate the mixture. Such an alternative may be used particularly in the manufacture of cottage cheese curd.

A still further improvement in the process for manufacturing cheese curd is put forth by H.G. Foster, Jr. and E.H. Cornwell; U.S. Patent 3,172,767; March 9, 1965; assigned to Swift & Company. In this process they find that it is possible to preadjust the acidity of the milk by direct acid addition prior to addition of enzyme. The natural acidity of milk (i.e., the titratable acidity of the milk) will vary generally from 0.13 to 0.20%, skim milk generally having a lower natural acidity than whole milk.

In practicing the process the minimum amount of directly added acid that will aid in development of a good cheese curd upon addition of enzyme is that amount of nontoxic acid which will increase the titratable acidity of the milk by at least about 0.015%. The maximum amount of acid that can be added will vary depending upon the type of milk, e.g., skim or whole, solids content of the milk and pH of the milk.

With skim milk an acid addition of 0.08% over the natural acidity is about the maximum practical amount of acid that can be added without danger of producing undesirable curd upon addition of rennet or other enzymes. When using whole milk, the maximum amount of acid will depend primarily upon the pH after adjustment of acidity and the solids content of the milk. As the solids content increases, more and more acid can be added without adverse results due to the buffering capacity of the solids.

For example, when working with a high solids whole milk, the titratable acidity has been raised by direct addition of acid to 0.34% (a value normally considered to be unsatisfactory because the milk would be sour and curdled), and yet, unexpectedly, the milk was not curdled and the curd subsequently prepared from this milk was of good texture and well adapted to manufacture of cheese. When making cheese curd by this method, the pH should not be lowered below the isoelectric point of the casein of the milk. This point is approximately at pH 4.6 to 4.7.

The acidity adjusted milk may be used in conventional batch or vat cheese processing providing good mechanical mixing of subsequently added enzyme to the milk is achieved. However, it is preferred to integrate the acid addition operation into the continuous cheese curd manufacturing process discussed in the previous patents so that an amount of milk is first acidified by direct addition of the desired amount of acid. The acid is added as a dilute solution, e.g., a 1 to 10 dilution of acid in water and is thoroughly mixed with the milk. The preferred acid is lactic acid.

The acidity adjusted milk is then heated to 85° to 180°F. and metered continuously into a curd-forming zone, such as a conduit at a predetermined rate. An enzyme capable of converting the milk protein from casein to para-casein is added to the stream of milk flowing in the curd-forming zone. Preferably the enzyme is continuously metered into the milk at a predetermined rate to supply sufficient enzyme to form the curd. A coagulum will form substantially immediately after introduction of the enzyme to the acidity adjusted milk thus giving a mixture of curd and whey.

Example: A 40 gallon batch of whole milk was divided into four aliquot lots A to D. The natural acidity of the milk was 0.19% and the pH was 6.9. Lots B to D were adjusted in acidity with a 10% lactic acid solution and to alternate lots cheese coloring was added to identify each lot passing through the line. The milk was heated to 108°F. and pumped through the continuous cheese curd-forming system at a rate of 400 lbs./hr. A 10% rennet solution was added at a rate of 7,200 cc/hr. Samples of the whey from each lot of milk was taken. The following table shows the analysis of this run.

Lot	Acidity	pH	Acid Solution Added, cc	Adjusted Acidity	Adjusted pH	pH Whey
A	0.19	6.9	None	--	--	6.7
B	0.19	6.9	50	0.22	6.85	6.65
C	0.19	6.9	150	0.235	6.75	6.52
D	0.19	6.9	250	0.265	6.55	6.38

Curd formed substantially immediately upon addition of rennet in the cases of lots B, C and D. The quality of the curd and clarity of the whey improved as the acidity was increased. There was no curd formation in lot A until the very end of the run and the whey was very milky.

Use of Enzymic and Nonenzymic Phases

A process for the continuous manufacture of curd is described by L. Radema, J. Ubbels, A.G.J. Arentzen and J.T. van der Linde; U.S. Patent 3,132,026; May 5, 1964; assigned to Nederlands Instituut voor Zuivelonderzoek, Netherlands. The process takes advantage of the fact that the clotting of milk caused by the addition of rennet may be divided into two phases, viz. one enzymic and one nonenzymic. In the enzymic phase casein is converted into paracasein. In the nonenzymic phase the coagulation of the paracasein with calcium takes place.

The temperature coefficient of the first phase was found to be much smaller than that of the second phase. Owing to this it became possible to study the two phases separately. In fact, if the milk which contains rennet is stored at 0°C., the first phase proceeds more slowly than at commonly applied coagulation temperatures, but still within a reasonable time.

At this low temperature, however, the rate at which the second phase is completed is decreased so much that this milk can still be stored for a long time without curdling. A subsequent increase of the temperature will cause flocculation within a very short time. If 130 minutes after the addition of the rennet the milk is heated from 0° to 37°C., its clotting time was 90 seconds. Longer storage of the milk at 0°C. caused only a slight further reduction of the clotting time.

The difference in temperature coefficient between the two phases may be utilized for the continuous production of curd. Rennet, starter and other required ingredients are added to the milk that has been cooled down to a temperature of -1° to +15°C., and after being kept at this temperature for 1 1/2 to 48 hours, this mixture is passed firstly through a heat exchanger, in which it is heated to 20° to 45°C., and then through a coagulation chamber, which near the inlet opening of the feed pipe comprises a device for preventing turbulences in the liquid mass and the outlet opening may be equipped with an apparatus for cutting the curd.

Preferably the process is started with milk having a temperature of 3° to 6°C., and after addition of the rennet, starter and the further ingredients, the mass is kept at the temperature for another 2 to 6 hours. The heat exchanger may consist of a conduit or tube heated by steam or some other heating medium, but preferably a tubular or a plate heat exchanger is used. The most favorable temperature for the coagulation in the coagulation chamber is 29° to 30°C. The cut curd thus obtained may further be converted into cheese.

A suitable apparatus for the conduct of the process is disclosed by L. Radema, J. Ubbels, A.G.J. Arentzen and J.T. van der Linde; U.S. Patent 3,210,845; October 12, 1965; assigned to Bedrijven van het Nederlands Instituut voor Zuivelonderzoek, Netherlands.

In Figure 5.11a, (1) denotes a tank in which the cold milk is stored and the required addition is made of rennet and other substances used in making cheese curd such as calcium chloride, saltpeter, starter culture and/or coloring agent. These ingredients are thoroughly mixed by stirring. After the desired reaction period, the mixture is pumped by pump (2) via lines (4) through heat exchanger (3) into the inlet opening in the center of bottom plate (6) of coagulation chamber (5). At a very short distance above the bottom plate, the flow of milk is interrupted by two perforated discs

FIGURE 5.11: USE OF ENZYMIC AND NONENZYMIC PHASES

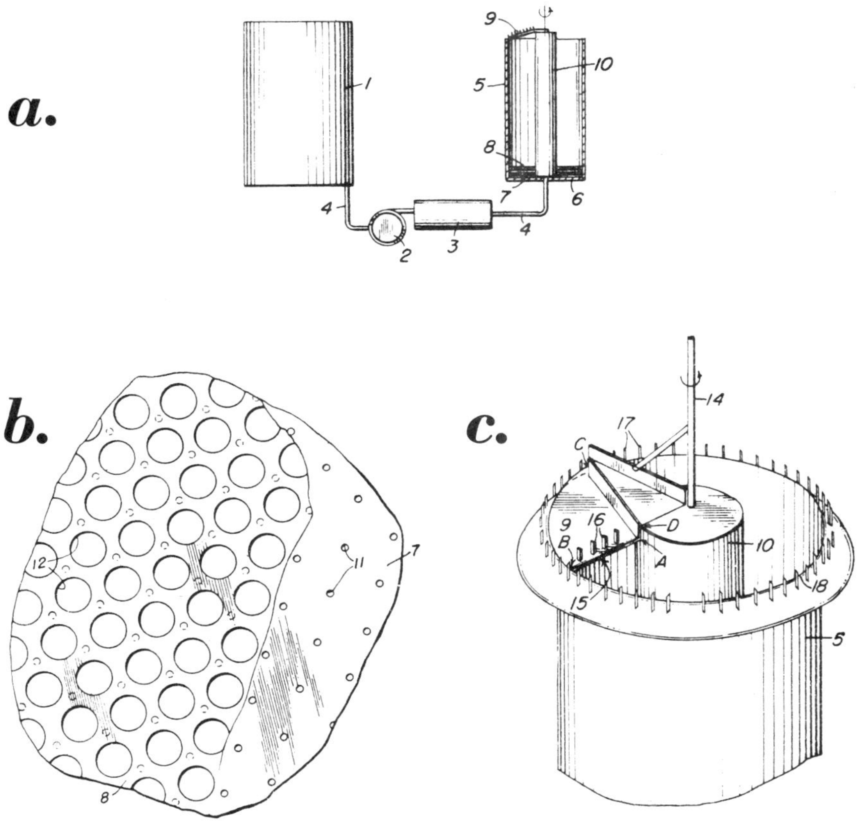

Source: L. Radema, J. Ubbels, A.G.J. Arentzen and J.T. van der Linde; U.S. Patent 3,210,845; October 12, 1965

(7) and (8) whose respective perforations (11) and (12) are staggered as shown in Figure 5.11b whereby the flow of coagulating milk becomes slow and uniform to the top of the coagulation chamber. In the center of the coagulation chamber is a smaller cylinder (10) which extends above the top of the coagulation chamber. Attached to the upper end of cylinder (10) is the cutting apparatus (9) which rotates in the direction of the arrow.

Figure 5.11b illustrates the size and staggered position of the respective perforations (11) and (12) in discs (7) and (8). The size and number of the perforations (11) in lower disc (7) are such that a pressure is set up beneath this disc which is large in relation to the pressure losses caused by friction of the warmed milk between disc (7) and bottom plate (6). Suitable perforations could be fifty 1 mm. diameter perforations per 100 cm.^{2} of disc (7). This ensures that the quantity of milk flowing through each of the perforations of disc (7) is about equal. The upper disc (8) contains an equal number of perforations (12) in the same pattern but they are much larger, i.e. fifty 10 mm. diameter perforations per 100 $cm.^2$ and are staggered with respect to the perforations (11) of the lower disc (7). The upper disc (8) prevents any secondary flow in the coagulating milk.

Figure 5.11c illustrates the cutting means of Figure 5.11a which is rotated about cylinder (10) by rotating shaft (14) so that cutting edge (15) and vertical blades (16) cut the curd into strips. The knife is part of a sector of a circle and has been mounted in such a way that the cutting edge or (A)(B) forms a small angle with the plane through the upper edge (18) of cylinder (5), so that point (A) lies higher than point (B).

The cutting surface (A)(B)(C) of the rotating knife may remain just inside the wall of the cylinder or it may overlap the wall of the cylinder as shown. The circular edge (B)(C) of the knife is inclined upwards in a helix in such a way that the underside of the plate (A)(B)(C) remains free of the continuously rising surface of the coagulated milk. The strips of curd formed by means of the cutting edge and blades slide over plate (A)(B)(C).

Thus, they remain intact but are straightened out. If (A)(B) were horizontal, the innermost strip would brush against the wall (C)(A)(D), the radius of curvature of wall (C)(A)(D) being larger than the curvature of the inner cylinder, and at the same time would have to go up the slope of plate (A)(B)(C), for (C) lies higher than (A) and (B). Actually, as already mentioned above, (A) lies higher than (B), so that the slope up which the central strip has to go is smaller.

Another object thus attained is that the cut strips will tend to slide off sideways, in consequence of which the pressure of the innermost strip against the wall (C)(A)(D), among other things, will diminish. At an adequate slope of (A)(B), the innermost strip may even be free from this wall. The firmness of the cut strips also plays a part in this. The less firm the strips, the more readily they tend to slide off sideways with the same position of plate (A)(B)(C). The position of plate (A)(B)(C) is preferably adjusted in such a way that the strips only just do not slide off sideways. As they arrive at the edge of the outer cylinder the strips are cut up by the vertical blades (17).

Example: In a tank, pasteurized milk was stored at a temperature of 3°C. To this milk the following ingredients were added: 0.6% (by volume) of starter, 0.03% (by volume) of rennet, 0.03% (by volume) of $CaCl_2$, 0.02% (by weight) of KNO_3.

Five hours after the addition of rennet and starter, the milk was begun to be pumped from the tank and to be forced through a plate-heat exchanger to the coagulation chamber. The milk was heated to 30°C. in the heat exchanger. Within 14 minutes after the temperature of 30°C. had been reached the milk had risen to the top of the coagulation chamber (height more than 1 m.), where it had a consistency such that it could be cut. The curd particles thus formed were regularly shaped and were converted into cheese.

Compression of Curd Granules

D.R.M. Lambert and J.M. Sharkey; U.S. Patent 3,404,009; October 1, 1968;

assigned to National Dairy Products Corporation describe a method and apparatus for conditioning cheese curd, in which curd granules are introduced into a first vessel, compressed to form a curd mass and transferred to a second vessel in such a manner as to enable the curd mass to expand uniformly.

The method comprises placing curd granules in a first vessel or conditioning tower, compressing the curd granules in the conditioning tower to cause knitting of the curd granules into a curd mass and then causing the curd mass to flow and to stretch. The stretched curd mass should not rupture but should generally flow. Drainage of whey from the curd mass is effected during the compression of the mass.

The curd mass is subjected to a controlled change of dimension or spreading while flowing from the lower end of the conditioning tower. The control substantially prevents rupture of the curd mass while permitting stretching or spreading. Thus it flows from one conditioning tower having one dimension into a second vessel such as a forming tray or confining receptacle having another dimension.

The change of dimension may be effected by progressively raising the conditioning tower relative to the forming tray which has a configuration different from that of the tower (or by progressively lowering the tray relative to the tower, or by both raising the tower and lowering the tray) in such a manner as to cause the curd mass to flow from the lower end of the conditioning tower and at the same time to slowly stretch the curd mass in desired manner and to a predetermined extent in the space between the lower end of the conditioning tower and the forming tray. The space is increased at a controlled rate during the change of dimension of the curd to permit flow without rupture of the curd mass.

The curd granules are confined or held in the conditioning tower during the draining, knitting and compressing operation, and this may be effected by maintaining the forming tray or confining receptacle in contact with the lower end of the conditioning tower during such operation. The curd should be drained at points throughout the height of the conditioning tower so that drainage is not limited to the bottom of the conditioning tower. Compression of the curd granules in the conditioning tower is preferably effected by a pressure plate which exerts downward pressure on the curd in the tower. A vacuum may be applied to the curd granules in the conditioning tower prior to the compression of the curd granules.

The method may also include applying heat and/or pressure to the curd mass after its removal from the conditioning tower to cause further stretching of the curd mass. The curd mass may be retained in the forming tray for a controlled period after heat and/or pressure have been applied to effect further knitting and acid development. The application of heat and pressure may be effected by placing a heated pressure plate in contact with the upper surface of the curd mass in the forming tray, the curd mass being subjected to increasing deformation at a gradual controlled rate to spread or stretch without substantial rupture of the curd mass in the forming tray.

Heating of the curd mass, particularly the top layer, after its removal from the conditioning tower increases the plasticity of the curd and enables further spreading or stretching of the curd mass without rupture. When subjecting the curd granules to compression in the conditioning tower, it has been found desirable, in order to effect the required knitting of the curd and expulsion of whey, to subject the curd granules first to low pressure, preferably about 0.6 lb./sq. in. for a period of about 3 minutes and second to higher pressures, preferably about 2.5 lbs./sq. in., for a similar period. The removal of the curd mass from the conditioning tower and the compression of the curd mass in the forming tray beneath the conditioning tower to cause stretching of the curd mass without causing rupture are effected in a controlled manner. Preferably the removal of the curd mass from the conditioning tower is effected by raising the conditioning tower relative to the forming tray in accordance with a predetermined cycle, the curd mass being caused to flow outwardly towards the boundaries of the forming tray, the cross-sectional area of the tray being greater than that of the conditioning tower.

The cycle of lifting operations of the conditioning tower relative to the forming tray may comprise a relatively rapid initial lift, a slower intermediate lift and a relatively rapid final lift. Satisfactory results have been obtained by a lifting cycle which comprises raising the conditioning tower a distance of 3 inches in the first 1/2 minute, 17 inches in the next 4 minutes and the remaining 15 inches in 1 minute.

Apparatus for carrying out this method of conditioning cheese curd comprises a conditioning tower into which curd granules are placed, means for compressing the curd granules in the conditioning tower to enhance knitting of the mass, means for removing the curd mass from the conditioning tower, and means for permitting stretching of the curd mass during or after its removal from the conditioning tower. The tower preferably is provided with spaced perforations for whey drainage.

Example: 18,000 pounds of milk was introduced into a cheese vat and inoculated with 180 pounds of a lactic acid starter culture. After a period of 1 hour ripening time, 50 ounces of rennet was added to the milk to coagulate the milk in the usual 30 minutes. When the curd was cut with curd knives the whey had an acidity of 0.10. The curd was agitated in the whey and cooked at a temperature of 100°F. until the acidity of the whey reached 0.155%, at which time the mixture of cheese curd and whey was introduced into a rotating trommel and the whey was separated from the cheese curd.

The cheese curd granules were then introduced into a star-shaped conditioning tower and were compressed by means of a pressure plate applied to the top surface of the cheese curd granules in the conditioning tower in order to effect knitting of the cheese curd into a curd mass. An initial pressure of 0.6 psi was applied to the cheese curd granules for a period of 3 minutes followed by application of pressure of 2.5 psi for an additional 3 minutes.

The curd mass was then delivered from the star-shaped conditioning tower into a rectangular forming tray positioned beneath the conditioning tower. The delivery of the curd mass from the conditioning tower into the forming tray was effected by raising the conditioning tower with respect to the forming tray a distance of 3 inches in the first 1/2 minute, 17 inches in the next 4 minutes and 15 inches in the next 1 minute, the lifting cycle being such that the curd mass underwent a controlled change of dimension without substantial rupture of the curd mass. The acidity of the whey expelled from the curd mass after delivery of the curd mass into the forming tray was 0.23%.

The curd mass was then further compressed by application of a pressure plate to the upper surface in the forming tray. The pressure plate weighed 2,500 lbs. and was lowered intermittently until the curd mass was caused to fill the forming tray, the lowering of the pressure plate being controlled to avoid substantial rupture of the curd mass. The curd mass was held in the forming tray until the acidity of the whey expelled reached 0.48 at which time the curd mass was cut into strips and milled in a curd mill. The milled curd was then salted, hooped and pressed and cured in accordance with conventional procedures. An American cheese was obtained which had a characteristic Cheddar cheese flavor and which was considered to be identical to Cheddar cheese prepared in a conventional cheddaring process.

Diverging Annular Screen Enclosure

In the apparatus designed by D.D. Winnie; U.S. Patent 3,633,276; January 11, 1972; assigned to Stoelting Brothers Company for a cheese cheddaring method, prepared whey and curd are transferred to the cheddaring machine to place the curd in the perforate downwardly diverging annular screen enclosure within the vat (Figure 5.12a). The screen is rotated and a mixer keeps the curd moving upwardly to obtain uniform dispersion of the curd and facilitate removal of air and gases.

After filling, the mixer is removed and whey drained while rotation continues at a slower rate. The vat is tilted (Figure 5.12b) so the now-knitting curd is manipulated to aid whey expulsion and work the cheese as required in cheddaring. Tilting to the Figure 5.12c position compresses the curd previously on the upper surface and knits the curd while the converging annulus further compresses and works the curd. Finally, (Figure 5.12d), the ring closure is pulled back and a knife is fixed across the opening to cut the curd as it moves down and out. By reason of the rotation of the screen, the curd comes out in a continuous ribbon for conveyance to further process stations.

FIGURE 5.12: DIVERGING ANNULAR SCREEN ENCLOSURE

a.

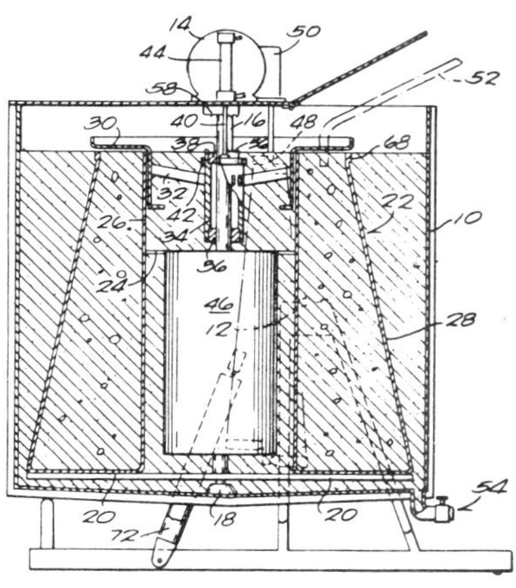

Apparatus During Filling Process

(continued)

FIGURE 5.12: (continued)

b.

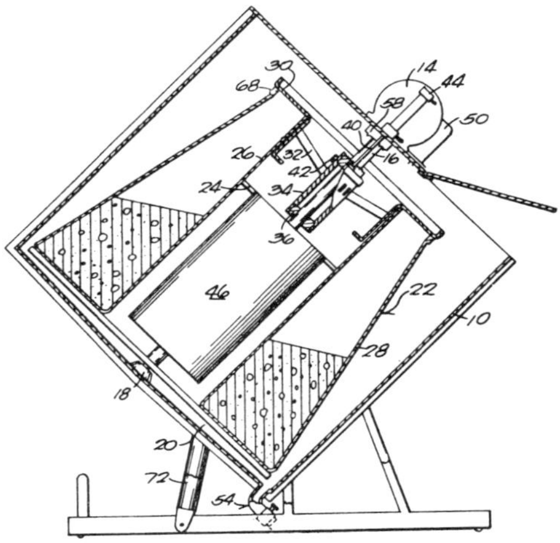

Vat Tilted After Whey Is Drained Off

c.

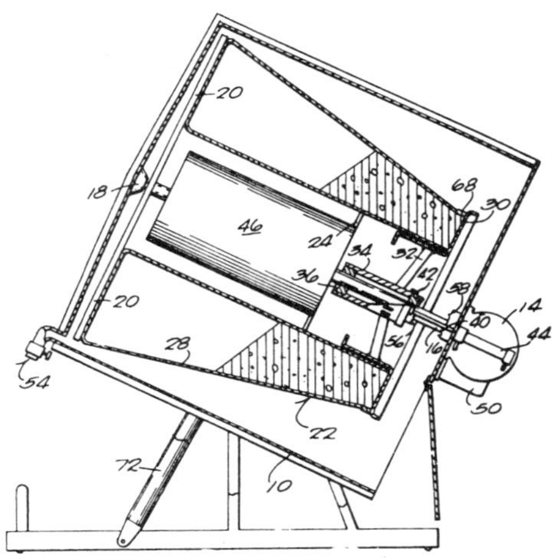

Vat Further Tilted

(continued)

FIGURE 5.12: (continued)

d.

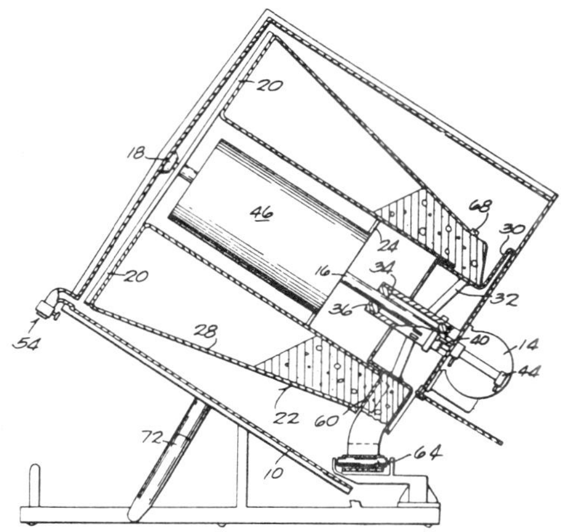

Vat Tilted to Discharge Position

Source: D.D. Winnie; U.S. Patent 3,633,276; January 11, 1972

The ability to tip the vat and the rotating screen enclosure is a key factor in construction in that it insures manipulation of the knitting curd since experience indicates that it is important to both rotate the screen enclosure and to provide agitation during the filling process, both to achieve a uniform filling within the enclosure and to expel air and trapped gases.

Without the agitation gas-like openings in the end product will result while the desired Cheddar cheese has a nearly solid structure. Possibly the agitation could be supplemented by or replaced by application of a partial vacuum during the filling step. An important feature is the ability to discharge the curd from the machine in a continuous ribbon which lends itself to further processing by automated equipment.

With this equipment hand labor is removed from the cheddaring step. It is usually necessary to have one worker per vat during the cheddaring step. With this equipment one worker can handle six machines. Furthermore, the cheese produced by this machine is at least equal to that produced by the traditional hand method at its best.

There is also some thought that the cheddaring time is reduced with the machine in that the desired breast-of-chicken consistency of the curd and acidity apparently are reached sooner than in the traditional method.

Compressing Curd in Containers

J. Czulak, N.H. Freeman and J.R. O'Connell; U.S. Patent 3,695,893; Oct. 3, 1972 describe an apparatus and method for processing cheese curd and whey in the manufacture of Cheddar cheese by passing the curds and whey over a screen to drain off the bulk of the whey. The separated curd is fed to a conveyor so that it forms a layer and the layer is stirred to facilitate further drainage of whey. Blocks are cut from the layer and compressed in respective containers each of which is large enough to permit its block to flow under compression to produce fiber development of the curd.

The process can be understood with reference to Figure 5.13 which is a diagrammatic sketch of the apparatus which functions in the following manner. Supported within frame (10) is a pipe (12) by means of which a mixture of curds and whey can be fed to a first separation stage (14). This separation stage comprises a perforated screen (16) which receives the curds and whey mixture from the pipe so that part of the whey passes through the screen and the remaining mixture is fed to a perforated drum (18). The bulk of the whey is separated from the curd by passage through the screen and drum and is passed to whey tanks (20) which extend along the bottom of the frame for its full length. The whey is removed along pipes (22) from the tank.

The separated curd after passing around the upper portion of the drum is passed by way of a guide plate (24) to a fusion conveyor (26). The curd forms a layer upon the fusion conveyor and travels along the upper run of the conveyor from one end to the other. While the curd is on the conveyor, it is gently stirred by sets of curd lifting means, (28) and (30).

Each lifting means gently opens up the curd to permit the remaining whey to drain off the sides of the conveyor from which it also passes to the whey tanks. Four sets of curd lifting means are provided, but two sets are mounted for pivotal movement about respective horizontal axes (31). These sets can be swung upwardly away from the conveyor as indicated by the arrows. Accordingly, the amount of stirring imparted to the curd can be adjusted. At the end of the conveyor away from drum (18), the curd layer will be almost completely drained and will be partially consolidated by gravity.

FIGURE 5.13: CURD PROCESSING APPARATUS

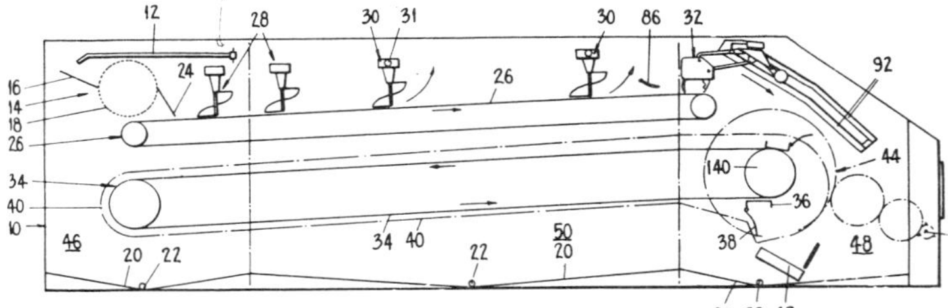

Source: J. Czulak, N.H. Freeman and J.R. O'Connell; U.S. Patent 3,695,893; October 3, 1972

As the layer reaches that end, blocks of curd are cut from it by a cutting and transporting assembly (32). Having cut a block of curd from the layer on the conveyor, the assembly transports it to one end of a second conveyor (34) which is located below conveyor (26) and rotates in the opposite direction. Conveyor (34) comprises a plurality of compression boxes (36). The assembly places the block of curd it has cut from the layer into one of the compression boxes which is just beginning to move along the upper run of conveyor (34).

Each box has a lid (38) pivotally mounted at its trailing edge, the lid being provided with cam followers which cooperate with cam tracks (40). The cam tracks and followers control the disposition of the lid relative to its box at all locations around the conveyor path. The lid is pivoted wide open so that the block may be freely placed in the box as the latter begins its travel along the upper run of the conveyor. As each box moves along the conveyor path from the block receiving position, the tracks and cam followers first pivot the lid to a position substantially parallel with the open top of its box and then move the lid until it closes the box.

Each box is dimensioned in relation to the block it receives so that the block can be placed in the center of it to be free of all side walls and to project from the open top at the receiving end of the conveyor (34). As the lid converges with the box, the block within is compressed and caused to flow outwardly to the side walls. During this step any remaining whey in the curd block is expressed by the positive pressure applied by the lid in cooperation with the cam tracks and the flow of the curd within the box is arranged to provide the required fiber development of the curd. If necessary, arrangements may be made to control the temperature of each box to maintain it at the desired value suitable for this processing step.

When a box reaches the receiving end of the lower run of the conveyor, the cam tracks diverge from the conveyor and permit the lid of the box to open wide again. Thereupon the block of compressed curd within the box is permitted to slide down the lid under the effect of gravity and onto a slide (42) positioned to receive it. From there the block can be passed under a guillotine mechanism (43) and the smaller blocks issuing from the guillotine can be salted and hooped.

Conveyor (34) is driven at its receiving end and coupled to it is a gear train diagrammatically illustrated and indicated by (44) which drives a shaft (45) upon which is arranged a plurality of cams to control operation of the cutting and transport assembly (32) in synchronism with the conveyor (34) so that a block of curd cut from the layer is always accurately placed in a box (36) on conveyor (34) which can be added to in order to increase production.

SEPARATING CURDS AND WHEY

Curd Removal with Tapered Tube

The object of the cheese process described by R.T. Budd and H.R. Chapman; U.S. Patent 2,982,655; May 2, 1961; assigned to National Research Development Corp., England is to produce the cheddaring or similar actions in a simplified manner, to reduce the amount of manual and/or mechanical effort required and to facilitate the replacement of the batch process by a continuous one. The process comprises feeding curd to the upper part of an upright column in which form the curd is confined and down which it flows by its own weight, supporting the curd at the lower end of the column and removing curd from the lower end at such a rate that cheese of a required body, texture, moisture content and acidity is produced. The curd may be fed continuously to the top of the column while treated curd is removed continuously from the bottom of the column.

The column is preferably tapered with the larger end lowermost and may be confined by means of an imperforate vertical tube which may be lagged against loss of heat. The free whey escapes at the lower end of the tube mainly between the tube and the column of curd. The curd is forced outwards towards the walls of the tube and it is thought that the whey forms a lubricating film on the walls of the tube and escapes with the curd at the bottom. The tube when used for true cheddaring requires a cross-sectional area which increases from top to bottom.

The ratio of top to bottom cross-sectional areas is dependent on the amount of spread of the curd which is required. During traditional cheddaring, a piece of curd will increase its superficial area by a factor of three or four. The factor will depend on a number of conditions, viz. the nature of the curd itself, the size of the blocks into which it is cut, how deep it is piled, whether extra weight is applied and how much, and the length of time for which the cheddaring is continued. In practice, a figure from four to five to one increase of area has been found convenient. An increase of area of less than 3 or greater than 7 may introduce undesirable difficulties.

The maximum pressure to which the curd is subjected depends on the height of the tube and the density of the curd and the latter depends on the curd preparation. In the traditional process, the maximum pressure is of the order of 1/10 to 1/4 psi. In some mechanical cheddaring machines where the sideways spread is restricted, pressures of 8 psi are used.

The condition of restricted sideways spread is made because there is a danger with high pressure on unrestricted curd of shearing the curd particles with resultant loss of fat. Experience has shown that a tube 6 ft. high producing a maximum pressure of about 2 psi is satisfactory. It is considered that a tube of height 3 to 10 ft. or generally 5 to 10 ft. and top area to bottom area ratio of 1 to 4-6 would be suitable and convenient for general use.

The volume of the tube is governed by the required rate of throughput and the length of time it is required to continue the cheddaring action. Having fixed the height and top/bottom cross-section area ratio, the volume determines the actual cross-sectional dimensions. There may be a maximum limit to these dimensions determined by the ability of the whey to drain out of a large mass of curd and difficulty of cutting. The internal surface of the tube should be smooth and clear of any projections which might impede the flow of curd. If desired, the tube may be surrounded by a heating means instead of or in addition to the lagging.

Separation of the whey from the curd should be carried out as near as possible to the top of the tube. There are existing pumps which will handle curds and whey mixture satisfactorily without damage to the curd which could be used to raise the contents of the curd making vat to the required level. The actual separation could be carried out using an agitating wire mesh, e.g., a Russell separator or an inclined rotary cylindrical sieve. Curds and whey are fed into the upper end of the cylinder, the whey drains out through the perforations and the curd particles eventually tumble out the lower end.

Initial filling of the tube at the start of a run presents a problem. Each discrete quantity of curd must take the same time to travel through the tube and be subjected to the same rates of pressure increase and sprad for the curd to be properly and uniformly cheddared. This presents no problem when the tube is full and curd is being fed to the top at the same rate at which the treated curd is being removed at the bottom. There is the problem at the start, when the first curd is presented to the empty tube, of filling the tube to produce a continuously moving column of curd.

One solution would be to have the curd delivered into the top of the tube, fall down to the bottom and be supported by the stationary cutter. When the tube is full the cutter is set in motion and the curd is removed at the normal rate. The first curd to come through will not be cheddared but subsequently the curd will become increasingly better cheddared until the last of the initial filling has passed through. At that point properly treated curd will start to issue. The early curd can be dealt with as desired but need not be wasted and need form only a small proportion of the total curd handled.

Having run the process continuously, the stage will be reached when no further separated curd is available to feed to the tube. There is, therefore, no further weight available to provide the cheddaring force to the curd remaining in the tube. The problem may be overcome by the addition of weight in some form or other at a rate equivalent to that at which the curd enters the tube. One such method is to seal off the tube at the top and supply compressed air at a suitable low pressure.

Inclined Tubular Body

J. Ubbels and J.T. van der Linde; U.S. Patent 3,193,927; July 13, 1965; assigned to Bedrijven van Het Nederlands Instituut voor Zuivelonderzoek, Netherlands describe an apparatus for separating whey from cut coagulated milk rapidly. The apparatus consists of an inclined tubular body which rotates about its longitudinal axis. Within the tubular body and attached to its inner walls is a perforated screw shaped member and associated stirring rods.

With milk that has been coagulated and cut in the conventional way the time needed for the separation of the whey is reduced to one-fourth of the normal time. The separation of the whey from milk that has been coagulated and cut according to U.S. Patent 3,132,026 (see page 170) is reduced to one-tenth of that required time.

An understanding of the principles of operation can be obtained by referring to Figure 5.14a and 5.14b. A power driven cylinder (1), not fully horizontally supported by the rollers (6), is equipped at the receiving end with a cone (2) and at the discharge end with a smaller cone (3). Mounted inside the cylinder is a screw shaped body (4).

The screw shaped body (4) is provided with holes (8) so that the whey can freely flow in the longitudinal direction of the cylinder, but the curd cannot. When the cylinder rotates, the curd particles are conveyed between the walls of the cylinder and the helical blades of the screw shaped body in the longitudinal direction of the cylinder. In Figure 5.14a, it can be seen that the longitudinal axis of the cylinder slopes upwards from the curd receiving end to the discharge end. By varying the angle of slope it is possible to control, in cooperation with the cones (2) and (3), the thickness of the liquid layer in the cylinder.

The liquid layer is thicker at the beginning of the cylinder than at the end. Due to the exudation of whey the curd volume decreases from the receiving to the discharge end. Owing to this construction the ratio of curd volume to whey volume as the curd approaches the discharge end is greater than with a horizontal cylinder. In consequence, the contact between the particles is more intensive, and accordingly, drying is more rapid than with a horizontal cylinder.

Extending through the screw shaped body in the longitudinal direction of the cylinder are a number of rods (7). During the rotation of the cylinder these rods move through the curd-whey mixture and produce the desired stirring effect. The rate at which the whey is separated from the curd is affected not only by temperature but also by the number of stirring rods, the circumferential velocity of the cylinder and the ratio of curd volume to whey volume.

FIGURE 5.14: INCLINED TUBULAR BODY

a.

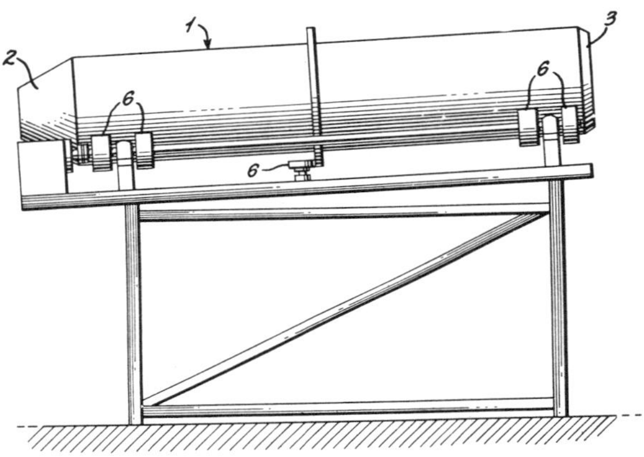

b.

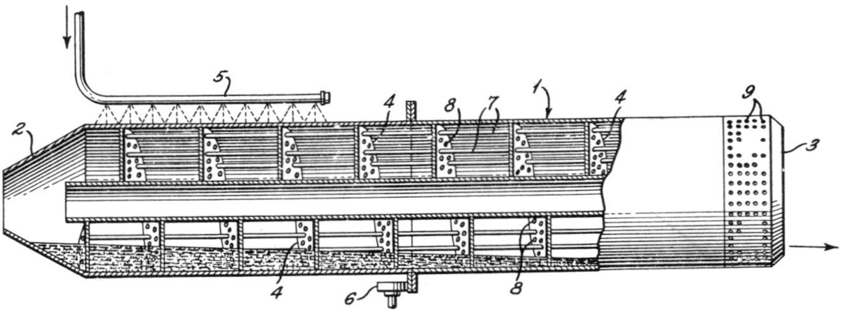

Source: J. Ubbels and J.T. van der Linde; U.S. Patent 3,193,927; July 13, 1965

Tiltable Rotatable Drum

D.R.M. Lambert and J.M. Sharkey; U.S. Patent 3,292,259; December 20, 1966; assigned to National Dairy Products Corporation describe an apparatus for separating whey from curd in the manufacture of cheese and is concerned particularly with that stage in the process of cheese-making in which a mixture of curd granules or particles and whey has been removed from the cheese vats in which the curd was formed and cooked and is subjected to a separation treatment to substantially remove the whey from the curd granules.

Apparatus of the kind described comprises a tiltable rotatable trommel or drum having less than one-half of its surface provided with elongated circumferentially extending slots. The trommel is cylindrical in shape and is provided with slots in a generally axial segment extending over approximately one-third of its surface. The segment containing the slots extends from end to end of the trommel and over one-third of its circumference, the remainder of the trommel being imperforate. The elongated slots have a width sufficient to permit whey to drain through, but not sufficient to allow curd granules to pass through or to block up the slots.

Of course, a small proportion of fines may pass through these slots. Preferably the width of each slot is less than the minimum dimension of the curd granules, other than fines. The length of each slot is several times greater than the width of the slots. The slots are preferably arranged in rows spaced axially from one another, the slots in each row being preferably opposite to the spaces between the slots in the adjacent rows.

The shape of the slots is important to minimize harm to the curd, particularly during rotation of the trommel. The elongated shape not only minimizes harm to the curd when the slots extend circumferentially, but they also permit improved drainage of whey. Also, the limited area and positioning of the area has been found to give improved results in the manufacture of cheese by limiting drainage of the curd to a certain sector of the trommel.

The ends of the trommel are partially closed to enable the curd and whey to be retained in the trommel during the separation operation. Means are provided for delivering a mixture of curd granules and whey to one end of the trommel and means are also provided for tilting the trommel in order to discharge curd granules from the other end of the trommel after the separation operation has been completed.

The trommel is provided internally with lifting members which comprise spaced parallel rods which are disposed to extend helically from end to end of the trommel and are secured to the internal surface of the trommel so as to be spaced a short distance inward. The rods serve to lift the curd as the trommel rotates and facilitate the separation of curd and whey and the drainage of the latter through the slots. Fingers or projections may be provided on some of the rods, for example, on every third rod, to assist in preventing the formation of lumps or large aggregations of curd.

A tray is provided beneath the trommel to collect the whey which drains through the slots. The trommel is rotated by suitable mechanism according to a predetermined cycle, both during and after the delivery of the mixture of curd and whey to the trommel. Means are provided to vary the speed of rotation of the trommel, which is preferably rotatable at either a low speed or a higher speed, as required.

The cycle of turning operations of the trommel is varied according to the condition of the curd and whey mixture supplied to the trommel and may comprise the following:

(a) four complete revolutions at low speed during the delivery of the mixture of curd and whey to the trommel, the trommel being held with the imperforate segment at the lowermost position for a period between successive revolutions;

(b) a one-half revolution of the trommel whereby the perforate segment is brought to the lowermost position, followed by a draining period of one minute while the drum is stationary with the perforations in a position to permit drainage;

(c) one complete revolution followed by one minute drainage period with drum stationary;

(d) one complete revolution followed by one minute drainage period with drum stationary; and

(e) continuous rotation of the trommel at high speed with the trommel in tilted position while curd granules are being discharged from the trommel.

Operation of the apparatus can be described by referring to Figure 5.15. The trommel (22) comprises a cylindrical stainless steel drum which is partially closed at its entry end by an annular plate (75) and is provided with a central entry aperture (76) through which a mixture of curd granules and whey is charged into the trommel through pipe (79) at the commencement of each cycle of operations (Figure 5.15a). The mixture of curd granules and whey is pumped from the vats in which the normal procedure of curd formation has been carried out.

In the drawing, the trommel is imperforate over two-thirds of its peripheral area and is perforated over the remaining one-third of its peripheral area. The trommel is thus provided with a generally axially extending imperforate segment and with a generally axially extending perforate segment. The perforations comprise a series of rows of slots (77) (Figure 5.15b).

These slots are disposed to extend in a peripheral or circumferential direction. The width of each slot is sufficient to permit whey to drain through them but is not sufficient to permit curd granules or particles to pass through or to block the slot, except for small quantities of fine particles which may pass through. This arrangement of slots has been found to effect efficient separation of whey from the curd granules, and a minimum discharge of fines with the whey from the trommel. The width of each slot is about 0.065", the length of each slot is about 1" and the ends of the slots are preferably rounded.

The slots are spaced relatively close together to permit maximum drainage and preferably the slots in each row are disposed opposite to the spaces between slots in the adjacent rows. The peripheral distance between adjacent slots is about 0.25", and the axial distance between rows of slots of about 0.12". Lifting spaced apart parallel rods or bars (78), preferably 16 in number, are mounted within the trommel and are disposed to extend from end to end of it in a helical direction, the angle of inclination of each rod to the axis of the trommel being about 15°. Each rod is secured to the internal surface of the trommel by suitable means and is spaced from the

FIGURE 5.15: TILTABLE ROTATABLE DRUM

a.

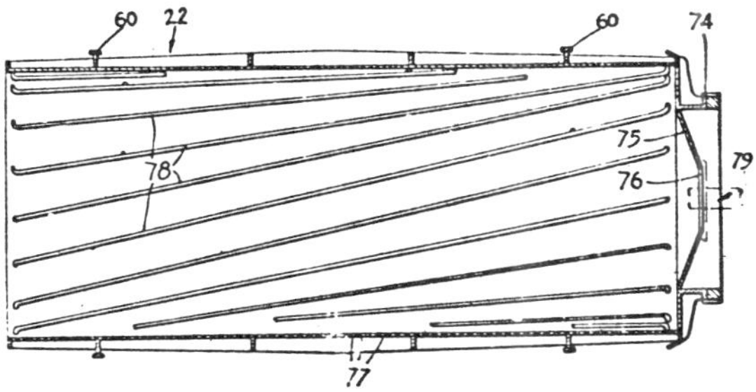

b.

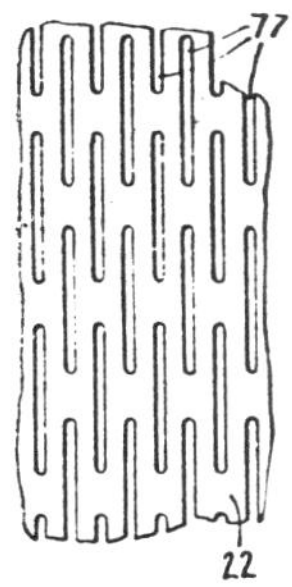

c.

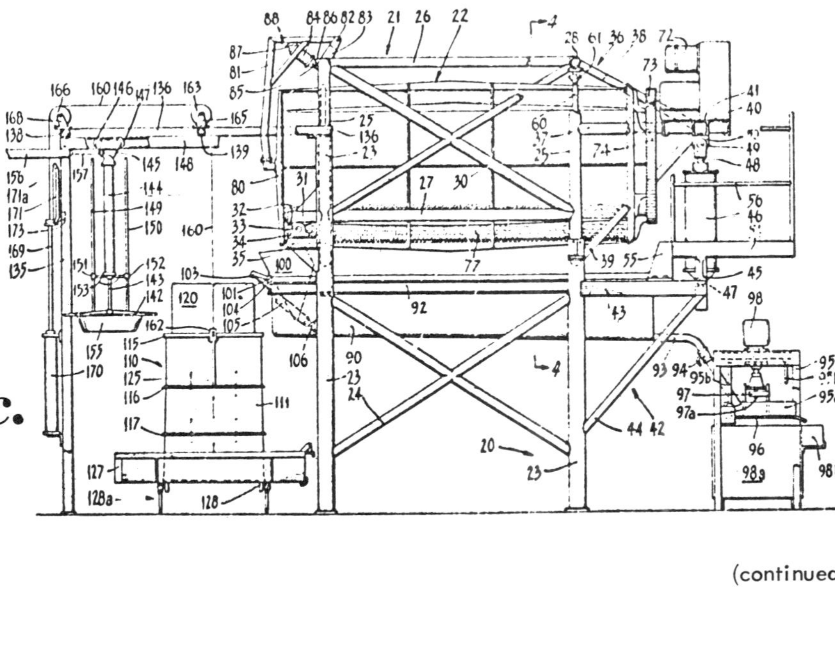

(continued)

FIGURE 5.15: (continued)

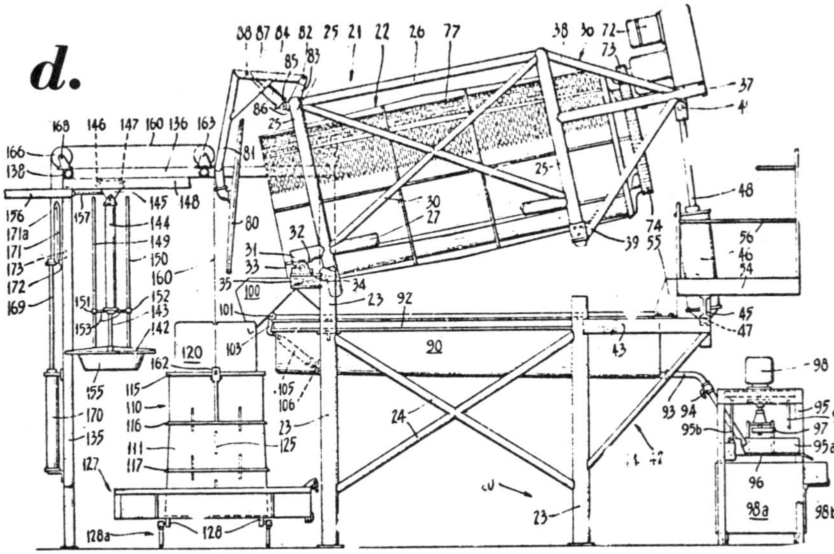

Source: D.R.M. Lambert and J.M. Sharkey; U.S Patent 3,292,259; December 20, 1966

internal wall by a relatively short distance. The lifting rods serve to lift the curd granules when the trommel is rotated and also serve to facilitate separation of the curd and whey and the drainage of the whey through the slots. Fingers may be provided on the rods or on some of such rods to assist in the prevention of formation of lumps in the curd.

The discharge end of the trommel is normally closed by a door (80), Figures 5.15c and 5.15d, which is secured to a frame (81) which is pivoted at pivot points (82) to brackets (83) secured to the transverse end member (28) of the trommel frame (21). An air cylinder (84) is pivoted at (85) to a bracket (86) secured to the end member (28) of the trommel frame and the piston (87) of the air cylinder (84) is pivoted at (88) to the frame (81). Actuation of the piston by the air cylinder effects opening or closing of the discharge end of the trommel by the door. This opening and closing movement is automatically controlled in accordance with a predetermined cycle.

Preferably the cycle of operation of the trommel is as follows. The trommel is moved to a position in which the imperforate segment of the trommel is lowermost and the slots are uppermost and the mixture of curd and whey is delivered into the trommel while it is stationary in this position. When the level of whey in the trommel reaches a predetermined level, the trommel is rotated through one revolution at slow speed, preferably about 2 1/4 rpm, to effect drainage of whey through the slots, delivery of the curd-whey mixture to the trommel being continued during rotation of the trommel. The trommel is then held stationary again for a period, and when the whey level again reaches the predetermined level the trommel is again rotated one revolution. This procedure is repeated four times during the filling operation. When delivery of the curd-whey mixture to the trommel has been completed, the trommel is rotated one-half turn to bring the slots into the lowermost position. Drainage of whey through the slots is then effected for about 1 minute. The trommel is rotated one revolution, at slow speed, followed by drainage for 1 minute, and this procedure repeated at least once.

It will be understood that the rate of separation of whey from the curd varies from one batch to another, and is dependent on the nature of the milk product from which the curd is formed and on the seasonal conditions, and consequently it may be necessary for the cycle of operations of the trommel to be varied according to the characteristics of the particular material being treated. Thus, the number of revolutions of the trommel, and the period of drainage of whey between successive revolutions, may be varied.

After separation of the whey from the curd and drainage of the whey from the trommel into the tray (90) have been completed, the trommel is tilted by means of the air cylinder (46) to the position shown in Figure 5.15d, the chute (100) is moved by the air cylinder (105) to the operative position as shown in Figure 5.15d and the door is moved away from the discharge end of the trommel by the air cylinder. The curd granules are then discharged from the trommel through the chute (100) into the conditioning tower (110), the trommel being rotated at higher speed, e.g., about 9 rpm, during the discharge operation to facilitate discharge of the curd granules from the trommel.

When the curd granules have been discharged from the trommel, the trommel is returned to the horizontal position shown in Figure 5.15c, the door is returned to close the discharge end of the trommel, the chute is returned to its inoperative position, and the filling cycle is recommenced.

MOLDS

To Prevent Bulging Rim

H.A. Rossen; U.S. Patent 3,095,647; July 2, 1963 describes a cheese mold intended for use in the production of pressed cheese. Its purpose is to prevent bulging or a disfiguring rim which may later have to be cut away. The apparatus described provides a groove about the circumference at one end, typically the bottom, of the mold in order to accomplish the stated function.

Figure 5.16 is a sectional view of the cheese mold. The mold has a mold wall (1), an outer end-bottom (2) and a loose insert (3) which together with the mold wall forms a groove (4) in which the lower edge of a slit single piece shell (5) may be introduced when pressing the cheese mass (6) present in the mold.

The insert is provided with reinforcing ribs (7) to prevent the insert which is made of comparatively thin plate, from bending during the pressing operation. The shell is provided with a lid (8) having a skirt (9) which grips round the uppermost edge of the shell so that a pressure exerted on the lid will push the shell inwardly into the

FIGURE 5.16: MOLD FORM

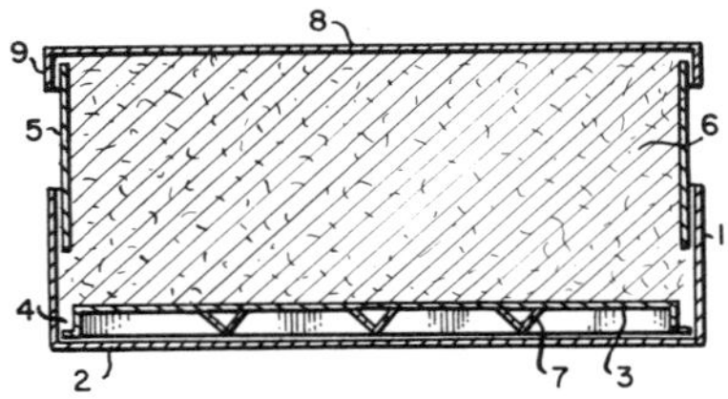

Source: H.A. Rossen; U.S. Patent 3,095,647; July 2, 1963

mold. The lid may be provided with a conventional reinforcement.

Removal of Gas by Suction

G. de Boer; U.S. Patent 3,199,195; August 10, 1965; assigned to NV Volma, Netherlands describes an apparatus for the manufacture of preformed pieces of curd from curd out of which the whey has been substantially eliminated. The apparatus comprises a stationary vessel providing an enclosed chamber having an inlet for the introduction of a mass of curd, means for receiving the mass of curd and for compressing the mass into a desired compressed shape. There exists a means for including an outlet conduit coupled to the vessel for establishing suction to enable the mass to be compressed while simultaneously being subjected to suction for the removal of gas.

It is an object of the apparatus to provide a method in which this inconvenience of air inclusion, if any, is avoided to the best possible purpose, whereby it is feasible to make hard cheese. Figure 5.17a shows a longitudinal section of the apparatus and Figure 5.17b is a cross-section taken along the line (II—II) in Figure 5.17a.

The apparatus comprises a container (1) which is provided with an inlet (2) for the prepared curd and a connection (3) for a suction duct. Provided within the container is a device (4) for prepressing and preforming the material which consists of two complementary parts (5) and (6), of which part (5) is composed of two semicylindrical bows (7), which on either side merge into a base plate (8).

The part (6) consists of two complementary semicylindrical bowls (9) of perforated material. Part (6) is displaceable in an upward direction within a space (10) in the container which is limited by a perforated inner casing (11). This movable arrangement is achieved through the intermediary of a threaded rod (12) which by way of a packing (13) protrudes from the container and which on its extremity is provided with a handwheel (14).

The part (5) of the device (4) for prepressing and preshaping is arranged in the container to be capable of travel. For that purpose part (5) is provided with a number of wheels (15) which can roll on rails (16), secured on the bottom of the container. On its front side the container is provided with a discharge opening (17) which can be closed by a cover (18).

FIGURE 5.17: REMOVAL OF GAS BY SUCTION

a.

b.

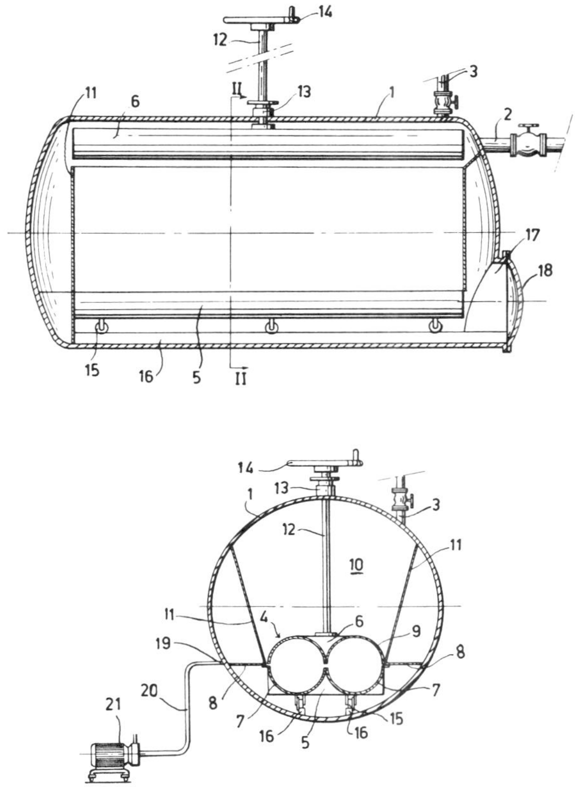

Source: G. de Boer; U.S. Patent 3,199,195; August 10, 1965

The opening is adapted to the dimensions of the movable part (5), so that this part, in a filled condition, can be wheeled outside of the container through the opening (17). The container is further provided with a discharge aperture (19) for the whey, which by way of a duct (20) is drained by a pump (21). The operation of the installation is described as follows.

The already prepared curd is introduced into the container by way of the inlet. The part (6) of the device (4) for precompressing and preshaping is in its elevated position, as represented in Figure 5.17a. After the space (10) has been filled to a definite level, the inlet is closed, whereupon by way of the suction connection an underpressure is generated in the interior of the container.

The gases in the curd are thereby expelled via the connection (3). When this process of expelling the gases has continued for a sufficiently long time, the part (6) of the precompressing and preforming device (4) is lowered by means of the hand wheel while the curd mass in the space (10) is simultaneously compressed. This compression is continued until the semicylindrical bowls (9) of the part (6) join the complementary bowls (7) of the part (5) (Figure 5.17b). The quantity of whey produced during this phase can be discharged by the pump (21) via the perforations in the bowls (9) and the inner casing (11) and via the aperture (19). In this way a compact curd cake, into which the air cannot penetrate, is obtained between the bowls (7) and (9).

After the eduction of the whey from the container the part (6) is again elevated by means of the hand wheel. The underpressure is neutralized and the cover is removed, after which the preformed pieces of curd, resting on the part (5), can be wheeled outwardly through the opening. The pieces are then cut up to a convenient size and conveyed to a stuffing trough.

Nondenting Construction

H.A. Rossen; U.S. Patent 3,514,857; June 2, 1970 designs a cheese mold which eliminates denting and bending of the mold which occurs during the various handlings of the mold such as turning and cleaning which takes place during the course of the cheese-making process. This mold is characterized in that the shell of the mold comprises two juxtaposed and rigidly interconnected plates having intermediate evenly distributed low elevations projecting from at least one of these plates, and of which plates at least the one facing the interior of the mold is finely perforated.

Without the advantage of the thin plate for the fine perforation being waived, this structure affords protection of the finely perforated plate against the rough mechanical influences to which it may be exposed during service in the cheese factories. It would even be possible to make the perforated plate thinner than would otherwise be normal and thus facilitate the production of the very fine perforation holes.

The reinforced shell may be carried out in different ways. Thus, it may for example be made up of a smooth perforated plate facing the interior of the shell, and a plate which surrounds the perforated plate and which at the side facing the perforated shell is provided with low elevations. The latter outer plate as such may be perforated but is preferably not perforated. However, the shell consists preferably of a combination of an inner finely perforated plate the outer face of which is provided with evenly distributed low elevations, and an outer nonperforated plate. Thus, the shell is made up of a shell of the known type and the described outer nonperforated plate.

Figure 5.18a is a front view of this cheese mold with the shell pushed only partly down into the mold and with raised covering plate and pressing lid; Figure 5.18b illustrates in cross-section part of a shell for use in the cheese mold of Figure 5.18a and Figure 5.18c illustrates a section, taken along the line (A—A), of the part of shell shown in Figure 5.18b, with part of the front plate broken away as indicated by (B).

The mold in Figure 5.18a has an outer wall (1) and a bottom (2) which may be made of stainless steel. The outer wall has at the upper edge a reinforcing ring (4), and in the edge formed by mold wall (1) and bottom (2) there is provided whey discharge holes (29) having a diameter of about 5 mm. The shell (3), the construction of which is seen in detail in Figure 5.18b and 5.18c, comprises an inner perforated plate (13) and an outer smooth nonperforated plate (14) which at top is bent to form an edge (15). The plate (13) is provided with perforation holes indicated at (5) and regularly

FIGURE 5.18: CHEESE MOLDING APPARATUS

a.

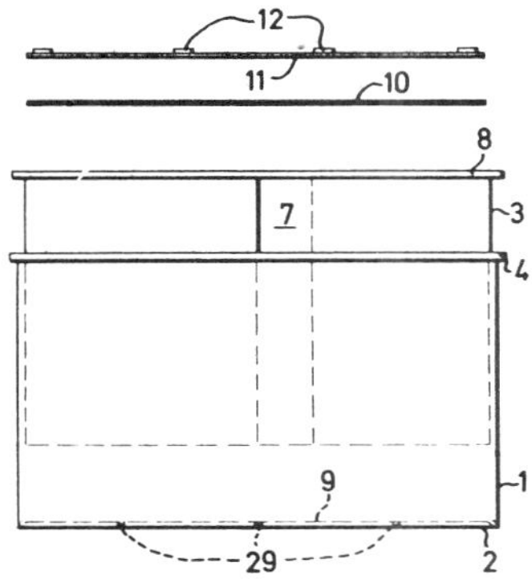

b.

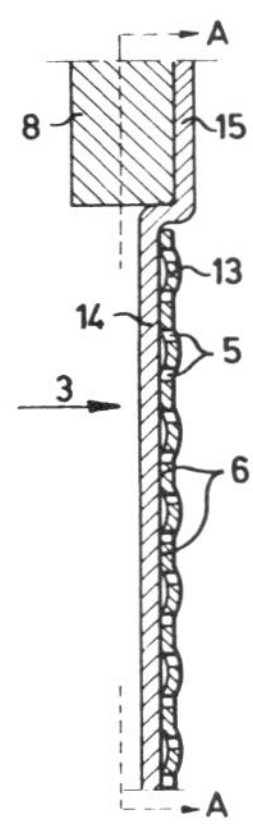

c.

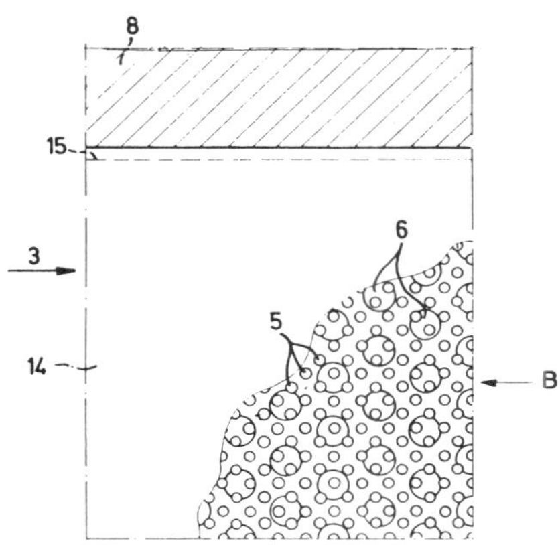

Source: H.A. Rossen; U.S. Patent 3,514,857; June 2, 1970

distributed elevations, indicated at (6), having been imparted to the outer side of the plate by embossing. The diameter of the perforation holes may be about 0.8 mm. or less, and the number of such holes may be 20 to 30 or more per $cm.^2$ of the plate, and the elevations may be of a height of from 1/2 to 1 mm. The ends (7) of the shell overlap each other when the shell is inserted into the mold. The shell is provided with an upper reinforcing split ring (8) which, when the shell has been pushed fully into mold, can rest on the reinforcing ring (4) of the outer wall.

The shell is provided with a perforated loose bottom plate indicated at (9), and a similar covering plate (10) which may slide down into the interior of the shell when this is inserted into the mold. The perforation of the bottom and covering plate is preferably similar to the perforation provided in the perforated plate (13) of the shell, as mentioned above. They are also preferably embossed in a similar manner.

The mold is furthermore provided with a pressing lid (11) of material similar to the material of the outer wall and provided with reinforcing ribs (12). This pressing lid can also during pressing slide down into the upper part of the shell in close contact with the interior surface of the shell.

The plate (14) shown in Figures 5.18b and 5.18c is connected by spot welding with the upper edge of plate (13) and similarly with the lower edge of the same plate. The reinforcement ring (8) is disposed aroung the edge (15) and preferably has been secured to the edge by welding. The total thickness of the plates (13) and (14) may be from 1 to 1 1/2 mm. In Figure 5.18b the elevations are shown to be round in shape, but they may have any other shape. The holes may also have any other shape than round.

Rectangular Hoop

The cheese mold apparatus described by D.A. Fowler; U.S. Patent 3,700,373; October 24, 1972; assigned to Cache Valley Dairy Association, in addition to providing an improved procedure for producing horn cheeses, also makes it practical to press a great variety of specially shaped cheeses without the waste associated with cutting such shapes from block cheeses. Many of the shapes previously impractical for most cheesemakers to produce a, now feasible products.

According to this process, cheese horns are pressed in rectangular horn cheese hoops shaped and dimensioned externally approximately the same as conventional rectangular block cheese hoops. These horn cheese hoops may include parts which are interchangeable with the block cheese hoops used in the same cheese plant, but ordinarily it is more convenient for the hoop structures used for the production of horns to be maintained and stored as complete sets.

It is possible to load the rectangular horn hoops of this process interchangeably with rectangular block hoops in a press to produce simultaneously both horns and blocks. With this process the automated washing equipment conventionally employed for washing block hoops is satisfactory for washing these rectangular horn cheese hoops. Horns are produced with much less loss of curd during the pressing procedure.

The apparatus is used in accordance with procedures generally followed in the production of block cheeses in a rectangular block cheese hoop, such as the standard 40 pound hoop. Such a standard rectangular block hoop is modified by the insertion of adapters. These adapters may be formed as insertable platens, one of which is positioned in the bottom of the hoop prior to insertion of the cloth and sleeve portions of the conventional hoop and the other of which is inserted in opposition to the first insert atop the cheesecloth after it is folded over the curd. The top of the rectangular hoop is then positioned in conventional fashion. Each platen has a pressing surface comprised of at least one, usually two, concave molds.

The molds of most interest are longitudinal cylindrical sections extending the full length of the pressing surface and dimensioned to produce horns of approximately the diameter produced in cylindrical horn hoops.

Many aspects of quality control and uniformity of product are enhanced by the method. The horns are produced by pressing a predetermined quantity of curd into an approximately cylindrical shape of predetermined volume by applying pressure across a much greater surface area than is the case when curd is pressed in a cylindrical horn hoop. When the platen members are urged together, pressure is applied in opposing directions normal the longitudinal axis of the pressed horn, rather than parallel to the longitudinal axis. It is generally desirable, from the standpoint of uniformity of quality and shape, to press curd in this fashion.

Although the platen inserts may include any convenient number of concave molds, rectangular horn hoops corresponding in dimensions to the standard 40 pound rectangular block hoops are appropriately dimensioned to receive platens containing two side-by-side concave molds. Each concave mold is shaped as a longitudinal cylindrical section of no more than semicylindrical volume. That is, each mold is shaped as the volume defined by a plane intersecting the cylinder parallel its longitudinal axis.

Each such longitudinal section should be no larger than, and preferably somewhat smaller than, a section defined by a plane intersecting the longitudinal axis of the cylinder. The precise shape of the molds used to produce horns need only approximate a cylindrical section. Subsequent aging procedures permit smoothing of the surface of the pressed horn, due to the natural flow characteristics of the cheese. The patent gives detailed diagrammatic illustrations of the apparatus.

PACKAGING

Process Cheese Slices

W.V. Green; U.S. Patent 2,931,729; April 5, 1960; assigned to Challenge Cream & Butter Association describes a process the object of which is to provide a method and apparatus for preparing or treating slices of process cheese so that when they are assembled together in direct contact with each other and in stacked relationship and are wrapped into packaged form the slices will not coalesce but will retain their individuality so that they may be readily separated from each other even after a considerable shelf life of the package.

The apparatus designed for this method consists of a pair of opposed upper rolls which are spaced from each other slightly less than the thickness of the slice so that as the slice passes between it will be engaged by the surfaces of these rolls. The spacing of the rolls should not be so much smaller than the thickness of the slice as to crush it. The rolls are rotatably mounted on standards and are hollow, enabling hot water or other hot fluid to be passed through by means of conduits. The temperature of these rolls is maintained at approximately 180°F. The function of these rolls is to heat the side faces of the slice sufficiently to melt and draw some of the butterfat to the surface.

The rolls are rotated at a speed so that the cheese slice remains in contact for approximately one-fourth of a second, the purpose being not to remelt the entire slice or even to heat it through but merely to heat the cheese adjacent the side faces sufficiently so that the butterfat constituent will be drawn to the surface. Thus as the slice passes from between the rolls each slice will have on its side faces a thin film of butterfat.

The slices are prevented from adhering to the rolls and from following around either of the rolls during their rotation by means of scrapers pivotally mounted beneath the rolls and urged into engagement with the holes by tension springs. The treated slices may be allowed to fall directly onto a stainless steel endless belt and on cooling to room temperatures the slices may be collected, stacked and wrapped.

The butterfat film seems to prevent the slices from adhering to each other excessively when they are ultimately packaged. There is some adhesion between the individual slices in the final package despite the presence of the butterfat film but such adhesion is not of such character that it prevents the slices from being separated from each other at the time of use without breaking of the slices or resorting to reslicing.

Preparing Cheese for Shipping

J. Meeker and W.H. Tate; U.S. Patent 3,100,712; August 13, 1963; assigned to Clearfield Cheese Company, Inc. are concerned with shipping bulk quantities of cheese and accordingly disclose the preparation and packaging of such bulk quantities for shipping purposes.

The initial step involves pouring of fresh cheese curd into a metallic cheese mold cylinder within which a cheesecloth bag is fitted and overlaps the open top end of the cylinder, which is supported in a vertical position. After the fresh cheese curd is poured into the mold cylinder the cloth bag is closed. A drain plate is placed in the bottom of the cylinder in a spaced relation which is maintained by a rim portion on a drain plate. In the center of the bottom member of the cylinder is a drain opening through which cheese whey is drained.

After the cheese is poured into the cylinder, pressure is applied to reduce its volume and force the moisture from it through draining apertures in the drain plate. Pressure is applied to the cheese by placement of a massive weight on top of the cheese through the open top of the mold cylinder.

To assure that the pressure is applied uniformly and to prevent development of surface irregularities in the molded cheese, a pressure plate is provided which fits within the open top of the cylinder. The weight is permitted to remain on top of the cheese for at least 10 hours during which time moisture will drain through the apertures in the drain plate and the drain hole in the bottom of the mold cylinder. After the molding process has been completed, a mold inverting mechanism is applied to the mold cylinder after the cloth bag has been opened so that the compressed and dry cheese may be directly deposited onto a stainless steel plate. The shipping container includes an outer tubular casing made of fiber-type cardboard having exposed on one side a corrugated reinforcement. The outer casing is polygonal in cross-sectional shape and forms an octagon. In view of the flat sided cross-sectional shape of the outer casing, it may be folded for more convenient storing and also will provide a more rigid closure. Furthermore, since the outer casing is made of cardboard, its weight will be no problem at all.

Disposed within the confines of the outer tubular casing is an inner cylindrical casing made of the same fiber-type cardboard having vertically disposed reinforcing corrugations. Thus, the inner and outer casings although made of cardboard, by virtue of their corrugations form a container having the requisite strength and forming an interior cylindrical surface for receiving the molded cheese after being encased within a plastic liner that is air and moistureproof.

Each of the flat sides of the outer tubular casing is provided with a flat extension at each axial end which cooperates with a foldable flat extension connected to a corrugation reinforced cardboard lid provided for the purpose of sealing both axial ends of the container. The sealing flaps from the lids are folded downward and upward behind the flat extensions of the outer tubular casing. The assembly is then sealed or bound with a metallic strap or band.

Thus the fresh cheese curd may initially be formed into an externally smooth surfaced mold and pressurized by a weight to reduce its volume and remove the moisture without developing any surface irregularities in the molded cheese. Development of surface mold on the cheese will be avoided by a more complete removal of the moisture and avoidance of surface crevices. The molded cheese may then be encased in plastic and enclosed within the cardboard container constructed to have the requisite strength and moistureproofing.

ADJUVANTS

MICROBIAL RENNETS

Starter Culture Apparatus

R.R. Boelter; U.S. Patent 3,117,009; January 7, 1964; assigned to R.R. Boelter Company, Inc. describes both the method and apparatus used for producing a starter culture in a small cabinet, preferably which has a separate cooking cover and hood used during the complete culture forming process. The culture bottle with the water and the skim milk mixture is immersed in a pasteurizing and sterilizing water bath within the cabinet which is releasably closed with a top cover. The water is raised to the proper temperature to sterilize and pasteurize the mixture. The bottle is then maintained immersed within the bath and is allowed to cool gradually to the temperature at which the bacteria media or mother culture is added. The cover is removed and the mother culture placed within the cabinet and a hood placed over the cabinet.

A chlorine or other suitable atmosphere is established within the cabinet to insure sterilization, after which the water bath level is dropped to allow inoculation of the culture. Means are provided to allow inoculation without contaminating the atmosphere within the cabinet which are, preferably, releasably sealed openings in the hood through which the operator's hands and arms are projected for transferring the bacteria media from its container into the culture bottle. After inoculation, the culture bottle is again tightly sealed and may or may not again be immersed. The latter immersion is desirable to insure complete sealing against possible entrance of bacteria or other contaminating foreign matter. Automatic timing means are provided for heating the culture and other such steps.

Figure 6.1a is a perspective view of this culture processing cabinet. Figure 6.1b is a vertical section through the cabinet. Figure 6.1c is a small fragmentary view of the processing cabinet with a cover replacing the hood of Figure 6.1a. Figure 6.1d is an exploded view of a front opening sealing structure shown in Figure 6.1a.

A culture processing unit is illustrated including an open top water tank (1) secured to vertical supporting legs (2). A bath of water (3) is provided in the tank with the upper level of the water being either somewhat below or completely over the top of a series of releasably sealed bottles (4) having removable top caps (5). The bottles each contain a liquid mixture (6) of skim milk and water which is to be treated and inoculated to form a starter culture. During the initial portion of the process, a relatively flat cover (7) is disposed over the top of the tank to prevent evaporation of the water from the bath into the culture processing room. During the inoculating period, an upper hood (8) is sealed to the top of the tank and defines a totally enclosed inoculating chamber. The hood includes a sealed side opening (9) and a pair of adjacent sealed front openings (10) for placing and removing items from the inoculating chamber and for working in the chamber.

FIGURE 6.1: METHOD AND APPARATUS FOR PRODUCING A STARTER CULTURE

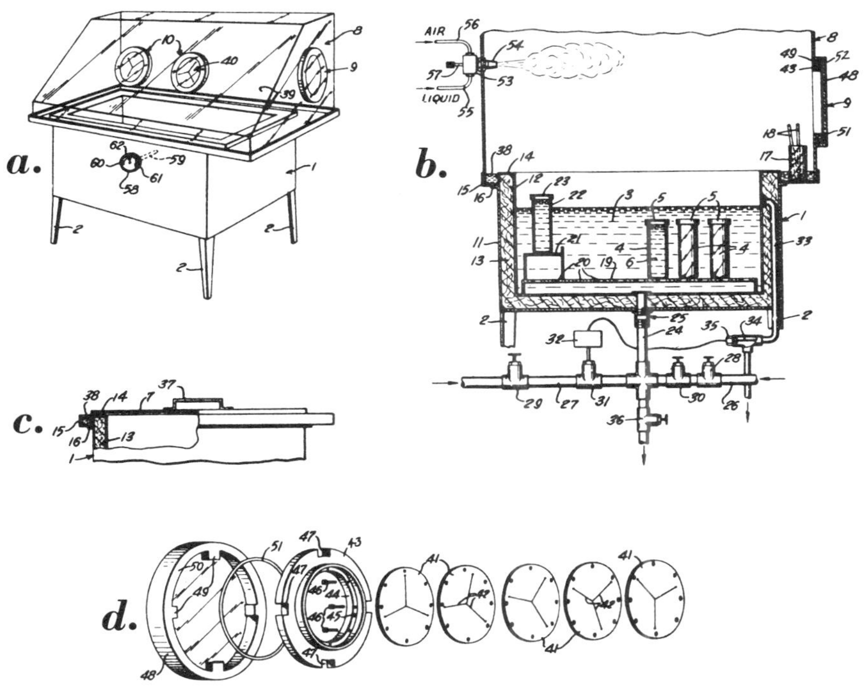

Source: R.R. Boelter; U.S. Patent 3,117,009; January 7, 1964

The tank is constructed of superimposed, rectangularly shaped, inner and outer base members (11) and (12) with a thermal insulation (13) disposed between them to eliminate heat loss from the water bath within the tank. An upper ledge (14) is welded to the upper end of member (11) and extends outwardly and terminates in an integral trough section (15) of rectangular cross-sections encircling the member (11). The trough section rests upon and is welded to a ledge (16) on the upper end of the member (12) on the front, back and left side. To the right side of the tank, trough section is extended outwardly to provide a table area for supporting any operating implements such as the container (17) for pipettes (18) for inoculation to the liquid. A water distributing plate (19) having a plurality of openings (20) rests on the bottom of the tank to support the bottles and to distribute evenly the incoming water which is introduced through the bottom of the tank. A small shelf (21) is shown to the left side of the tank on the plate (19). A bottle (22) generally corresponding to bottles (4) and containing a previously formed starter culture liquid is placed on shelf (21) after removal of cover (7) and before positioning the hood on the tank. The shelf is of a proper height to hold the upper end (23) of bottle (22) above the uppermost level of the water bath.

A steam and water pipe (24) is connected to the center of the bottom of the tank by a suitable coupling unit (25). A water pipe (26) and a steam pipe (27) are connected respectively to the steam and water conduit (24). Manually operated on-off valves (28) and (29) are provided respectively in lines (26) and (27) for connecting the tank to sources of water and steam. A needle control valve (30) is connected in water line (26) to establish a preset flow of water into the tank. A modulating valve (31) is connected in steam pipe (27) and includes an operating drive unit (32) to continuously adjust the steam-water mixture for automatic regulation of the temperature of the bath.

A high-level overflow pipe (33) is secured to the inner member (11) adjacent the upper level of the bath and extends downwardly between the members (11) and (12) and out of the tank. A temperature sensing unit (34) is mounted in the overflow pipe (33) by coupling (35) and is connected to drive unit (32) for controlling valve (31) for automatic temperature regulation. A drain valve (36) is connected to the pipe (24) to allow lowering the level of the bath and access to the caps of the bottles.

During the initial processing of the culture liquid, the cover is disposed over the tank. The cover is generally a flat rectangular metal cover having depending side walls slightly deeper than the trough section (15). A handle (37) is provided on the top of the cover for placing and removing the cover on the tank. The side walls rest in trough section (15) to support the cover in proper position close to the tank. A chlorine water (38) substantially fills the trough section (15) and establishes a water seal which prevents contaminating bacteria and other foreign matter from entering the tank.

When the liquid has been sterilized and is in condition for inoculation, the cover is removed and the hood placed on the top of the tank. The hood is open at the bottom with the lower end of the hood resting within the trough of the tank contiguous to the outer side wall of the trough. The chlorine water (38) in the trough section maintains a liquid seal between the hood and the tank. The hood is formed of clear plastic and includes an inclined front wall (39) having the openings (10) provided and sealed with soft rubberlike diaphragm units (40).

Referring particularly to Figure 6.1d each of the diaphragm units includes a plurality of stacked individual round soft rubberlike discs (41) having 3 radial slits (42) equicircumferentially spaced and extending from the center of each opening. A slotted ring (43) includes a central spacing hub (44) which is secured to the hood at each opening and spaces the ring (43) from the front wall (39). The hub (44) includes an internal clamping shoulder (45) having a diameter corresponding to the discs which are disposed between the shoulder and the front wall of the hood. Small cap screws (46) pass through the shoulder and discs and thread into suitable tapped openings in the hood to clamp the discs over the openings. The slits in the adjacent discs are angularly offset, and when the operator places his hands and arms through the diaphragm units to inoculate the liquid, the soft rubberlike discs tightly cling to the arms and maintain a seal at the openings. Foreign matter is thus prevented from entering the cabinet.

Ring (43) includes 4 equicircumferentially spaced edge recesses (47). A cap (48) includes a plurality of radial projections (49) integrally formed on the inner edge of circular cap rim (50) and spaced in accordance with slots (47). Cap (48) is placed over the corresponding ring (43) and turned to releasably close the opening (10). An O-ring seal (51) is disposed between the adjacent faces of cap (48) and ring (43) to seal the opening. The side opening (49) includes a cap (52) which is secured to the hood generally in accordance with the securement construction of the cap (49) to releasably close the opening for removing the bottles from the cabinet.

A chlorine fogger (53) is secured to the side wall of the hood opposite the opening (9) and includes a nozzle (54) projecting into the hood. A chlorine line (55) and an air line (56) are secured to the fogger (53) for supplying a chlorine fog within the cabinet. A small needle valve control (57) is provided to control the mixture of air and chlorine introduced into the cabinet.

A thermometer (58) is secured to the front of the tank and includes a sensing unit (59) projecting into the bath. Switches (60) and (61) are mounted within the path of a temperature indicator (62) of the thermometer. The switches are connected to actuate indicating means, such as a lamp (63), if the unit malfunctions during a cycle of operation.

Although the temperature of the bath may be automatically controlled, the apparatus may be manually controlled as follows. The culture liquid is placed in the bottles (4) which are then sealed with a suitable leakproof cover or lid (5) and placed within the tank (1). The water valves (28) and (30) and the steam valve (29) are opened, and the tank (1) filled to the level of the overflow pipe (33) which will automatically maintain this level. The valves (29) and (30) are manually adjusted to set the temperature of the bath (3) at 200°F., at which temperature the bath is held for one hour for pasteurizing and sterilizing of the liquid. At the end of one hour, valve (29) is closed and the temperature of the bath is allowed to slowly decrease to 70°F. Valves (29) and (30) are then adjusted to maintain this temperature for 30 minutes, after which these valves are closed to allow the liquid to be inoculated.

The cover (7) is removed and the mother culture bottle (22) and the container (17) with pipettes (18) are then placed on the supporting ledges as shown in Figure 6.1b and the hood placed on the tank with the lower edge resting in trough section(15). The chlorine fogger (53) is actuated to establish a chlorine fog within the hood which sterilizes the atmosphere. The bath is then lowered by manually opening the valve (36) until the lids of bottles are exposed. Caps (49) are removed from the opening (10) and the operator places his arms through the sealing discs (41). The lids (5) and (23) are removed and the required quantity of the bacteria media or mother culture in bottle (22) transferred to the bottles (4) by the use of the pipettes. Bottles (22) and (4) are then again sealed and bottle (22) is placed on the ledge adjacent the opening (9).

The water valve (29) and steam valve (30) are opened to raise the level of the bath to that of overflow pipe (33) and again immerse the bottles (4). The temperature of the bath is set at 72°F., and then the valves (29) and (30) are closed. The insulated tank will maintain the temperature during the propagation period. After the culture is completely formed, cap (52) is removed and the bottles are removed from the cabinet and stored in a suitable refrigerator for subsequent use or they may be used directly.

Pediococcus cerevisiae

According to a process described by T.W. Humphreys and D.S. Titus; U.S. Patent 3,086,866; April 23, 1963; assigned to Merck & Co., Inc., the time required for the coagulation of milk can be shortened considerably by including in the milk *Pediococcus cerevisiae* together with suitable sources of carbon and nitrogen for the microorganism. Suitable carbon sources are carbohydrates assimilable by the microorganism, such as dextrose, fructose, sucrose, mannose, galactose and maltose. Dextrose is preferable. The sugars may also be added in the form of solutions such as corn syrup. Alternatively, the milk itself can be used as a source of carbon by adding an enzyme such as lactose which is capable of converting the lactose of the milk to dextrose and galactose.

As suitable sources of nitrogen are yeast products such as yeast extract, yeast autolysate, solubilized yeast, food yeast, etc; amino acids such as hydrolyzed proteins (enzymatic digest of casein); and proteins such as peptone. The amount of the source of carbon is not critical, but is preferably not in excess of 2% by weight of the milk. When dextrose is the source of carbon, it is preferably used in an amount between 0.5 and 2.0%. Usually, 1% dextrose is sufficient and gives good results. The quantity and the source of nitrogen used is also not critical and can vary depending in part upon the particular source being utilized. In general, it is preferred to use an amount equivalent to 0.5 to 2.0% by weight of the milk.

The coagulation of the milk can be accelerated further by including in the milk, in addition to the carbon and nitrogen sources, a small amount of a purine such as guanine. An amount of purine of 0.5 to 2 mg. per 100 ml. of milk will accelerate the formation of lactic acid and therefore the coagulation of the milk. When the milk is thus supplemented with carbon, nitrogen, and purine, an incubation period of from 16 to 24 hours at about 37°C. will bring about complete coagulation of the milk. These purines are also present in various yeast products, and when these materials are used as the source of assimilable nitrogen, it is not necessary to add additional purines.

In coagulation of milk, the formation of the lactic acid can be allowed to continue until complete coagulation occurs, or the coagulation can be effected earlier when a particular desired level of titratable acidity is reached by the addition of rennet extract.

Example: The milk samples in the examples were prepared by reconstituting commercial nonfat dry milk solids using 10 g. of the milk solids in 90 ml. of distilled water. The milk was dispensed in 10 ml. quantities in cotton-plugged Pyrex test tubes, sterilized by autoclaving at 250°F. for 10 minutes, cooled quickly by immersion in tap water, inoculated with the *Pediococcus cerevisiae* cells and incubated at 37°C. for 20 to 24 hours. At the end of the incubation period, the relative amount of lactic acid formed was determined by titrating the incubated tubes directly with N/10 sodium hydroxide to the phenolphthalein end point. The nutrient source of carbon was 1% by weight of dextrose. The dextrose and the nutrient source of nitrogen, which is indicated as the additive, were added either from stock solutions or as solids and the pH of the milk was brought to 6.6 to 6.8 by the addition of sodium hydroxide or hydrochloric acid before adjusting the final volume.

The inoculum of Pediococcus cerevisiae was prepared by growing this microorganism at 37°C. for 20 to 24 hours in the following medium. A two liter solution of the following composition was formed:

Tryptone	20 grams
Yeast extract	10 grams
Sodium citrate	10 grams
K_2HPO_4	10 grams
Dextrose (anhydrous)	20 grams
Tween 80	2 ml.
$NaNO_2$	0.2 grams
Salt solution*	1,000 ml.
Distilled water, q.s.	2,000 ml.

*1.4 g. $MnCl_2{\cdot}4H_2O$, 0.4 g. $FeSO_4{\cdot}7H_2O$ 8.0 g. $MgSO_4{\cdot}7H_2O$, and 1,000 ml. distilled water.

After preparing the medium, the pH was adjusted electrometrically to 7.12 with concentrated hydrochloric acid and 50 g. of sodium chloride were dissolved in the medium. The pH was then 6.8. Prior to inoculation with the microorganism, the medium was sterilized by autoclaving at 120°C. for 30 minutes.

The broth culture at the time of use contained about 10^8 to 10^9 cells per ml. and 0.1 ml. of this broth culture was used to inoculate the 10 ml. milk samples.

Reproducible Method

The process developed by E.J. Tynan and M.W. Hales; U.S. Patent 3,142,575; July 28, 1964; assigned to Chas. Pfizer and Co., Inc. provides a uniform and reproducible method for the preparation of a starter culture. This process enables cheese to be made with a minimum development of off-flavors, poor texture, bad odors and gas pockets. The strains of this culture may be conveniently maintained in one group or divided into only two stable groups, thus decreasing the possibility of introducing wild, deleterious microbial strains during transfer operations. The process also allows culture upgrading, that is, freeing contaminated cultures of undesirable strains of microorganisms, which also tend to produce poor cheese.

The process first plates out a multistrain cheese starter culture, selecting from the strains one rapid-rate acid former, one moderate-rate acid former, one slow-rate acid former, and one nonacid-former. A nutrient medium is inoculated with the selected strains, which is then incubated until a peak state of viability is reached. The medium is then dried to isolate the improved culture. Drying is widely used as a method of preserving bacterial cultures and, while removal of the volatile elements in a high vacuum at normal room temperatures (20° to 25°C.) is a convenient method for drying, it is preferred, because of a tendency to diminish mortality, to dry the cultures in vacuo from the frozen state. This preferred freeze drying method is known as lyophilization.

The multistrain cultures used as sources for the individual strains used can be obtained from a number of sources. However, three mixed cultures, ATCC 14871, ATCC 14872 and ATCC 14873, serve as useful starting materials for this process. These cultures comprise commercially available cheese starter cultures, each of which contains many morphologically distinct strains. The cultures are, in the main, lactic acid-producing Streptococci.

To separate the strains, each culture is transferred to an agar plate by spreading techniques and the plate is incubated until well formed colonies of growing microorganisms can be observed. Then, with the aid of magnification, a number of the individual colonies are aseptically transferred to separate batches of nutrient media and the strains thus isolated are propagated. The nutrient media can be any synthetic or natural medium suitable for growth; a sterile reconstituted skim milk solids medium is preferred, however. When a peak state of viability is reached, it is convenient to preserve the strain by freeze drying the medium and to store the strain as the solid residue, the so-called "lyophile." Although neither critical nor necessary to this process, the best microbiological technique requires at least 3 series transfers during the propagation of the pure strains although it is suitable to use only one transfer.

To classify the strains in terms of their ability to effect the rate of acid formation, they are grown in a reconstituted milk medium under standardized conditions of time and temperature, and a lactic acid titer is determined. The technique comprises transferring cells of the viable strain to be classified into a flask containing a quantity, conveniently about 100 ml., of a sterile medium of dry skim milk solids reconstituted with distilled water. Although the concentration is not critical, it is preferred to use from 9 to 10% by weight of milk solids.

The inoculated medium is incubated for 16 hours at 21°C., in a quiescent state, then is treated with phenolphthalein indicator and is titrated with 0.1 N sodium hydroxide solution to determine the percent of acid, calculated as lactic acid, present. The amount of acid developed can also be determined by measuring the pH of the resulting medium.

This procedure is repeated for all members of the strains isolated as described and the strains are then grouped with respect to the amount of lactic acid produced under these conditions. Those strains which do not cause the production of significant amounts of lactic acid, i.e., do not cause an increase in the normal 0.17 to 0.20 lactic titer of sweet milk, are grouped with the nonacid-formers contemplated here. The fast acid formers have a lactic titer of 0.70 to 0.85; the moderate acid formers have a titer of 0.40 to 0.69; and the slow acid formers have a titer of 0.20 to 0.39. As previously stated, sweet milk itself, the control, has a titer of 0.17 to 0.20.

It is also possible to classify the strains with respect to acid-forming rate by observing the rate at which they curdle milk. The fast acid formers always form curds in 72 hours; the intermediate acid formers form curds in substantially all cases; and the nonacid formers never do. It is preferred, however, to rely on the quantitative titers rather than on curd formation as the criterion for selection.

After the strains have been isolated and classified, the cheese starter culture may be prepared by selecting at least one strain from each of the four groups - fast, moderate, slow acid formers and nonacid formers - and growing them to a peak state of viability. It is preferable, since it minimizes the effect of selecting strains with low phage resistance, to select about three strains from each group. The process need not be limited specifically to three since more than three may be used with equal effectiveness, although no particular advantage is derived in so doing. It is believed to be essential, however, to select at least one strain from each of the four groups for use in the culture to be prepared.

After the strains have been isolated, classified and selected, they are grown in a nutrient medium to a peak state of viability and are dried and stored for later use as starter cultures. This production procedure comprises inoculating a sterile milk solids medium of a convenient concentration, preferably from 9 to 14% by weight of milk solids in water, and incubating for a convenient time, preferably for 8 to 24 hours, at about 21°C. Although not necessary, it is preferred to transfer and inoculate through three series of pure cultures. The final pure strains are then combined conveniently in a flask of the same medium. Here it is convenient to use 0.1 to 1 volume of each strain per final volume. This mixed culture is then used to inoculate a milk solids medium, conveniently from 9 to 14% by weight of milk solids at a rate of about 1% of inoculum based on the growth medium.

This inoculum is then incubated, at 20° to 21°C., under quiescent conditions, and periodic observations of curd formation and determinations of pH and lactic acid percentage are made. When the broth has reached a pH of 4.3 to 4.6 and the titrable acid has reached 0.8 to 1.3%, no appreciable further increases in acidity will generally occur, no further visual changes are observed, and a peak state of viability has been reached. This usually requires from 12 to 24 hours. The suspension is freeze dried and the dry, solid residue from freeze drying is preferably stored in evacuated, sealed tubes until it is ready for use as a starter culture.

Since, as has been previously mentioned, differences in the reproduction rates among some strains make it desirable to propagate and to store them in separate groups, a preferred form is to produce the culture in 2 groups. In so doing, the strains which tend to be incompatible, i.e., which tend to grow at the expense of others, may be kept separate until used. Such a method is illustrated in the following diagram:

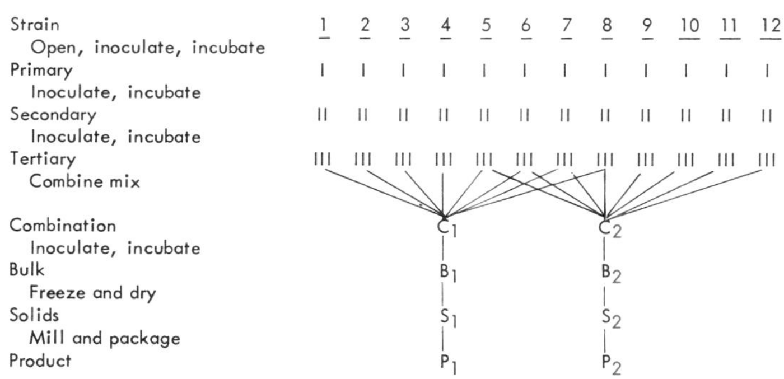

Lyophiles of the selected strains are opened and used to inoculate a sterile 10% milk solids medium, Primary I. This is incubated at about 21°C. for 24 hours, or until good growth is obtained, then is used to inoculate a fresh portion of the same medium, Secondary II. After incubation of Secondary II at 21°C. for 24 hours, it is used to inoculate Tertiary III, a third series of fresh sterile medium of the same composition. After incubation at 21°C. for 24 hours, each III is pooled by transfer in a predetermined manner to the combinations C_1 and C_2. These are flasks of sterile milk which are well mixed, then used to inoculate the sterile bulk propagation medium, preferably from 9 to 14% skim milk solids dissolved in water.

The individual bulks are incubated at about 21°c. for 14 to 24 hours, then are freeze dried and the amorphous solids are milled and packaged. This procedure is employed in preparing the lyophilized milled cultures designated ATCC 14874 and ATCC 14875.

Example: A commercial multistrain cheese starter culture, deposited in the American Type Culture Collection, Washington D.C., ATCC No. 14873, is suspended in distilled water and aliquots of the mixed culture are spread on the surfaces of sterile solidified beef lactose agar (Difco Laboratories), and on sterile Triple "S" media in sterile petri dishes. The Triple "S" medium has the composition:

Triple "S" Ingredients	Grams
Tryptone	20
Yeast extract	5
Gelatin	2.5
Dextrose	5
Lactose	5
Sucrose	5
Sodium chloride	4
Sodium acetate	1.5
Distilled water, q.s., 1,000 ml.	

The plates are incubated at 21°C. for 7 days during which time many well-separated colonies with different growth habits are observed to form. Eighteen single colonies, which appear to be morphologically distinct, are then individually transferred, under aseptic conditions, on the point of a sterile culture needle to tubes containing 10 milliliters of a sterile culture medium of the following composition:

Powdered skim milk	90 grams
Distilled water, q.s.	1,000 ml.

The inoculated tubes are incubated at 21°C. for 72 hours, whereupon single pure cultures are obtained comprising the selected individual strains originally present in the starter. The strains are isolated by freezing the cultures and desiccating them under a high vacuum and are maintained as lyophiles in a sealed, evacuated tube at 5°C.

Each strain is then classified with respect to the rate at which it promotes acid formation. Each lyophile is opened and is transferred to 100 ml. of a sterile 9% reconstituted skim milk medium of the preceding composition. The 18 flasks are incubated at 21°C. for 24 hours, and then 1 ml. of each is used to inoculate 18 new bottles of sterile 9% milk medium. After the new series has been incubated for 24 hours at

21°C., 1 ml. of each is used to inoculate a third series of 18 bottles, each containing 100 ml. of sterile 9% milk medium. After incubating for 18 hours at 21°C., the formation of curds is noted and approximately 20 ml. of suspension is aseptically removed from each, is weighed and titrated with 0.1 N sodium hydroxide solution using phenolphthalein indicator, and the acidities, expressed as percent lactic acid, are determined. The expression used is:

$$\text{Percent of free acid (as lactic)} = \frac{\text{ml. NaOH} \times \text{Normality} \times 0.09008 \times 100}{\text{Sample weight (g.)}}$$

The following data are obtained:

Strain No.	Colony Morphology[1]	72 hr. Curd Formation	Titer,[2] Percent Acid, as Lactic	Acid-Former Designation
C-1	Small, glistening	No curd	0.22	Non.
C-2	Large, glistening	Slight Curd	0.34	Slow.
C-3	Small, raised	Curd	0.71	Fast.
C-4	Dull, flat	No curd	0.23	Non.
C-5	Flat, dull; circular border.	do	0.3	Slow.
C-6	Large, glistening	Slight curd	0.38	Do.
C-7	Flat, dull; circular border.	No curd	-----------	
C-8	Raised, glistening	Curd	0.72	Fast.
C-9	Flat, dull	do	0.81	Do.
C-10	Flat, dull; circular border.	No curd	0.21	Non.
C-11	Large, glistening	do	0.33	Slow.
C-12	Large, raised, glistening	Slight curd	0.38	Moderate.
C-13	do	do	0.34	Slow.
C-14	do	No curd	0.38	Moderate.
C-15	Fairly large, flat, dull.	Curd	0.83	Slow. Fast.
C-16	do	do	0.74	Do.
C-17	Small, raised; with slight border.	do	0.79	Do.
C-18	do	do	0.73	Do.

[1] Strains 1-11, incl., isolated on Triple "S"; medium; strains; 12–23, incl., isolated on beef lactose agar medium.
[2] Titer of sterile medium (control), 0.19% acid, as lactic.

The procedure is repeated substituting a second commercial starter culture, deposited and designated ATCC 14871, for the first culture. After isolation of 18 single colonies and propagation of these, observations of curd formation and determination of lactic acid formation rates are made. The data obtained are shown as follows:

Strain No.	Colony Morphology[1]	72 hr. Curd Formation	Titer,[1] Percent Acid, as Lactic	Acid-Former Designation
A-1	Large, flat, dull, etched.	Curd	0.34	Slow.
A-2	Smaller, raised, glistening.	do	0.71	Fast.
A-3	Large, flat, dull etched.	do	0.35	Slow.
A-4	Smaller, raised, glistening.	do	0.73	Fast.
A-5	Very small, glistening.	No curd	0.21	Non.
A-6	do	do	0.22	Non.
A-7	do	do	0.21	Non.
A-8	Intermediate size, raised, glistening.	Curd	0.73	Fast.
A-9	do	do	0.73	Do.
A-10	do	No curd	0.37	Slow.
A-11	Large, flat, dull etched.	Curd	0.51	Moderate.
A-12	do	do	0.41	Do.
A-13	do	do	0.36	Slow.
A-14	Large, raised, glistening.	No curd	0.34	Do.
A-15	Large, flat, slightly glistening.	Curd	0.43	Moderate.
A-16	Very small	No curd	0.25	Non.
A-17	Small, raised	Curd	0.78	Fast.
A-18	Raised, glistening	do	0.81	Do.

[1] In contrast to 18 hr. C series, 16 hr. incubations made are before titration.

The procedure is repeated substituting a third commercial starter culture, deposited and designated ATCC 14872, for the first culture. After isolation of 20 single colonies and propagation of these, observations of curd formation and determination of lactic acid formation rates are made. The following data are obtained:

Strain No.	Colony Morphology[1]	72 hr. Curd Formation	Titer,[1] Percent Acid, as Lactic	Acid-Former Designation
B-1	Large, raised, slight glistening.	Curd	0.35	Slow.
B-2	Slightly smaller, dull.	do	0.31	Do.
B-3	do	No curd	0.35	Do.
B-4	Large raised, slight glistening.	Curd	0.68	Moderate.
B-5	Small, raised	do	0.79	Fast.
B-6	Large, raised, glistening.	do	0.61	Moderate.
B-7	Flat, dull, milky	do	0.34	Slow.
B-8	Flat, glistening	No curd	0.32	Do.
B-9	do	do	0.19	Non.
B-10	Large, glistening; ragged border.	Curd	0.79	Fast.
B-11	Large, glistening; very ragged border.	do	0.68	Do.
B-12	Raised, glistening; slight border.	do	0.70	Do.
B-13	do	do	0.74	Do.
B-14	Small, glistening	do	0.77	Do.
B-15	Small glistening; etched border.	No curd	0.30	Slow.
B-16	Very small, glistening.	Curd	0.74	Fast.
B-17	Flat, fairly large, glistening.	do	0.71	Do.
B-18	Flat, glistening; slight irregular border.	do	0.60	Moderate.
B-19	do	do	0.77	Fast.
B-20	Small, raised, glistening, etched.	do	0.41	Slow.

[1] 16 hr. incubations were made before titration.

Lyophiles of strains C-3, A-4, B-10, C-2, A-11, B-2, C-11, A-14, B-7, B-9, C-10, C-12 are opened and used to inoculate individual flasks of 100 ml. of 10% skim milk medium which has previously been sterilized at 10 pounds steam pressure for 10 minutes. The flasks are incubated at 21°C. for 24 hours, then 1 ml. of each is used to inoculate a second series of flasks containing 100 ml. of the same sterile medium. The second series is incubated at 21°C. for 24 hours, then is used to inoculate (1 ml./100 ml.) a third series of flasks containing 100 ml. of the same sterile medium.

After incubating the third series for 24 hours at 21°C., 0.1 ml. from each of the 12 flasks is transferred aseptically to 100 ml. of a sterile 10% skim milk solids medium. Ten milliliters of this medium, which now contains all 12 strains, is used to inoculate each of 3 Roux bottles, each of which contains 1 liter of 10% skim milk solids medium which has been sterilized at 6 pounds steam pressure for 10 minutes. The Roux bottles are incubated at 21°C. for 15 hours. The pH of the medium reaches a constant 4.6 and the percent acid, as lactic, rises to 0.82. The media are then freeze dried and the solid residue is stored in small tubes. The culture which is obtained is eminently suitable for the preparation of Cheddar cheese and cottage cheese.

Swiss Cheese Culture Paste

R.V. Hussong and M. Purko; U.S. Patent 3,159,490; December 1, 1964; assigned to National Dairy Products Corporation describe a process for manufacturing Swiss cheese using *Propionibacterium shermanii* which relates to a method of preparing a bacterial cell paste and to a method of cheesemaking incorporating such a cell paste.

The bacteria suitable for cheesemaking can be prepared in concentrated form and on a small or large scale, batchwise or continuously. A system is schematically shown in the patent which is adapted to provide a bacterial cell paste product as a result of continuous generation of bacteria in the vessel. A nutritive bacterial culture solution can be continuously passed into the vessel, and so can also a pH-adjusting solution, and a portion of bacteria-containing nutritive solution can be continuously passed from the vessel to a centrifuge for separation into a low solids medium and desired bacterial cell paste.

The system is particularly adapted for the preparation of a cell paste consisting essentially of Propionibacterium shermanii or similar propionic acid-producing bacteria for use in Swiss cheese manufacture. The same system could also be used in the production of Streptococcus lactic bacteria, or other various bacterial cultures used in the manufacture of cultured dairy products. The following is an example of the production of Propionibacterium shermanii cell paste on a batch basis by this method.

Example: A nutritive bacterial growth medium which includes: 45 g. of soy peptone; 60 g. of extractables from brewers' yeast; 75 g. of enzyme hydrolyzed milk protein; 300 ml. of a salt solution comprising 50 g. of sodium chloride, 25 grams KH_2PO_4, amd 50 g. $(NH_4)_2HPO_4$ per 500 ml. water; 300 ml. of a second salt solution comprising 5 g. $MgSO_4 \cdot 7H_2O$, 0.95 g. $MnSO_4 \cdot H_2O$, and 0.25 gram $FeSO_4 \cdot 7H_2O$, per 500 ml. water; 300 ml. of a third salt solution comprising 50 grams of tripotassium citrate and 125 g. of sodium acetate per 500 ml. water; and water sufficient to make up to 14 liters.

The pH of this solution was adjusted to about 6.7 with aqueous sodium hydroxide solution, and then the solution was autoclaved for one hour at 250°F.

To this solution was added 1 liter of water containing 150 g. of lactose, thereby providing a bacterial growth medium. The lactose solution was autoclaved for 15 minutes at 250°F., before addition to the solution.

The prepared bacterial growth medium was placed in a vessel and was then inoculated with a culture of Propionibacterium shermanii at a level of 200 ml. of culture per 15 liters of growth medium. The growth medium containing the bacterial culture was then held at 86°F. in the vessel for 48 hours, with the pH being adjusted at least every 12 hours with aqueous sodium hydroxide solution through a suitable line to maintain it within the range of 6.7 to 7.0. Once the optimal bacterial concentration was obtained in the initially introduced inoculated growth medium, that is, at the end of the 48 hour period, the entire batch of bacteria-containing growth medium was then passed to a centrifuge to separate out the propionic acid-generating bacteria therein in the form of a cell paste.

The system was put on a continuous basis utilizing the same constituents. Thus, the initial 48 hour growth period was used and aqueous sodium hydroxide was then continuously run into the vessel to maintain the Propionibacterium shermanii at a pH of 6.7 to 7.0, and nutritive bacterial growth medium was continuously fed to the vessel. An equal amount of bacteria-laden growth medium was continuously passed from the vessel through an outlet line to the centrifuge in which the finished cell paste was obtained by separation from the growth medium. The volume of growth medium in the vessel and the rate of overflow of bacteria-containing growth medium from that chamber were controlled so that maximum development of bacteria was assured. For example, with Propionibacterium shermanii, the average residence time for bacteria within the growth chamber in the continuous treatment was from 20 to 30 minutes. Since the bacteria were maintained at optimal concentrations and under optimal growth conditions throughout the residence time in the growth chamber, there was optimal continued growth of the bacteria so that relatively large yields of bacteria per unit of time were obtained.

This continuous process yielded from about 15 to 18 grams of Propionibacterium shermanii cell paste per liter of growth medium, the cell paste having a moisture content of about 80% by weight and a bacterial population of the order of about 10^{10} per gram, which is 50 to 150 times as great as the bacterial population in other bacterial starters.

The cell paste was then stored, in the as-produced condition, at about 35°F., for up to 4 months without losing appreciable viability. It was found that the paste could also be stored at room temperature (70°F.), and would retain adequate viability for a week or more. The cell paste could be further concentrated by lyophilizing (freeze drying), resulting in a product of enhanced storage stability.

In the case of Streptococcus lactis bacteria, it has been found that 100 liters of the nutritive growth medium will yield, under conditions of continuous operation in the described process and utilizing a residence time of about 20 to 30 minutes, a total of about 400 g. of bacterial cell paste having about 100 times greater bacterial population than the usual Streptococcus lactis, starter culture. The per gram bacterial population can be further increased, for example, 1,000 to 10,000 fold by lyophilization of the paste.

With this system it is possible to provide on a commercial scale a cell paste production center and to ship economically the cell paste to particular points of use in cheesemaking. Due to low bulk and good storage stability of the cell paste, it is no longer necessary to provide culturing facilities at each cheese factory.

In making Swiss cheese, this propionic acid-generating bacteria cell paste is incorporated into milk before setting of the milk in place of the usual bacterial culture. The remaining steps of the Swiss cheese make are then carried out in the conventional manner.

In this connection, 2 g. of Propionibacterium shermanii cell paste is added to approximately 200 ml. of milk, and is blended in a high-speed blender. The resultant blend, comprising 2 g. of cell paste in suspension is then added to 10,000 pounds of whole milk contained in a conventional Swiss cheese vat. Following the addition of this blend to the milk, the normal methods of Swiss cheese manufacture and cure are employed. The resultant Swiss cheese is a high-quality product, is of uniformly high grade, color, texture, pH, moisture, appearance, and taste.

Enzyme from Fungi

K. Arima and S. Iwasaki; U.S. Patent 3,212,905; October 19, 1965; assigned to Meito Sangyo, K.K., Japan, found it possible to provide milk coagulating enzyme "microbial rennet" having very low protease activity but remarkably high milk coagulating activity by cultivating fungus selected from the group consisting of Mucor pusillus Lindt belonging to the genus Mucor, genus Rhizopus, genus Monascus, genus

Ascochyta, genus Sclerotium, genus Collectotrichum, their varieties and the natural and artificially induced mutants thereof that possess most of the characteristic properties of the parent fungi concerned.

The medium for cultivating the fungi may be any suitable artificial or natural medium so long as it contains at least a carbon source, a nitrogen source and inorganic salts. As the carbon source, the monosaccharides, disaccharides and polysaccharides which are assimilable to the fungi, such as sucrose, lactose, glucose, starch, etc. can be used. Similarly, in case of the nitrogen source, so long as they are assimilable, a wide range of nitrogen sources can be used such as inorganic ammonium salts, amino acids, and a wide variety of other protein substances. The inorganic salts may also include, for example, phosphates, magnesium salts, calcium salts, etc. Wheat bran is the preferred medium. It is to be understood, of course, that the amounts of carbon and nitrogen sources and the composition of the medium itself will vary depending upon the adaptability of the fungus that is used for the production of "microbial rennet."

While the temperature of cultivation will vary depending on the fungus, between 28° to 40°C. is suitable. It is usual to carry out the cultivation for 2 to 7 days by means of liquid culture or solid culture at a pH of 5 to 7, and preferably under aerobic conditions. Thus, while in case of liquid culture, surface culture may be employed, better results are obtained by means of shaking culture or submerged culture. As a result of cultivation, the milk coagulating enzyme, "microbial rennet," is produced in the medium. The "microbial rennet" is collected from the culture medium, in case of solid culture, by extracting with water or an aqueous solution, while in case of liquid culture it is done by removal of the mycelium whereby in both cases microbial rennet-containing solution can be collected. Examples of aqueous solvents are the solutions containing salts such as sodium chloride, potassium chloride, etc., the solutions containing acids such as hydrochloric acid, citric acid, acetic acid, etc. as well as various other types of buffer solutions.

The "microbial rennet"-containing solution can then be concentrated or solidified by carrying out purification treatments such as precipitation with organic solvents, salting out, concentration under reduced pressure, purification by means of ion exchanges etc.

Microbial rennet exhibits characteristics clearly different from proteases of animal or vegetable origin or those proteases produced by other microbes. Microbial rennet possesses powerful milk coagulating activity, but its protease activity is very weak. The milk coagulating activity of microbial rennet as in the instance of rennet obtained from the fourth stomach of suckling calves, shows powerful activity in the acid side, and as the pH rises its potency decreases, no activity being manifested when the pH is in the alkaline side. Hence, in making cheese, microbial rennet can be used as in the conventional method.

When curds are formed, the nitrogen that transfers into the whey after 150 minutes was about 20% determined by the Kjeldahl method. Thus, since it resembles substantially the instances resulting from the use of rennet, the casein in the milk is advantageously utilized. When the protease activity of "microbial rennet" was measured by the Anson method, using respective amount of enzymes which show the same milk coagulating activity, the absorption at 660 mμ in case of the acid protease produced by Aspergillus saitoi was 0.641 (O.D.) and of rennet, 0.059 (O.D.), while that of the microbial rennet was less than 0.102 (O.D.), indicating that this enzyme was a completely different type of enzyme than the usual protease, its milk coagulating activity being unusually powerful as compared with its protease activity.

Example: Seven parts of water was added to ten parts of wheat bran. After mixing, it was cooked for 30 minutes at 100°C. This was inoculated with a strain belonging to Mucor pusillus Lindt, and was incubated at 30°C. On the third day the milk coagulating activity reached its maximum. The wheat bran to which water was added in the proportion of 10 g. to 50 ml. was allowed to stand at room temperature for one hour, after which the mycelium was separated by filtration and 40 ml. of microbial rennet solution was obtained. The milk coagulating activity of this extract was 800 Soxhlet units per ml. for a 10% solution of powdered skimmed milk containing 1/100 M of $CaCl_2$. Consequently, from 10 g. of wheat bran 32,000 units of microbial rennet were obtained.

Next, by salting out this extract with ammonium sulfate of 0.22 to 0.66 saturation, practically all of its activity could be transferred to the precipitate portion. Furthermore, activity recoveries of 78%, 80%, 60%, and 75% were shown respectively by means of organic solvents of the following concentrations: ethanol, 70%; methanol, 75%; acetone, 66%; and isopropanol 70%. The activity of microbial rennet thus obtained was 100 Soxhlet units per mg., while that of the rennet tablets of Chr. Hansen's Laboratory, Inc., was 200 Soxhlet units per mg.

The microbial rennet was then used and cheese was made. To 55 kg. of raw milk mixed with 15 kg. of skimmed milk which were maintained at 30°C. were added 13 grams of KNO_3 followed by 8.5 g. of $CaCl_2 \cdot 2H_2O$ and 1 kg. of starter. Ten minutes later, 5 g. of microbial rennet was added, and the mixture was stirred. After 25 minutes' standing, the curd was formed and cutting was performed. Cooking was carried out in three consecutive steps at 33°C., 36°C. and 38°C., and after 1 hour and 20 minutes of cooking at 38°C., 120 g. of sodium chloride was added. The draining of the whey from the curd was normal, and the texture of the curd was fine and smooth. 6.3 kg. (yield 8.9%) of green cheese was obtained.

Endothia parasitica

J.L. Sardinas; U.S. Patent 3,275,453; September 27, 1966; assigned to Chas. Pfizer & Co. has found that strains of Endothia parasitica elaborate a desirable mildly-proteolytic, milk-curdling enzyme. This enzyme can be used in place of all or part of the animal rennin in standard cheese-making processes and is fully equivalent to rennin in clotting time, desired protease activity, and in freedom from the development of off-flavors and bitter taste in the cheese during storage. The enzyme has been found to accelerate aging of Cheddar cheese, when compared with rennin. For example, some of the Cheddar cheeses made with the enzyme have a tangy, but clean and mild flavor with long body after only 9 months of aging; in contrast, 18 months of aging is ordinarily required to reach the same stage of cure when animal rennin is used. The enzyme can be used advantageously with rennin also, if desired, the rennin providing the clotting activity and the enzyme providing accelerated aging effect.

Example: Endothia parasitica, ATCC 14729, is rinsed from a potato dextrose agar slant, under sterile conditions, into a flask containing one liter of the aqueous inoculum medium in a 2.8-l. Fernbach flask (the medium has been previously sterilized in an autoclave for 45 minutes at 121°C.). The medium has the following composition in the quantities indicated per liter of medium.

	Quantity Per Liter
N-Z Amine B*	5.0 g
Soya protein digest**	2.0 g.
Yeast extract	2.0 g.
Soluble starch	10.0 g.
D-mannitol	5.0 g.
$FeSO_4 \cdot 7H_2O$	15.0 mg.
Microelement solution***	1.0 ml.

*Commercial hydrolyzed casein preparation.

**Commercial preparation, Soytone.

***Microelement solution: Fe (as $Fe(NH_4)_2(SO_4)_2$), 1.0 mg./ml.; Zn (as $ZnSO_4$), 1.0 mg./ml.; Mn (as $MnSO_4$), 0.50 mg. /ml.; Cu (as $CuSO_4$), 0.08 mg./ml.; Co (as $CoSO_4$), 0.10 mg./ml,; B (as H_3BO_3), 0.10 mg./ml.

The flasks are incubated in a rotary shaker at 28°C. for 96 hours during which time a considerable amount of growth occurs.

An enzyme production medium is prepared and is sterilized in two-liter portions in four-liter fermentation vessels by heating in an autoclave for one hour at 121°C. The medium has the following composition:

	Grams/Liter
Soybean meal	30.0
Cerelose (glucose)	10.0
$NaNO_3$	3.0
Skim milk	10.0
KH_2PO_4	0.50
$MgSO_4 \cdot 7H_2O$	0.25

The medium has a pH of about 6.8.

After cooling the growth medium to 28°C., 100 ml. of the inoculum is added to the vessel containing 2,000 ml. of sterile growth medium and fermentation is carried out at 28°C. while air is introduced at the rate of 0.5 volumes of air per volume of broth per minute and while the medium is agitated at 1,700 rpm. After 48 hours, it is found that sufficient enzyme is present to enable 0.5 ml. of broth to curdle 5.0 milliliters of sweet milk within 3 minutes. The final pH is about 6.0 to 6.5.

The enzyme is isolated by filtering the broth through a Buchner funnel precoated with a diatomaceous filter aid and the filter cake is washed with 100 ml. of deionized water. The combined washing water and filtrate are freeze dried and the freeze dried solids assay about 1:8,000. The freeze dried solids are then reconstituted in deionized water to five times the original broth concentration and the solution is cooled to 1°C. During the cooling period and in all subsequent steps, nitrogen is flushed over the solutions and solids to aid in the exclusion of oxygen. Acetone, -25°C., is then added slowly, with stirring, to the enzyme solution until an amount equivalent to twice the original volume has been added. Toward the end of the addition of acetone, the precipitated enzyme becomes completely congealed. The acetone-water supernatant layer is decanted from the vessel leaving behind the viscous, congealed precipitate. Excess liquid is expressed from the cold solid by pressing with a spatula and the enzyme is further dried in a rapid stream of nitrogen. The precipitate is then suspended in an equal volume of water and is rapidly freeze dried; these solids contain 85% of the activity.

A 15% (w./v.) aqueous solution of the purified freeze dried solids is prepared, cooled to 1°C. and a volume of acetone, cooled to -25°C., is added slowly, with stirring, until such an amount is added that congealing of the precipitate just begins. An amount of acetone is required approximately equivalent to 1.5 volumes of original solution. The precipitate is allowed to settle and the supernatant is decanted and to this is added slowly, with stirring, a volume of cold (125°C.) acetone equivalent to 0.75 volume of the solution originally taken.

The precipitate which forms is allowed to settle, the supernatant layer is decanted off, excess acetone is removed, the solids are dissolved in an equal amount of deionized water and are recovered by freeze drying. The freeze dried solids contain 40% of the original activity and assay about 1:45,000. This precipitate is dissolved in an equal volume of water and is dialyzed against cold tap water for 5 days. The dialysis apparatus itself is refrigerated during this time. The nondialyzables are then freeze dried. The solid product has an activity assay value of 1:100,000.

A Cheddar cheese is made substituting for the animal-rennet preparation ordinarily used in the setting step, an equivalent amount of the enzyme preparation. The temperature of 1,000 lbs. of pasteurized sweet milk is adjusted to 86° to 88°F. and 1 percent by weight of a cheesemaker's commercial lactic acid starter solution is added. The mixture is agitated for one hour then is adjusted to 88°F. and 3 ounces of the enzyme preparation dissolved in 120 ml. of deionized water is added and stirred in thoroughly.

Agitation is stopped and, after 30 minutes, the firmness of the curd is tested by inserting a blunt instrument and raising it slowly whereupon the curd splits ahead of the instrument. The curd is cut into 1/4-inch cubes with curd knives, then is stirred and, after 15 minutes, it is firmed by heating during 30 minutes from 88°F. to 100°F. The curd is stirred slowly during heating.

The whey, which separates from the curd, is drained 2.25 hours after the enzyme preparation has been added. After the whey has been drained, the curd is ditched and packed, then layered into slabs about 5 to 6 inches wide. The slabs are turned five times at 10 minute intervals then are repiled and the curd is milled. The temperature of the curd at milling is 98°F. The milled curd is spread evenly and is treated with 3 lbs. of flake cheesemaker's salt sprinkled over the surface. After the salt has dissolved in the curd, which requires about 30 minutes, the curd is transferred to hoops and the filled hoops are pressed for 30 minutes. The cloth covering on the outside of the hoop is adjusted and then the cheese is put back into the press and given a final pressing. This final pressing frees the rind of openings and of surface defects and requires about 24 hours.

The cheese is then dried for four days at 55°F., then is coated with paraffin, is boxed and is cured at 75% relative humidity and 45°F. The plug sample taken when the cheese is three weeks old is free of mechanical openings and the cheese is free of acid flavor and bitterness. Then the cheese is one month old, it has a firm, smooth waxy body and is free of mechanical openings, is free of unclean flavors and retains these properties until it is cured, which process requires six months to one year.

Exclusion of Alkaline Peptidase

E.D. Murray and M.P. Prince; U.S. Patent 3,507,750; April 21, 1970; assigned to John Labatt Limited, Canada, have found that if an asporulant mutant of a bacterium of the genus *Bacillus* is grown in a nutrient medium containing a carbon source, a nitrogen source and inorganic salts, acid and neutral peptidases are selectively grown to the substantial exclusion of alkaline peptidase, and the yield of acid and neutral peptidases is greatly increased over that obtained with other methods. Moreover, if the medium is enriched with a source of L-leucine, the acid and neutral peptidases are directly obtained in a ratio which is highly suitable for use in the production of cheese.

Thus, the process permits the direct production of acid and neutral peptidases without the necessity of either balancing the ratio of acid to neutral peptidases or removing, inhibiting or inactivating alkaline peptidase.

Example 1: Production of Culture Filtrate — An inoculum of *Bacillus subtilis* (parent) strain) was prepared by transferring organisms from an agar slant to a sterile seed medium of trypticase soy broth (3% w./v.) fortified with calcium chloride (0.01% w./v.) and L-leucine (0.01% w./v.) all adjusted to pH 6.2 with 1.0 N HCl. The seed medium was prepared by dispensing 100 ml. of medium into 300 ml. Erlenmeyer flasks. The inoculated seed preparation was shaken on a rotary shaker (2 inch throw at 125 revolutions per minute) for 12 hours at 36°C. and after this incubation period the seed was used on 2% v./v. basis to inoculate a production medium.

The production medium was prepared by using water extractable nutrients from wheat bran. To accomplish this, a slurry of wheat bran (3% w./v.) and water was boiled under pressure at 120°C. for 45 minutes. The mixture was then cooled and allowed to settle and the clear supernatant was drawn off and fortified with calcium chloride (0.01% w./v.) and L-leucine (0.01% w./v.). The medium was adjusted with 1.0 N HCl so that after sterilization the final pH was 6.2 ± 0.1.

The inoculated production medium was maintained at 36°C. and stirred at 300 rpm in a fully baffled fermentor of 5.0 liters working volume. Sterile air was added at the rate of 2 liters per minute, and normal liquid antifoaming agents were used to suppress foam. The system was allowed to grow until a maximum level of enzymatic activity was reached (judged by milk coagulation tests), this growth period being generally 12 ± 1 hours. The culture broth was then clarified by centrifugation and/or filtration.

The final yield of each enzymic component in crude culture filtrate is shown in the table on the following page. For comparative purposes, data are included for the nonamino acid fortified production medium. The table gives data for enzymic components in crude culture filtrate after growth. Activities are expressed as units per ml. of culture filtrate. The ratio is of acid peptidase to neutral peptidase.

Medium	Acid peptidase	Neutral peptidase	Ratio acid to neutral	Alkaline peptidase
Wheat bran+calcium chloride.	2,000	4,740	1:2.37	1,308
Wheat bran+calcium chloride+L-leucine.	2,300	4,140	1:1.80	987

Example 2: Manufacture of Cheeses — Whole pasteurized milk (16 Imp. gals.) was heated to 31° ± 1°C. When this temperature was reached, 0.5 to 1.0% cheese starter culture was added and the mixture was stirred thoroughly. The starter organisms were allowed to grow until the acidity of the milk reached 0.17% as lactic acid. At this time 15 ml. of concentrated enzyme complex were added. The mixture was stirred thoroughly and allowed to remain, covered and unagitated, in a jacketed vessel for 25 to 30 minutes.

The coagulated milk was then cut with a curd knife and the normal techniques of Cheddar cheese manufacture were followed. After pressing, the final cheese was placed in a ripening cabinet at a temperature of 12° ± 2°C. with a relative humidity of 70 to 75%. The cheeses were allowed to ripen for periods of one to eighteen months and evaluation of the cheeses at various time intervals showed them to be of normal Cheddar flavor and quality.

Earlier work done by E.D. Murray and M.S. Kendall; U.S. Patent 3,482,997; December 9, 1969; assigned to John Labatt Limited, Canada, also discusses making cheese using a modified bacterial enzyme complex from the genus Bacillus. The enzyme complex contains 1 part acid peptidase to 0.5 to 5.0 parts neutral peptidase and is free from alkaline peptidase.

Using the peptidase complex obtained in the previous two patents, E.D. Murray; V.E. Gruetzner and D.M. Irvine; U.S. Patent 3,542,563; November 24, 1970; assigned to John Labatt Limited, Canada, have determined that in the production of Cheddar cheese using a milk coagulating enzyme complex as clotting agent, the moisture level of the cheese can be reduced by applying a flash of additional heat of short duration during the holding stage. The temperature of the curd is raised to 105 to 115°F. during application of the flash of additional heat.

During application of the flash of additional heat, the temperature should be raised from the regular holding temperature as quickly as possible, maintained at the elevated temperature for a short time sufficient to stimulate syneresis and then depending on the starter organisms used, it may be quickly cooled to the normal holding temperature. The elevated temperature is preferably about 108° to 112°F. and is maintained for not more than about 15 minutes, usually about 1 to 10 minutes. Particularly good results have been obtained at 110°F. for 3 minutes. The main point in applying the flash of additional heat is that the temperature of the batch should be raised to the desired level and returned again to the normal level in order to obtain a desired degree of stimulation of syneresis with the least possible input of heat since this minimizes heat inactivation of the lactic acid producing bacteria.

The flash of additional heat is preferably applied during the first 15 minutes of the holding stage, after the batch reaches the maximum cooking temperature. However, the selection of the actual time during the holding stage for the application of the flash of additional heat can be varied considerably depending on the size of the batch, the size and capabilities of the equipment, etc.

Example: Whole milk containing about 3.4% fat was pasteurized and cooled. 160 pounds of this milk were placed into a two hundred pound capacity steam-jacketed vat and the temperature of the milk was adjusted to 88°F. One and one half percent (by weight) of conventional commercial starter culture was added and the milk was allowed to ripen until there was an increase in acidity of about 0.01 to 0.02%, this being obtained in about 30 minutes. After acid development 15.0 milliliters of a peptidase complex were added to the milk. The peptidase complex was prepared according to the preceding processes and had an activity such that 3 fluid ounces coagulated 1,000 pounds of milk. Coagulation resulted and a firm coagulum was produced.

When the milk was properly coagulated, it was cut into cubes with regular 1/4 inch curd knives. The cut curd was agitated slowly to prevent breaking and matting of the soft particles. The cooking stage was started about ten minutes after cutting, and the curd and whey were heated to 103°F. within a period of about 30 minutes. During the holding stage a temperature of 103°F. was maintained until there was an increase in acidity of about 0.01 to 0.015%. This required an additional time of about 25 minutes.

After the increase in acidity, steam was reapplied to the jacket and the temperature of the curd and whey was raised to 108°F. within about 5 minutes. The temperature was maintained at 108°F. for about 10 minutes, after which cooling water was applied to the jacket and the temperature of the vat was adjusted to 103°F. within a period of about 5 minutes. When the curd was firm and dry, acidity had increased to 0.03% and then it was allowed to settle. The whey was then drained off and the normal cheddaring procedure was followed. The resulting cheese was of excellent texture and possessed a moisture content of about 35%.

Treatment to Change Cell Permeability

W.M. Cort and L.K. Riggs; U.S. Patent 3,295,991; January 3, 1967 have found that cheese of high uniformity may be produced in a relatively short time by the direct addition of suitable enzymes to milk or milk curd which catalyze chemical reactions in the milk, which reactions produce the desired effect, e.g., curd formation, cheese flavor, etc. Appropriate enzyme preparations, when added to milk, catalyze the formation of lactic acid, and assist in the production of curd or when added to curd, catalyze the reaction which produces the curing of curd which in turn develops the characteristics of the cheese, such as the desired texture, bouquet, flavor, etc.

Enzymes are organic catalysts which trigger chemical reactions. "Appropriate" and "suitable" enzymes, as used here, mean those enzymes which trigger specific types of chemical reactions, i.e., development of lactic acid in milk, development of curd and development of flavor, texture, bouquet, etc., of specific types of cheeses from milk curd. Each chemical reaction in the milk requires its own catalyst.

The choice of enzyme preparation, therefore, depends upon the desired reaction. Despite the long evolution of the cheese-making industry, relatively little is known about the chemical reactions which produce the characteristics of the various cheeses;

even less is known about the catalysts which bring about these chemical reactions. Accordingly, the suitable enzymes used here cannot be specifically defined by their chemical formula but rather by the results of their catalytic action, such as enzymes which induce the formation of curd in milk, enzymes which trigger the ripening or curing of curd, producing the characteristics of the desired cheese. Suitable enzymes are selective; those enzymes used to trigger the development of blue cheese will differ from those which induce the formation of Cheddar, Swiss, etc.

These enzymes may be obtained from any available source. For example, it has been found that suitable enzymes can be obtained from microorganisms commonly used in the cheese industry. S. lactis and S. cremoris, for example, are commonly used as starters or curd producers. Enzyme extracts from these are suitable starter enzymes for this process.

Many of the microorganisms from which these enzymes can be obtained are of known strains and can be obtained from commercial sources. Others are not known, known only in part or are not commercially available; in such instances cultures may be isolated from ripened cheeses and plated on agar, picked and transplanted. The growth of such cultures is accomplished by submerged fermentation or moldy bran fermentation, etc. The nutrient media as well as the method of growth vary with the cultures.

These enzyme preparations may be obtained from the various microorganisms by known methods. In general, the microorganisms are treated in order to change the cell permeability, i.e., by freeze drying, ultrasonics, etc. The broken cells may then be extracted or extracted and precipitated. The extraction is usually accomplished by a mildly alkaline solution, such as sodium tetraborate; the precipitation by acetone, alcohol or other suitable solvents.

Any of the ripened or unripened cheeses may be produced by adding suitable enzymes to milk, which are capable of catalyzing the chemical reactions in the milk which develop the cheese characteristics. This direct addition of suitable enzymes directly to milk or milk curd enables the distribution of the catalyst uniformly throughout the curd and the control of the amount of enzyme added. This procedure produces a cheese in very little time, produces a more uniform cheese, eliminates involved, complicated and time-consuming procedures, enables the reproduction of the quality of the cheese and eliminates the ring and mold growth usually found on the ripened cheeses. The process is particularly useful for the development of ripened cheese.

Example: An isolant from a good quality of English Cheddar cheese is plated on agar, colonies are picked and transplants made. The culture is inoculated into tryptic digest of casein, glucose and soybean protein (TGE) and incubated for several days at 84°F. The cells are extracted with pH 7 phosphate buffer for 20 hours at 40°F. The extracts are centrifuged with cold acetone. The precipitate is solubilized in water and freeze dried. 10 mg. of this enzyme powder is added to 50 g. of curd and mixed uniformly with it. The mixture is allowed to stand at 40°F. for three days. The Cheddar cheese characteristics are developed.

Filbert Nut Extract

J.P. Cox; U.S. Patent 3,375,118; March 26, 1968; assigned to Robert W. Beach describes a process designed to expedite the ripening of cheese, particularly of the

Cheddar variety, by the action of enzyme cheese-ripening material which will not impair the flavor of the cheese, but which may actually enhance its flavor.

In general, the raw cheese, such as Cheddar cheese, is prepared in accordance with customary procedure and when it is ready for the ripening step, the raw cheese is subjected to the accelerated ripening action. Such action can be accomplished by wrapping the cheese in an absorbent wrapping material, such as paper towel, impregnated with the cheese-ripening material. It is preferable for the thus wrapped cheese to be further covered with an outer wrapper of air impervious material. The cheese is then placed in storage for ripening for a period of 12 to 15 days. The ripening action can be terminated simply by removing the impregnated wrapper, after which the blocks of cheese are covered for market by an air-impervious covering such as a coating of wax or a transparent air-impervious polyethylene film.

Even for the production of mild Cheddar cheese ordinarily the minimum ripening period should be 2 or 3 months; for medium Cheddar cheese the ripening period will be 6 months to a year and for sharp Cheddar cheese ripening period can be from 9 months to 18 months. Even sharp Cheddar cheese will be produced by the ripening action in this process in a period of 12 to 15 days. At that time the ripening process can be interrupted simply by removing the impregnated wrapping. The cheese is washed in water and then sealed either in wax or in an air-impervious wrapper. The completed cheese can then be stored until ready for use or shipped to stores for sale.

The cheese-ripening agent which has been found to be satisfactory for this process is an extract of filbert nuts and, alternatively or supplementally, enzymatic material of yeast designated *Saccharomyces cerevisiae*. Preferably the filbert nut enzyme cheese-ripening material is extracted from such nuts by leaching and then the extract is concentrated. Various procedures can be employed for preparing the enzyme cheese-ripening concentrate.

A satisfactory method of preparing the enzyme cheese-ripening extract or concentrate involves crushing filbert nuts and then comminuting them in a blender. A small amount of salt, such as approximately one-tenth of one percent by weight, should be added at this time. This comminuting operation develops heat and it is important to limit the rise in temperature to preserve the cheese-ripening effectiveness of the enzyme material. Consequently, during the comminuting process a heat-absorbing liquid such as water is added to the nut material from time to time in sufficient quantities and sufficiently often to prevent the temperature of the nut material from rising to a value higher than 50°C. The amount of water thus added during the comminuting operation should be from 2 to 3 times as much by weight as the weight of the nutmeat.

To extract enzyme cheese-ripening material most effectively, the cell structure of the nutmeat material should be ruptured as much as possible. This can be accomplished by mixing with the comminuted nutmeat material in the blender pulverized glass, preferably Pyrex glass. The amount of such pulverized glass added to the nutmeat may be up to about 10% by weight of the nutmeat material.

The comminuting of the nutmeat material alone, and mixed with the pulverized glass, is continued until a milky suspension is formed which will contain enzyme cheese-ripening material liberated from the nutmeat. At this time a small amount of *Saccharomyces cerevisiae* is added, preferably in the proportion of one-tenth of a gram per pound of nutmeat material. This yeast will be in powder form and can be mixed into the suspension by the blender in a few seconds. Such addition is, however, optional.

The next step is to separate the enzyme cheese-ripening material from the suspension which can be accomplished by filtering the suspension through filter paper, of the number 2 type. The liquid filtrate contains the enzyme cheese-ripening material and if such solution is sufficiently concentrated, it may be used directly to impregnate the absorptive wrapper for the cheese. Alternatively, such solution can be concentrated with ether or by natural evaporation in a dry atmosphere. The solution cannot be concentrated by heat above 50°C. because such heat will destroy the cheese-ripening effectiveness of the enzyme material.

Other procedures may be used for extracting the enzyme material from the suspension, such as ion exchange, electrophoresis, electrodialysis, dialysis, ultracentrifugation, pervaporation, solvent extraction, adsorption by an adsorbent material such as kaolin, precipitation or crystallization. If the enzyme material is finally extracted in concentrated liquid or solid form, it will be necessary to dilute such material before it is used to impregnate the absorptive wrapping for the cheese.

As applied to the cheese the absorbent wrapping material impregnated with enzyme cheese-ripening material should be somewhat damp, but the whey from the cheese will moisten such wrapping material even if it is not damp initially, to activate the ripening process. The ripening action can be terminated by removing the impregnated wrapper from the cheese. If ripening action should continue to some extent, the enzyme cheese-ripening material can be deactivated by placing the cheese in a container made of a heavy metal, such as zinc or lead, or by irradiating the cheese or subjecting it to ultrasonic vibration.

Stabilized Mixed Bacteria Concentrate

S.M. Farr; U.S. Patent 3,420,742; January 7, 1969; assigned to Dairy Technics, Inc. describes a method which relates to a concentrated bacterial product useful for fermenting milk in order to make cottage cheese and cheese.

The method involves separately incubating in separate culture media two or three types of bacteria. The first type is selected from *Streptococcus lactis*, *Streptococcus cremoris*, *Lactobacillus bulgaricus* and *Streptococcus thermophilus*. The second type is selected from *Streptococcus citrovorous* and *Streptococcus paracitrovorus*. The third type is *Streptococcus diacetylactis*. The respective media are concentrated to obtain separate concentrates of the two or three types of bacteria. The two or three types of bacteria are mixed together in the desired proportions to produce a mixed concentrate without permitting further growth of the bacteria. The mixed concentrate is then frozen so that it can be stored for a long time without major loss in the viability of the bacteria. What is produced is a stabilized mixed bacteria concentrate consisting essentially of a substantially neutralized mixture of two or three types of bacteria which has been stabilized by the addition of a stabilizing agent and a nutrient medium.

In describing the process, "acid-producing bacteria" refers to bacteria capable of fermenting lactose or other, similar carbohydrates rapidly at temperatures of 70° to 110°F. with the production of principally lactic acid, namely, *Streptococcus lactis*,

Streptococcus cremoris, Lactobacillus bulgaricus and Streptococcus thermophilus. The term "flavor-producing bacteria" refers to bacteria capable of fermenting citric acid or citrates at a favorable pH with the production of biacetyl, acetylmethylcarbinol, volatile acids and carbon dioxide, namely, Leuconostoc citrovorum (Streptococcus citrovorus) and Leuconostoc dextranicum (Streptococcus paracitrovorus) and subspecies of these. Streptococcus diacetylactis produces some lactic acid and also produces substantial amounts of flavor constituents.

A critical feature of this method is that the acid-producing bacteria, the flavor-producing bacteria and the Streptococcus diacetylactis be grown and concentrated separately from each other. The three types of bacteria are separately grown or propagated in their respective culture media and then they are concentrated by centrifuging to produce separate concentrated bacterial products. About five cubic centimeters of each concentrate contain approximately the same number of bacteria as would be present in one quart of milk which had been fully fermented by the same bacteria. Because of the variability of bacteria strains, the particular conditions under which they are fermented and other variables, it is impossible to give precise bacteria counts but, in general, each milliliter of the concentrates will contain 10 to 60 x 10^9 cells per milliliter. The two or three types of concentrated bacteria are then mixed with each other in the proper proportions depending on usage and with glycerol and with a nutrient medium and then the mixture is frozen to produce the stabilized, concentrated, bacterial product. The bacteria must be grown separately in culture media and care must be taken to insure that the bacteria are phage-free.

Example: Preparation of Cottage Cheese — 100 gallons of milk were pasteurized by heating to 200°F. for about a half-hour or longer. The milk was cooled to 70°F and then 60 cubic centimeters of the mixed bacterial concentrate containing about 98% by weight of acid-producing bacteria and about 2% by weight of flavor-producing bacteria was added to it and thoroughly mixed. The mixture was incubated at 78°F. for about 14 hours and there was obtained a cottage cheese product of pleasing flavor and proper consistency.

Cooling to Stop Cultivation

V.W. Christensen; U.S. Patent 3,483,087; December 9, 1969; assigned to Miles Laboratories, Inc. provides improved cheese starter cultures by cultivating lactic acid-producing bacteria in a buffered culture medium comprising an aqueous solution of a mixture of non-fat milk powder, ammonium orthophosphate and an alkali metal orthophosphate. The cultivation of the bacteria is stopped when the change in acidity of the culture medium falls within the range of 0.85 to 1.05%; the cultivated bacteria culture is cooled to stop rapidly further growth of the bacteria. The so cultured lactic acid-producing bacteria is then maintained at a temperature below -100°F., preferably below -300°F.

The uniformity and stability of lactic acid-producing bacteria prepared and stored in accordance with this process is shown by the following example for preparing a seed culture by the process described .

Example: A specific variety of lactic acid-producing bacteria composed of equal blends of Streptococcus cremoris, strains #9625, #144F, and #11E obtained from the stock of Oregon State University was prepared. One thousand milliliters of this culture was prepared in a phage-resistant buffered medium and cultured for 14 hours at 71°F. using a 1% inoculation. The buffered culture had the following composition.

	Percent
Pretested nonfat milk powder	67.6
Demineralized whey powder	16.9
Diammonium phosphate	5.1
Monoammonium phosphate	5.1
Disodium phosphate	5.1
Dried extract of pancreas powder	0.2

When the change in acidity reached 0.96%, it was cooled immediately to 40°F. in 1/2 hour. Then special plastic vials were filled with 1.2 ml. of the cooled culture using a sterile syringe. These vials were then immediately immersed in liquid nitrogen and the temperature lowered to the temperature of the coolant (-300°F.) in less than 1 minute. It was then stored. Subsequently in use, a vial was then removed and thawed and transferred under aseptic conditions into a 100 ml. sterile phage-resistant buffered medium. This is Transfer 1, T_1. It was incubated at 71°F. for 14 hours reaching a developed acidity of 0.97%.

The T_1 culture above was used to inoculate a larger amount of the same culture medium. One gallon of the culture medium was prepared and sterilized and cooled to 71°F. It was inoculated using 38 ml. of T_1 culture above. The medium was incubated at 71°F. for 14 hours reaching a developed acidity of 0.97% and is identified as Transfer 2, T_2.

Subsequently, 100 gallons of bulk starter medium were prepared by dissolving the culture medium at 11.8% solids in water. After sterilization it was cooled to 71°F. and inoculated with the previous 1 gallon, T_2. It was ripened for 14 hours and attained a developed acidity of 0.95%. It was then cooled immediately to 45°F. within one hour.

One hundred and twenty pounds of this bulk starter was taken to a cheese plant and used to inoculate 15,000 pounds of milk. The manufacturing acidities were: 0.23 when the whey removed, 0.33 when cheese was packed, 0.62 when cheese milled.

The cheese was stored at 38°F. and graded after 3 days and showed good cheese characteristics. The cheese was graded 6 months later by a Federal Grader and graded Grade A, or Wisconsin State Brand. Flavor and body texture were excellent.

Similarly, 30 additional vats of cheese were made with this culture obtaining normal manufacturing acidities grading Grade A. Subsequently, this culture was checked for storage characteristics in liquid nitrogen. Results are shown in the following table:

CULTURE STORED AT −300° F

Time	Plate count (billions of organisms/ml.)	Activity test 6 hrs. at 89° F.
0	2.8	0.70
2 weeks	2.8	0.73
1 month	2.6	0.77
4 months	2.8	0.73
1 year	2.6	0.74

The plate count and activity tests of the starter culture (after one year's storage) were substantially the same as when the samples were prepared and stored initially.

Mucor miehei

R.L. Charles, D.P. Gertzman and N. Melachouris; U.S. Patent 3,549,390; December 22, 1970; assigned to Miles Laboratories, Inc. have found that a milk-clotting enzyme can be prepared by growing under aerobic conditions a culture of Mucor miehei NRRL 3420 or mutants of it in a medium containing appropriate nutrients and then recovering the enzyme. This enzyme has a desirably high ratio of clotting activity to proteolytic activity and can be employed under usual calf rennet milk-clotting conditions to produce cheese of good flavor and having desirable body and texture.

Example: A culture of Mucor miehei NRRL 3420 was transferred to several one-liter Erlenmeyer flasks each containing a sterilized medium consisting of 10 g. corn starch, 12.5 g. ground corn, 7.5 g. soy meal, 1.25 g. calcium carbonate and 250 ml. water. This medium also contained 100 mg. bacterial amylase to liquefy the starchy components of the medium. The inoculated flask contents were then shaken in a reciprocal motion shaker with a two-inch stroke at 176 strokes per minute for 6 days while maintained at a temperature of 32.5°C. The resulting fermentation beer was then filtered from the mycelium and the combined filtrate from all the flasks was assayed to contain enzyme activity of 1,233 Soxhlet units per ml. A portion of the filtrate was frozen and dried from the frozen state. The dried powder was assayed to contain 114,823 Soxhlet units per gram and 727 proteolytic units per gram. The ratio of Soxhlet units to proteolytic units was 157.9.

An enzyme prepared in the above manner was then used to make cheese. A 240 lb. quantity of pasteurized whole milk was placed in a cheese vat and heated to 88°F. (31°C.). A 2.4 lb. quantity of lactic acid starter culture was then added to the milk. After an hour a quantity of the above prepared liquid enzyme filtrate containing a total enzyme potency of about 800,000 Soxhlet units was added to the milk in the cheese vat. The milk clotted, and a satisfactory curd was prepared in 28 minutes. It was then cut into desired shapes and heated to 102°F. (39°C.)

The whey was drained off and the usual steps of cheddaring, milling, salting and hooping were carried out to obtain green cheese. A control cheese was prepared at the same time using a calf rennet under similar conditions. The cutting quality of the curd and the texture of the green cheese prepared from the microbial enzyme were found to be comparable with those prepared from calf rennet. The cheeses were ripened at 45°F. (7°C.) and 90% relative humidity for 60 days. The organoleptic evaluations, such as smelling, tasting, breaking, and the like, indicated that the cheese made from the microbial enzyme had normal body and texture and normal pleasing flavor completely free of any bitterness, rancidity or other off flavors and was substantially the same as that produced from calf rennet.

Purification with an Absorbing Silicate

H.C.T. Moelker and R. Matthijsen; U.S. Patent 3,591,388; July 6, 1971; assigned to Organon Inc. describe a process for the purification of microbial rennets contaminated by proteolytic enzymes, which give rise to a bitter flavor during the aging process of the cheese. The process comprises contacting the impure rennet with an absorbing silicate at a pH between 3 and 9, by which the taste spoiling enzymes are selectively absorbed.

An aqueous solution of the impure rennet is contacted with an absorbing silicate at a pH between 3 and 9, by which the enzymes are selectively absorbed, after which the absorbent is separated from the solution. In this manner unspecific proteases and other undesired products are removed from the rennet preparation by absorption, a purified rennet remaining behind in solution.

It was quite unexpected that in a solution of microbial rennet a selective separation could be effected by silicates between two substances that are so related as microbial rennet and unspecific proteolytic enzymes. In this process the unspecific enzymes are for the greater part absorbed, the rennet remaining behind in the solution without having lost its activity. The absorption of the contaminating enzymes can be effected by adding an absorbent to the solution of the rennet and stirring the mixture for some time, followed by the removal of the absorbent, by filtration or centrifugation, or by passing the solution to be purified through a column of the absorbent.

Finally an enzyme concentrate or, if desired, a dry enzyme preparation, can be obtained by evaporation of the water or precipitation of the milk-clotting enzyme by salting out or precipitation with an organic solvent.

From rennet preparations, in which the ratio between clotting activity and proteolytic activity was about 6,000 to 11,000, purified rennets were thus obtained, in which this ratio had improved to about 14,000 to 20,000.

The choice of the pH during the absorption is determined by the stability of the microbial rennet at different pH's and by the absorption properties of the absorbent. The rennet is stable at a pH between about 3 and 9 dependent upon the temperature. On the other side the absorption of proteolytic enzymes occurs within this whole pH range. The absorption properties decrease as the pH rises. Below pH 4 some microbial rennet, too, is absorbed as a rule, while outside the pH range 3 to 9 the rennet denaturates. The purification is preferably performed at a pH of about 5, that is to say between 4.5 and 6, at which pH the best results are obtained.

As starting material for the purification, a powdery rennet can be used which is dissolved in water or a buffer solution at a pH preferably about 5. Suitable absorbents are bentonite such as the types BIV 350 bianco, BIV 220 alvorio and Micro BIV, and Permutit such as Decalso F.

Example: Ten grams of a solid microbial rennet originating from *Mucor pusillus* was dissolved in 250 ml. of distilled water. To the resulting solution was added 10 grams of bentonite BIV 350 bianco, after which the pH was adjusted to 5.0 by means of 1 N acetic acid. The resulting suspension was stirred for 15 minutes and then filtered over a fluted filter. The clotting and proteolytic activities were determined before and after the absorption step.

The purified rennet obtained was used in the production of cheese in a comparative experiment with unpurified rennet. To 20 ml. of uncooked new milk was added 3 milliliters of the purified rennet solution, 4 g. of saltpeter and 1% starter culture. The temperature at which coagulation occurs is 30°C. Coagulation time is 31 minutes. In the same manner 20 liters of milk was coagulated with 3 milliliters of the unpurified rennet preparation. Clotting time was 30 1/2 minutes.

The curdled milk was cut into slices, during which operations excretion of whey occurred. To further this excretion process, the mass was regularly kept in motion for 20 minutes with a cheese-harp. Then, after removal of 10 l. of whey, 6 l. of water of 45°C. was added to the curdled milk, after which the temperature was 37°C. The mass was kept in motion for about 50 minutes with a cheese-harp. Then the curdled milk was separated from the whey by means of a pressing plate fitted with perforations. Finally the curdled milk was divided into two portions. These portions were put into warm cheese-molds, which were placed under a press. After 6 weeks and 6 months respectively, a group of 8 experienced tasters submitted the cheeses to an organoleptic examination by a comparative taste test.

The following result was obtained. The cheese prepared with the purified rennet is stated under A and the cheese prepared with the unpurified starting material under B.

	Cheese A		Cheese B	
	After 6 weeks	After 6 months	After 6 weeks	After 6 months
Bitter	0	0	1	2
Somewhat bitter	1	0	7	6
Not bitter	7	8	0	0

Streptococcus diacetilactis for Blue Cheese

W.T. Williamson and M. Purko; U.S. Patent 3,635,734; January 18, 1972; assigned to Kraftco Corporation describe a method for producing blue-veined cheese in which a suitable culture of a gas-producing micro-organism is selected and is grown in a suitable media. The culture is then concentrated and added to the milk from which the blue-veined cheese is to be manufactured. The gas-producing micro-organism is selected to provide a desired level of gas during curing of the cheese.

A suitable *Streptococcus diacetilactis* culture is used to inoculate a growth medium composed of 1.0 weight percent hydrolyzed casein, 1.0 weight percent dextrose, 1.0 weight percent yeast extract and 0.5 weight percent dibasic potassium phosphate. The inoculated medium is then incubated at a suitable temperature until an optimum level of *Streptococcus diacetilactis* cells are produced. The yield of *Streptococcus diacetilactis* cells may be increased by neutralizing the growth medium with an alkaline material such as potassium hydroxide at periodic intervals. It is desirable to maintain the pH of the inoculated medium above 5.0 so that acid produced during growth of the *Streptococcus diacetilactis* does not interfere with further growth.

Optimum growth at a temperature of 72°F. is usually reached after about 9.5 to 12.0 hours. After optimum growth is reached, the inoculated medium is cooled to below 60°F. to arrest decay of viable organisms. At this time the medium usually has from 10^8 to 10^9 cells per milliliter. To provide the desired level of cells for inoculating batches of milk to be made into blue-veined cheese it is desirable to provide a cell concentrate having about 10^{11} cells per milliliter. A preferred method for preparing the cell concentrate is to pass the inoculated media through a centrifuge. In general, a desired cell concentrate is obtained when one liter of the inoculated medium is reduced to from 5 to 25 grams.

The cell concentrate is then combined with a suitable storage medium and frozen. One such storage medium may be prepared by combining 15 parts of sterile glycerol with 100 parts of sterile skim milk. The cell concentrate is combined with about 5 times its weight of the sterile storage medium and frozen in suitably sized lots. The concentration and freezing of the gas-producing microorganism enables the microorganism to be stored for future use without loss of its producing properties.

Example: A Streptococcus diacetilactis culture is used to inoculate 150 gallons of a growth medium of the above composition. The growth medium is first sterilized at 121°C. for 15 minutes. Thirty-five hundred milliliters of a citrate broth culture of the Streptococcus diacetilactis which has been tested for gas-producing ability is then added to the 150 gallons of growth medium. The pH of the inoculated medium gradually decreases to 5.0 after 8 to 10 hours and is then adjusted to a pH of 7.0 with 3.6 liters of sterile potassium hydroxide. After an additional period of 1 1/2 hours, the pH has again decreased to a level of 5.0 and is again neutralized to a pH of 7.0 with potassium hydroxide. The inoculated medium is then cooled to 40°F. and the contents are conducted to a centrifuge. The entire contents of the inoculated medium are centrifuged without stopping. A yield of about 4.0 to 4.5 pounds of a cell paste of Streptococcus diacetilactis is obtained.

A milk medium is then prepared by heating a suspension containing 10 percent by weight of nonfat milk solids to 180°F. for 60 minutes and thereafter cooling to 40°F. Fifteen milliliters of sterile glycerol are then added to each 100 milliliters of the cool milk medium. Twenty grams of the cell paste obtained by the above method are then added to each 100 milliliters of the milk-glycerol mixture to provide a cell paste suspension. The cell paste suspension thus obtained is then divided into 250 ml. lots and each lot is frozen immediately in a dry ice-alcohol bath. Each lot is then stored at a temperature of -20°F. until required for use. Forty to forty-five pounds of a suspended cell paste suspension are obtained.

The cell paste suspension is suitable for direct inoculation of milk to be used in making blue cheese. In this connection 500 ml. of the frozen cell paste mixture are used for each 20,000 pounds of milk. When used at this level, the Streptococcus diacetilactis cell paste concentrate provides sufficient carbon dioxide gas during curing of the blue cheese to provide an improved bein structure in cured blue cheese.

The cell paste produced by the above method may also be used to prepare a bulk starter by adding 15 ml. of the cell paste to about 225 pounds of milk. The inoculated milk is then incubated overnight at a temperature of 72°F. The bulk starter is then added to the vat of milk at the same time as the lactic starter cultures. The gas-producing ability of the cell paste is not impaired by this single transfer.

Bacillus polymyxa

T. Imai, Y. Irie and Y. Kanazawa; U.S. Patent 3,661,594; May 9, 1972; assigned to Godo Shusei K. K., Japan describe a milk coagulating enzyme preparation by a process comprising culturing a strain of Bacillus polymyxa on a nutrient medium containing an assimilable carbon source and inorganic salts. The enzyme "bacterial rennet" produced by the process is then used as a milk coagulating enzyme useful for making cheese.

Milk coagulating enzyme having high activity is obtained from the cultivated broth of a selected strain of bacteria belonging to Bacillus polymyxa identified as (FERM-P) No. 412 cultivated on several kinds of artificial or natural nutrient media containing carbohydrates, nitrogen sources, growth factors and inorganic salts. As the carbon sources, assimilable carbohydrates such as the flours of corn, soybean or other crops, powdered skimmed milk and various sugars can be used. Similarly, as the nitrogen sources, a wide range of organic or inorganic compounds can be used. The inorganic salts may also include phosphate, magnesium, calcium and other salts.

The cultivation of the bacterial strain is carried out in a conventional manner in a liquid broth at a temperature of 25° to 37°C. and at a pH of 5 to 8 for 2 to 7 days. Thus, while in the case of liquid culture, surface culture may be employed, better results are obtained by means of shaking or submerged cultures.

As a result of the cultivation, the milk coagulating enzyme "bacterial rennet" is produced in the nutrient medium.

To separate the crude enzyme from the culture broth, a refining process such as filtration or centrifugation (removal of the bacterial cells and insoluble matters) followed by salting out or precipitation with organic solvents or concentration under reduced pressure is carried out. In the case of salting with ammonium sulfate of 0.3 to 0.8 saturation, the milk coagulating activity can be almost quantitatively transferred to the precipitated fraction. Also, in the case of the precipitation by using organic solvents, the recovery of activities of 88%, 80%, 70%, or 75% were shown respectively by adding aqueous solutions of methanol, 75%; ethanol, 70%; acetone, 60% or isopropanol, 70%. In this manner the crude "bacterial rennet" is obtained.

In the separation process using organic solvents, calcium salts such as calcium acetate, calcium chloride, or the like, are added to cultured broth in an amount such that the final concentration reaches 0.01 to 0.1 mol and the supernatant liquid is separated by centrifugation. Then the supernatant liquid is able to be fractionated very easily by following the treatment with the abovementioned organic solvents. If desired, the supernatant liquid is mixed with aqueous ethanol having a concentration of up to 40%. The resulting precipitate is removed and then to the remaining ethanol solution is added an additional amount of ethanol until the concentration of ethanol reaches 70% by volume of the solution. This produces a precipitate and the precipitate is then recovered. It is found that the resultant precipitate gives the bacteria-made rennet high water solubility.

It was found that the precipitation of impurities occurs when calcium salt is added to the cultured broth in the amounts of 0.01 to 0.1 mol. The resultant precipitate contains the impurities and therefore the precipitate is discarded. It is, however, practical to use the calcium salt in an amount up to 0.1 mol for avoiding the excessive cost due to using a large excess of the calcium salt.

The crude product maintains a high milk coagulating activity at a pH of from 4 to 9 when it is stored at 5°C. for 7 days or at 27°C. for 24 hours. However, it becomes unstable at pH value of less than 3 or more than 10. Furthermore, it is observed that the crude enzyme exhibits a strong coagulating activity in acidic solution. It decreases in activity in solution as the pH of the solution increases.

To better explain the merits of this process, reference is made to Figure 6.2.

FIGURE 6.2: NITROGEN IN WHEY

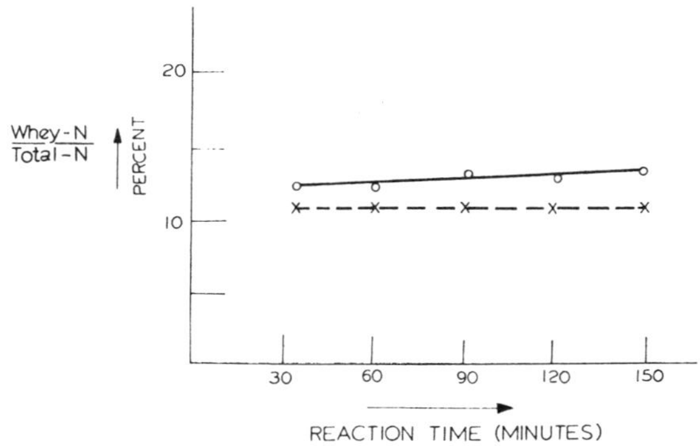

Source: T. Imai, Y. Irie and Y. Kanazawa; U.S. Patent 3,661,594; May 9, 1972

In the drawing, the solid line shows the amounts of nitrogen (Kjeldahl-N) transferred to the whey part after milk was coagulated by adding the enzyme (obtained from Bacillus polymyxa). The dotted line gives those amounts transferred in the case of commercial rennet under the same conditions. From the drawing it is observed that the enzyme exhibits substantially the same efficiency as in the commercial rennet because of nitrogen being as low as about 14% of total amount of nitrogen constantly being transferred to the whey during the incubation periods of from 30 to 150 minutes. This shows that the enzyme can coagulate the majority of casein in milk without much proteolysis.

Thus, this enzyme is confirmed to have very high milk coagulating activity and is, consequently, considered to be useful "bacterial rennet" in cheese manufacture.

In a process of any cheese making, it is necessary to produce a curd by adding to milk a coagulating enzyme. The rennet, as a milk coagulating enzyme, is considered to be indispensable for this process because it is well-known and powerful in activity. According to this process, however, it is possible to produce cheese of excellent quality very readily on a commercial scale without the use of rennet.

Example: A nutrient medium containing 10% corn flour, 0.1% powdered yeast extract and 1/200 mol of $CaCl_2$ was inoculated with a strain of bacterium belonging to Bacillus polymyxa No. 3-8, also identified as (Ferm-P) No. 412, and after 4 days' shaking or submerged culture at 30°C., the milk coagulating activity of the bacterium reached its maximum. The cells were separated by filtration or centrifugation and an enzyme solution was obtained. The milk coagulating activity of this enzyme solution was 400 Soxhlet units per ml. for a 10% solution of powdered skimmed milk containing 1/100 mol of $CaCl_2$.

By salting out this enzyme solution with ammonium sulfate of 0.3 to 0.8 saturation, substantially all of its activity could be transferred to the precipitated fraction. Furthermore, the activity of enzyme recovered is on the order of 88%, 80%, 70% or 75%, respectively by adding organic solvent of the following concentration;

methanol, 75%; ethanol, 70%; acetone, 60% and isopropanol, 70%. Thus the crude enzyme can be obtained. The activity of the crude enzyme precipitated by the methanol 70% was 20 Soxhlet units per milligram.

This "bacterial rennet" was then used to make cheese. To 100 kg. of raw whole milk maintained at 30°C. was added 0.1 kg. of $CaCl_2 \cdot 2H_2O$ followed by 1 kg. of starter. When the acidity of the milk reached 0.18% (as lactic acid), 8 g. of bacterial rennet obtained as described above was added, and the mixture was stirred. After 25 minutes' standing, the curd with elasticity was formed and cutting was performed. Cooling was carried out in two consecutive steps, at 34°C. and 38°C., and after 80 minutes of cooking at 38°C., casting into molds was performed with compression. The whey was then drained from the curd and 8.6 kg. of green cheese was obtained in the yield of 8.6%. This green cheese was salted by soaking in a sodium chloride solution. Test of taste and flavor after 4 months' ripening showed no sign of bitterness and no difference between this cheese and commercial rennet cheese.

MICROBIAL RIPENING

Accelerated Cheddar Cheese Process

In a process by A.J. Luksas; U.S. Patent 3,689,286; September 5, 1972; assigned to Beatrice Foods Co. fermented flavor of the Cheddar type is prepared by acidifying a protein, e.g., milk or recombined milk to form a curd and then inoculating the curd with micrococci and allowing the micrococci to grow. A flavor intensity of 100 times one year old Cheddar is produced. The acidification is preferably carried out by growing nontoxic enterococci bacilli on the protein medium. This can be accomplished by acidifying a water containing fat and protein and/or no lactose source to form a slurry and then inoculating the slurry with micrococci and then allowing the micrococci to grow under controlled environment until the desired Cheddar flavor is developed.

The initial media can be skim milk, whole milk or recombined milk. If recombined milk is employed, there can be employed sodium caseinate, soy protein, cottonseed protein or any food grade protein. The fat source is normally butterfat in the amount present in milk, i.e., 2.6 to 8.4%. It is also essential that lactose or other utilizable carbohydrate be present. The amount present in milk, (2 to 6%), is adequate although this also can be increased or decreased. The preferred medium contains a whey plus a source of casein dispersed in water.

The initial media for the Cheddar production can be acidified with nontoxic inorganic or organic acids, e.g., hydrochloric acid, acetic acid, maleic acid, tartatic acid, citric acid, phosphoric acid, lactic acid or the like to a pH of 4.8 to 5.4, preferably 5.2. However, the preferred method of acidifying the milk is the conventional procedure employing *Streptococcus lactis* or other lactic acid forming microorganisms until the desired pH is reached. Of course the microorganisms should be nontoxic. Suitable organisms are enterococci, lactobacilli, gram negative bacilli and aerobic spore formers such as *Bacillus subtilis*, *B. alvei*, etc.

Normally the milk is sterilized in an autoclave prior to addition of the acidifying agent, in this case microorganism. The microorganisms are allowed to grow until the proper pH is reached. At 72°F., this is normally not over 24 hours when using the Streptococcus lactis. The time will vary to some extent depending on the temperature, which of course affects the growth rate of the microorganism. Under optimum growth temperature conditions, generally not over one day is required.

After the desired pH is reached, a curd is developed either by raising the temperature to 113° to 122°F. or by adding rennet and incubating until a curd forms. The curd is centrifuged, the supernatant decanted and the curd washed with water. Up to this point the process is conventional for forming Cheddar cheese. The curd thus prepared is suspended in water at 5 to 50% solids, preferably 15%, and is then inoculated with a nontoxic micrococcus such as Micrococcus caseolyticus, (e.g., A.T.C.C. 13548), Micrococcus conglomeratus, (e.g., A.T.C. 401) or Micrococcus freudenreichii (either A.T.C. 407 or A.T.C. 8459).

The fermentation is then continued, preferably aerobically, with the micrococci at a preferred temperature of 89.6°F. for 5 to 7 days with vigorous agitation. The agitation is important to obtain rapidly the desired flavor. When less favorable growth temperatures are employed, the fermentation must be continued somewhat longer (up to 2 weeks) to get an equivalent amount of Cheddar flavor.

Example: Whole milk was pasteurized at 200°F. for 30 minutes, cooled to 72°F. and inoculated with Streptococcus lactis and incubated until the pH was 5.2. Rennet was then added and the milk incubated until a curd was formed. The curd was centrifuged, the supernatant decanted and the curd was then washed with water.

The curd was then sterilized in an autoclave to kill the Streptococcus lactis. Next the curd was suspended in water at 15% solids and inoculated with Micrococcus caseolyticus (A.T.C.C. 13548) and fermentation continued aerobically with strong stirring to insure vigorous agitation at 32°F. for 6 days. The product had a typical Cheddar flavor but had a Cheddar flavor intensity approximately 100 times that of one year old Cheddar even though the process took only a week from the first inoculation of the whole milk with the Streptococcus lactis.

The flavor thus produced was suitable as such for blending with other ingredients to impart a Cheddar flavor. Thus it was suitable for placing on apple pie or incorporating in cookie dough.

FLAVORINGS

Vegetable Hydrolysate and Yeast Autolysate

W.A. Rosenthal and H.A. Pinkalla; U.S. Patent 2,928,740; March 15, 1960; assigned to American Bio-Synthetics Corporation have found that a mixture of a vegetable protein hydrolysate and a torula yeast co-hydrolysate made by autolyzing the yeast in the presence of the vegetable hydrolysate, has a pleasing flavor and fixes and intensifies the flavor of other foods.

In the preparation of a vegetable hydrolysate, wheat gluten is mixed with excess concentrated hydrochloric acid and refluxed for 6 to 12 hours. During heating the ratio of amino acids to total protein content is periodically determined to obtain a maximum of amino acids. When the ratio of amino acids to total protein no longer rises, the end point has been reached and heating is stopped. The humins are then filtered off and the filtrate is neutralized with sodium hydroxide.

A concentration of sodium hydroxide is used to arrive at a 16% solution of sodium chloride in the hydrolysate. The ratio of addition of the sodium hydroxide to the hydrolysate is controlled so that the temperature of the hydrolysate, during the neutralization step, does not rise above 25°C. as there are changes in the protein and amino acids resembling the burning of such materials upon exceeding that temperature during the neutralization. After neutralization the hydrolysate is an opaque light brown liquid which may be again filtered if a clearer product is desired. Control of temperature during neutralization is critical.

The yeast is approximately one half protein. A total protein to sodium chloride ratio of 2.2 to 2.5 to 1.0 is desired. This is critical to making a nonfiltered nonsettling colloid which can be stored and used without further treatment. (The carbohydrates of the yeast do not affect the desired properties of the final product.) The vegetable hydrolysate is mixed with the torula yeast dried to no more than 5% moisture and at a temperature inactivating the yeast, preferably in the proportion by weight of 2.7 to 3.7 of the hydrolysate-sodium chloride solution to 1.0 of the dried torula yeast. The mixture is kept at 37° to 40°C. for 20 to 24 hours with stirring, to obtain substantially complete autolysis of the yeast without substantial change in the yeast proteins from their original form. The mixture of the vegetable protein hydrolysate and of the torula yeast autolysate is then ready for use.

Tests were made in the manufacture of process cheese spreads, using cheese of the American and Cheddar varieties to make lots of each cheese respectively containing 1, 2, 3 and 4% of the vegetable hydrolysate-torula yeast autolysate which is a smooth paste and acts as an emulsifier for the cheese so that the texture of the cheese spreads are improved. In addition to the qualities of making the cheese spreads easier to manufacture, the composition also produces creamier and smoother cheese spreads.

Blue Cheese Flavor

V.S. Bavisotto; U.S. Patent 3,034,902; May 15, 1962; assigned to Chas. Pfizer & Co. provides a blue cheese aroma and flavor composition by combining a plurality of pure or relatively pure chemical compounds, which can be added to salad dressings, mayonnaises, cheese dip preparations, crackers, rolls, sauces, etc. to impart a blue cheese flavor and aroma to them.

The process involves using two concentrates. Concentrate A contains 3.0 g. acetone, 3.0 g. 2-pentanone, 1.5 g. 2-heptanone, 1.0 g. 2-octanone, 1.0 g. 2-nonanone, 0.5 g. undecanone, 1.0 g. ethanol, 0.04 g. acetaldehyde and 80.0 g. stabilized cottonseed oil. Concentrate B contains 16.8 g. butyric acid, 24.1 g. caproic acid, 5.7 g. caprylic acid, 15.5 g. capric acid and 27.9 g. stabilized cottonseed oil.

The cottonseed oil used in concentrates A and B was an edible, refined grade, stabilized with the antioxidant NDGA #10A ((1:2000). Additional ingredients such as antioxidants, coloring agents, dispersing agents, etc. could, of course, be added to the above concentrations if desired.

The above concentrations can be added to any food to which it is desired to impart a blue cheese flavor and aroma. The intensity of the flavor and aroma can be varied by combining the concentrations in the relations shown on the following page to produce the indicated flavor and aroma. All parts are based on 100 parts by weight of the product to be flavored.

Flavor	Ratio
Sharp (fully-ripened blue cheese flavor)	0.9 part A + 0.45 part B
Moderately strong	0.55 part A + 0.27 part B
Medium	0.22 part A + 0.22 part B
Mild	0.036 part A + 0.22 part B
Very mild	0.018 part A + 0.22 part B

These ratios may be varied to achieve a preferred flavor and aroma characteristic. Once the concentrations have been mixed in the proportions indicated, the flavor and aroma will remain the same for an indefinite period of time. The stability of this formula makes the blue cheese flavor and aroma composition a very desirable ingredient for commercial application since the flavor and aroma will have the same intensity for an indefinite period of time.

Streptococcus diacetilactis

The process of E. Lundstedt; U.S. Patent 3,408,490; August 7, 1962; assigned to H.P. Hood & Sons, Inc. is based upon the discovery that the organism Streptococcus diacetilactis, by modified processing techniques, can be utilized to produce creamed cottage cheese of consistently good aroma and flavor without effecting the acidity of the finished cottage cheese and do so economically. Streptococcus diacetilactis cultures are assigned Accession No. 13675 by the American Type Culture Collection.

Milk starters containing Streptococcus diacetilactis have been used but, when added to milk according to conventional techniques, usually produce sizable amounts of carbon dioxide gas, therefore producing a gassy cottage cheese curd. Gassy curd breaks up in the cheese vats and also retains moisture to a degree which makes drainage of curd very difficult and nearly impossible in automatic draining systems. Such starters in conventional use produce less curd and undesirable curd particles without producing, on a consistent basis, better aroma in cottage cheese than the more conventional L. citrovorumlactis starters, consequently, they are avoided in the manufacture of cottage cheese.

In this process, cultures of or cultures containing Streptococcus diacetilactis are inoculated into citrate enriched cheese whey, preferably cottage cheese whey, and ripened for a suitable period of time. The resulting medium is added in small amount to conventionally prepared cottage cheese. This ripened cottage cheese whey culture is supersaturated with carbon dioxide gas and is of such taste and smell as to be useless according to conventional judgment. However, the addition of small amounts of such cultured material to creamed cottage cheese will, within the three days normally required to market such cheese and under reduced temperature conditions at which no bacterial growth takes place, produce a cottage cheese of consistently excellent flavor and aroma.

The initial step is to provide a cheese whey as the necessary medium for growth of the bacteria. While cottage cheese whey is greatly preferred, it has been found that other wheys can be employed. Useful wheys are the lactic acid wheys produced by bacterial souring of milk to a ripened pH of from 4.8 to 4.3. Useful examples also include cream cheese whey and Neufchatel cheese whey. Rennet whey and whole or skim milk are not satisfactory.

Whey contains soluble proteins which are denatured and coagulated by heat, as in the sterilization subsequently described. While the proteins cause no particular harm and may be left in the whey if desired, coagulation renders the whey cloudy and it is preferred to remove the proteins by heat coagulation and separation by filtering, centrifuging, or settling and decanting. This is preferably done prior to citration but can be done at any stage prior to inoculation.

Citrate is added to the base medium in an amount equivalent to 1.0 to 4.3 parts by weight of anhydrous citric acid per 100 parts of whey. This citrate addition can comprise citric acid or any of its monovalent cation salts suitable for use in food, with or without water of hydration. For example, commercial sodium citrate pentahydrate in the amount of 2 to 8 parts by weight per 100 parts of whey can be used and is equivalent to the amount of anhydrous citric acid stated above. 2.7 parts anhydrous citric acid or 5.0 parts sodium citrate pentahydrate are preferred.

The pH of the citrated whey is then adjusted to about 6.0. This can be done with sodium hydroxide, sodium carbonate, etc. by conventional neutralization procedures. While the citrated whey is not suitable for further processing, it can be spray dried for storage. The citrated whey powder can be reconstituted with water.

The citrated whey is then pasteurized or sterilized. Any suitable heat treatment can be used which will assure that no organisms are present to propagate in competition with the subsequently added bacteria culture. Suitable procedures are autoclaving at not more than 10 pounds steam pressure for 15 minutes, or heating for one hour at 190°F.

The citrated whey medium, after sterilization and cooling, is then inoculated with *Streptococcus diacetilactis* organisms either pure, as preferred, or in the presence of other starter bacteria. Strains should be employed with high citrate-fermenting and low lactose-fermenting abilities. This inoculation can comprise from a few parts to a fully developed culture per million parts of citrated whey to about 5.0 parts culture by weight per 100 parts citrated whey, with about 2.0 parts per hundred being preferred. The culture is sufficiently developed if it has or contains sufficient carbon dioxide to produce foam on agitation. The amount of this inoculation is not particularly critical since the bacteria multiply rapidly in the medium within 18 hours at 72°F.

The medium is then ripened at about 80°F. The time of ripening can vary from 15 hours to 7 days; 18 hours at 72°F. with a 2% bacterial culture inoculation has been found satisfactory. The ripening produces sufficient fermentation and bacterial growth to provide an adequate amount of flavor and aroma precursor materials in the culture.

Growth of a particular citrate fermenting culture should be continued until approximately the maximum pH, as measured on a CO_2-free basis, is achieved. At this time sufficient of the desired precursor materials are present. The ripened culture may now be refrigerated for storage or may be left at ripening temperature for approximately one week without undue deterioration or loss of precursor material. The maximum pH also coincides with approximately the maximum production of gas (CO_2)

so that sufficient ripening of culture can also be judged by agitation which will produce foam. Similarly, the titratable acidity of the ripening culture, calculated as lactic acid and on a CO_2-free basis will decrease from about 0.3% in the original inoculated culture mixture to less than 0.1% as the pH increases.

The culture mixture, prepared and ripened as described, can be used immediately or refrigerated for as long as one month. It is added to creamed or uncreamed cottage cheese prepared by otherwise conventional techniques, in an amount sufficient to produce about 1 to 4 parts per million diacetyl in the cheese after about 3 days. In practice it has been found that from 1/4 to 1% addition of the ripened culture mixture will give diacetyl within the desired range. It is conveniently added to the cold cream dressing prior to its addition to the cottage cheese curd in making the final product.

Example: Sodium citrate pentahydrate was added to cottage cheese whey, 5.0 parts by weight per 100 parts whey. The pH was adjusted with 2 N sodium hydroxide to 6.0 and the citrated whey heated at 190°F. for one hour and cooled to 72°F. The heat treated citrated whey was then inoculated with 2% of a developed *Streptococcus diacetilactis* culture and ripened for 18 hours at 72°F. The ripened culture was then refrigerated (38° to 42°F.) for storage. One-half percent by weight of this ripened culture was added to conventionally prepared creamed cottage cheese, mechanically agitated for five minutes and stored at 38°F. Excellent organoleptically-determined flavor and aroma were developed. Diacetyl, in parts per million, was developed in the cheese (at 38°F.) as follows:

	2 Days	6 Days	17 Days	pH (17 Days)
Test (total diacetyl)	2.68	4.08	6.16	5.0
Control	0.84	0.84	1.52	5.0

The cottage cheese was in marketable condition for the full 17 days.

Penicillium roqueforti

S.G. Knight; U.S. Patent 3,100,153; August 6, 1963; assigned to Wisconsin Alumni Research Foundation discusses the fact that *Penicillium roqueforti* and *Penicillium camemberti* could be grown rapidly under certain conditions in milk or mixtures of milk and cream and that the resulting products were adaptable for use, without aging, for flavoring foodstuffs. Homogenization of the butterfat in the milk and cream along with aeration and maintaining the medium free from unwanted bacterial contaminants were found necessary for proper growth and development of the flavoring compositions.

These fungi are characterized by their ability to convert fatty acids to ketones when fermented under submerged aerobic conditions. The fermentation period using vegetative inoculum (mold cells in mycelium form with nutrient growth medium) was relatively short, e.g., 2 days, compared to the fermentation period, e.g., 5 to 6 days, using mold conidia or conidiospores. The fermentation period using 7 day old spore containing vegetative inoculum or spores alone was longer than that required with 2 day old, substantially spore-free vegetative inoculum. In view of this, it is assumed that the spores as such were relatively inert and that vegetative inoculum and preferably young inoculum should be used. Investigations also disclosed that the fermentation mechanism apparently proceeded first by the hydrolysis or splitting of the fats and then by the conversion of the resulting free fatty acids to the desired highly flavored aromatic ketones. It was also noted that the development of the desired flavor could be hastened by the addition of small amounts of lipase or like fat splitting enzyme.

After lipase modified the milk fat, the resulting mixture containing free fatty acids derived from the milk fat was converted to the desired highly flavored composition much more rapidly by the old spore containing vegetative inoculum, in which the cells were relatively inactive, than by the young substantially spore-free inoculum, in which cells were known to be relatively active.

Milk fat was first modified with lipase to release free fatty acids. Conidia or conidiospores of Penicillium roqueforti were then added to the modified fat mixture (free fatty acids derived from the milk fat) and the mixture subjected to fermentation under submerged aerobic conditions. In a matter of minutes it was noted that the spores (heretofore considered inert) were converting the free fatty acids to ketones. Instead of days, desired flavor was obtained in this manner in hours.

Investigations based on the above have shown that the young active vegetative inoculum provides enzymes for splitting the fat and that the free acids are then converted to ketones when spores are produced in the old relatively inactive vegetative inoculum. In this action the spores introduce a ketone (C=O) group on the number 3 carbon atom and then decarboxylate the fatty acid. This can be illustrated as follows:

$$RCH_2CH_2CH_2COOH \xrightarrow{\text{spores}} RCH_2{-}\overset{\overset{\displaystyle O}{\|}}{C}{-}CH_3$$

where R is an alkyl group, e.g., containing 1 to 18 or more carbon atoms.

Free fatty acids can be used as such where available and where not available they can be made readily by chemical hydrolysis of fats or by modification of the corresponding fats, e.g., milk fat, by use of fat splitting enzymes such as lipase. While homogenization of the fat or fatty acid is not essential where the process is carried out with agitation or agitation accompanying aeration, the use of homogenized fat or aqueous fat emulsions such as in homogenized milk fat is preferred both for the conversion of fat to free fatty acid and conversion of fatty acid to ketone by the spores. The overall action of the enzymes and spores is speeded up when the fat and fatty acid are dispersed in minute particles to provide large areas of surface.

The process can be readily carried out in an aqueous medium, e.g., pasteurized, homogenized milk containing milk fat or fatty acids derived from milk fat, at about 25° to 30°C. in a fermenter of the type used for submerged aerobic fermentation. Sterilization or equivalent treatment such as pasteurization, which eliminates bacteria which deleteriously effect fermentation, is necessary. One of the preferred ways of making an inoculum is to grow or culture the mold for 2 to 3 days in pasteurized or sterilized whole or skim milk at about room temperature in a fermenter. The resulting vegetative inoculum (mold cells in mycelium form and growth medium) can be used in a pasteurized as well as a sterilized fat containing milk medium and provides a relatively short fermentation period of about 1.5 to 2.5 days.

After fermentation the broth can be used as such or be condensed with or without vacuum down to 50% or more in volume. This product can be used as such or be spray dried, vacuum drum dried, etc. Temperatures which cause darkening or caramelization of the product should be avoided.

Example: The fermenter is first charged with 1,000 lbs. of a homogenized mixture of whole milk and cream in a ratio of about 1 to 1 by volume to provide a mixture containing about 12% fat. The mixture is then sterilized with live steam at about 15 lbs. pressure for about 20 minutes. After cooling, the mixture is then inoculated aseptically with about 50 to 100 lbs. of a vegetative inoculum containing mold cells of a white mutant of *Penicillium roqueforti*. The mixture is then agitated with aeration of about 1 volume per minute and at a temperature of about 25° to 30°C. for about 2 days. The fermented mixture is then preheated to about 140°F. and spray dried in swirling air with an inlet temperature of about 225° to 325°F. and an outlet temperature of about 150° to 170°F. to yield a highly flavored product adaptable for use in the flavoring of foodstuffs.

J.C. Watts, Jr. and J.H. Nelson; U.S. Patent 3,072,488; January 8, 1963; assigned to Dairyland Food Laboratories, Inc. describe an improvement in the above process. They found that the yield of spores (which term includes spore-like entities) could be greatly increased in submerged culture with aeration, e.g., using mechanical shakers, by the use of sodium chloride and/or lactose, or equivalent nontoxic salts or sugars. Also, when homogenized milk, skim milk or whey (including reconstituted nonfat milk solids and reconstituted whey solids) and various blends of these were used as the aqueous culture media, the resulting high spore count products were essentially free from true vegetative growth material and could be used as such as the inoculum without purification, i.e., without separation of the spores.

The amount of added salt or sugar or mixtures of these used in the production of the spore inoculum can vary with optimum amounts under the culturing conditions employed. Ordinarily, it is preferred to start the preparation of the inoculum with a homogenized whole milk medium containing about 1 to 3% by weight of added salt and/or sugar and then after culturing the medium (submerged aerobic) for about 3 to 4 days, adding about 4 to 6% by weight additional salt or sugar and continuing the culturing of the medium for about another 1 to 2 days. The use initially of about 2% of salt and/or sugar followed by an additional 5% of salt or sugar, to provide the medium with a total of about 7% by weight of added salt and/or sugar, as described in the following example is in the preferred range.

It is advantageous at the start of the culturing period to add a small amount, e.g., about 3% by weight of lactic acid, to provide the culture medium with an acidic pH of about 4.5 to 5.5. The culturing times and temperatures can vary somewhat with periods within 4 to 6 days at room temperature, e.g., 21 to 25°C. giving good results and being within the preferred range.

The spore inoculum used to inoculate the culture medium containing the salt and/or sugar in the preparation of the high spore count inoculum used in the final fermentation step in the production of the flavoring composition can be prepared by washing the spores from a 5 to 15 day old salt culture of *P. roqueforti* grown on a mixture made up of 50% by volume of tomato-vegetable juices and 50% by volume of 3% aqueous agar solution. The resulting spores, suspended in water, which may contain 1% Tween 80 are then added to the culture medium and the resulting medium cultured as described. The amount of spores used in this inoculum can vary although to keep the culturing period within about 4 to 6 days, sufficient spores are preferably added to yield a level of about 5×10^7 to 5×10^8 spores per cc of substrate. Using a substrate of this type a finished culture with a spore count (including spore-like entities) in the range of about 50×10^9 to 150×10^9 spores per cc can be readily obtained.

Example: About 391 grams of homogenized whole milk are added to each of a series of 1,000 cc wide mouth Erlenmeyer flasks, and the flasks are plugged with cotton and the composite sterilized at about 121°C. for 12 minutes. When the sterilized milk is cooled to room temperature (21° to 25°C.), 1.2 g. of sterilized lactic acid and 8 g. of sterilized salt (NaCl are then added aseptically to each flask). The resulting culture medium has a pH of about 4.5 to 5.5. The flasks are then inoculated with a spore suspension of a white mutant of P. roqueforti (U.W.-12), obtained as described above, to provide the medium with about 5×10^7 to 5×10^8 spores per cc of substrate. The inoculated flasks are next placed on rotary shakers (approximately 160 cpm) and agitated continuously for about 96 hours at room temperature. At this time 20 g. of sterilized salt are added to the culture medium aseptically in each flask and the shaker agitation is resumed and continued for an additional 24 hours. The finished culture or inoculum contains around 100×10^9 spores per cc and is ready for use without further processing. These high spore count inocula can be used to prepare flavoring compositions in accordance with the fermentation process described in U.S. Patent 3,100,153 above.

Genus Trametes

S. Hori and H. Shimazono; U.S. Patent 3,348,952; October 24, 1967; assigned to Takeda Chemical Industries, Ltd., Japan describe a method for improving the flavor of cheese and for imparting supplemental cheese flavor to any other dairy product which comprises treating a starting material with flavoring enzyme produced by microorganisms belonging to the genus Trametes.

Animal milk is sterilized and 1 to 2% starter is added to the milk to produce lactic acid, followed by the addition of rennet to obtain curd. After whey is removed, curd is smashed. Then whey is completely removed by pressing, followed by heating to obtain green cheese. This prepared green cheese is subjected to ripening at 5 to 15°C. in a ripening room.

Ordinarily ripening requires about 4 to 6 months or, in some cases, about 2 yrs. In this process cheese-like flavor comes out by treating the green cheese with flavoring enzyme for a short time. Trametes sanguinea (L. ex Fr.) Lloyd, has been found to be especially useful here.

The microorganisms are incubated in a liquid or solid medium. In general, the use of a liquid medium is preferable. In most cases, it is preferable to subject these microorganisms to submerged cultivation. The culture medium employed desirably contains carbon and nitrogen sources which are assimilable by the microorganisms.

The incubation is desirably carried out at about 25 to 35°C. and the accumulated flavoring enzyme in the culture broth reaches a maximum usually between several tens of hours and several hundred hours. the pH of the medium is adjusted to around 3.0 to 6.0. Thus accumulated flavoring enzyme can be recovered from the culture broth. Generally known means can be applied to the recovery of the flavoring enzyme. This flavoring enzyme or material containing the flavoring enzyme is then allowed to act on the starting material which may be milk or green cheese. The procedure may vary depending on the kind or state of the starting material, etc. However, in general, the procedure may sufficiently be carried out by only mixing flavoring enzyme with the starting material, preferably under agitation. Temperature of the treatment, time required for the treatment, the amount of flavoring enzyme, etc. may also vary.

More concretely stated, when green cheese is treated with flavoring enzyme, it may be preferable to agitate a mixture of green cheese and flavoring enzyme at about 10° to 20°C. By this treatment, strong cheese-like flavor can be brought about in a short time. When flavoring enzyme is caused to act on animal milk or condensed milk, it may be preferable to adjust pH of the milk to the range of from acid to slightly alkali, desirably to pH 2 to 6, and then to agitate at about 20° to 40°C. for 1 to 3 hours. In this case, 0.1 to 1.5% by weight of flavoring enzyme fraction may be sufficient for treatment. Flavoring enzyme has an ability of coagulating milk as well as an effect of flavor enhancement on milk. Therefore, flavoring enzyme may serve as rennet, or when the pH is lowered with a suitable acid, e.g., lactic acid, it may serve as rennet and starter.

Example: *Trametes sanguinea* (L. ex Fr.) Lloyd (ATCC-14622) is inoculated to a sterilized aqueous medium of pH 6.0 containing 4% of sucrose, 2% of glucose, 3% of soybean cake, 1% of cornsteep liquor, 0.2% of ammonium sulfate, 0.2% of potassium dihydrogenphosphate, 0.1% of magnesium sulfate (7 hydrate) and 0.05% of copper sulfate. The medium is agitated with 160 rounds per minute under the aeration of 500 liters per minute. The incubation is carried out for 140 hours at 28°C. After adjusting to pH 4.0 with ammonium sulfate, the culture broth is filtered. Then 200 kilograms of ammonium sulfate is added to thus obtain 400 liters of filtrate to yield precipitates. The precipitates are collected by centrifugation, and dried at 35°C. to give 4 kilograms of crude flavoring enzyme.

15 grams of the thus obtained crude enzyme is dissolved in 150 ml. of distilled water to be dialyzed at 4°C. for 1 day. The whole volume of the so dialyzed solution is made up to 220 ml. and is then salted out with ammonium sulfate at 10 to 58% saturation. This flavoring enzyme is added to about 200 times of its weight of whole milk adjusted to pH 3.7 with lactic acid, and the mixture is agitated at 28°C. for 2 hours. Using the resultant mixture, cheese is prepared by means of a conventional method, whereupon cheese having a pleasing flavor can be obtained without being subjected to ripening. Use of skim milk in place of whole milk also gives cheese having a pleasing flavor without being subjected to ripening, though the flavor is weaker to some extent that that of whole milk. The flavoring enzyme is also added to green cheese prepared by a conventional method, and the mixture is agitated at 15°C. After 2 hours, a pleasing cheese-like flavor is achieved.

Aliphatic Fatty Acid, Phenol and Lactone

G.J. Henning; U.S. Patent 3,520,699; July 14, 1970; assigned to Lever Brothers Company provides Cheddar cheese flavor composition by combining at least one aliphatic fatty acid selected from the group consisting of butyric, caproic, caprylic and capric acids, at least one phenol and at least one lactone of hydroxycarboxylic acids having from 4 to 22 carbon atoms in the molecule or precursors of these. The addition of a ketone having a methyl group and an alkyl group with 2 to 15 carbon atoms provides a blue cheese flavor.

When imparting a Cheddar cheese flavor to a fat-containing foodstuff, the amount of each component that may be employed, expressed in mg. per kg. fat present in the foodstuff, are preferably as follows: from 0.1 to 5 mg. of each phenol; up to 100 mg. of each lactone or a precursor; up to 600 mg. of each fatty acid; and where the flavor required is that resembling blue cheese, up to 400 mg. of each methyl ketone or a precursor. The flavor composition can be incorporated into the foodstuff by admixture with a solution, dispersion or emulsion of the composition in an edible oil and/or water. When the foodstuff, for example pastry, that is to be flavored has already been shaped, the shaped foodstuff may suitably be sprayed with an oil solution or a water emulsion of the flavor composition. The flavor composition may be used in processed cheese.

Example 1: A composition for imparting a Cheddar cheese flavor to foodstuffs was prepared by dissolving the following components in 1,000 g. refined groundnut oil:

Lactone of delta-hydroxydodecanoic acid	20 g.
Lactone of gamma-hydroxybutric acid	4 g.
Lactone of gamma-hydroxyoctanoic acid	10 g.
Lactone of epsilon-hydroxydodecanoic acid	40 g.
Butyric acid	250 g.
Caproic acid	300 g.
Caprylic acid	200 g.
Capric acid	250 g.
Phenol	250 mg.
p-Cresol	250 mg.
Guaiacol	400 mg.

Example 2: A composition for providing a foodstuff with a flavor similar to that of blue cheese was prepared by incorporating in 1,000 g. refined groundnut oil the substances listed in Example 1, and in addition the following ketones:

Methyl propyl ketone	100 g.
Methyl pentyl ketone	250 g.
Methyl heptyl ketone	200 g.
Methyl nonyl ketone	40 g.
Methyl undecyl ketone	10 g.
Methyl tridecyl ketone	10 g.

Example 3: A processed cheese was prepared in the following manner. 540 g. purified coconut oil, to which 0.2% soybean lecithin and 0.67% anatto has been added, was emulsified at 50°C. in 12 l. skimmed milk by means of a centrifugal stirrer. 67 milliliters concentrated lactic acid, containing 80% monomeric lactic acid, was added to the emulsion at 35°C., the pH then being 4.65. The clotted milk was filtered through a cloth and pressed to a dry matter content of 48%. The following substances were then added to the curd:

Polyphosphate	2.5%
Potassium sorbate	0.2%
Casein hydrolysate	3.0%

The curd was mixed with the added substances in a Küstner melting tank and subsequently melted by heating at 80°C. for 10 minutes under a vacuum of 34 cm. mercury, while mixing with a high speed stirrer. After the melting operation, 0.04% of the solution in oil described in Example 1 was added, and on solidification, a processed cheese was obtained with a good consistency and an excellent taste, which was virtually indistinguishable in flavor from a processed cheese prepared from Cheddar cheese.

Cream Cheese Flavor

A.J. Luksas; U.S. Patent 3,535,121; October 20, 1970; assigned to Beatrice Foods Co. describes a process in which a cream cheese type flavoring is produced by growing a mixture of *Streptococcus lactis* and *Streptococcus diacetilactis* on milk, preferably skim milk, under aerobic conditions.

It is critical that the organisms be grown under aerobic conditions. To insure aerobic conditions, the fermentation is carried out with agitation while introducing air. The fermentation can be carried for for 1 to 10 days. Maximum cream cheese type flavor is obtained in 2 to 5 days. The organisms are preferably grown at 32°C. in the presence of citric acid. The product can be used as such but preferably it is dried. The resulting powder has a strong pleasant cream cheese type flavor and can be used to impart to a dip, cream cheese, butter or oleomargarine. The use of skim milk is preferable to the use of whole milk since the product obtained when skim milk is employed as the growth medium has a longer shelf life. After the cultures have grown to develop the maximum amount of flavor, the bacteria are killed by pasteurization using conventional pasteurization temperatures and times.

Example 1: Skim milk containing 0.1% added sodium citrate was inoculated with *Streptococcus lactis* and *Streptococcus diacetilactis* and agitated constantly for 3 days at 32°C. while passing air through the milk. The product was then heated at 175°F. for 20 minutes pasteurization to kill the organisms and then chilled to 40°F. The product obtained had a strong cream cheese flavor and was suitable for use as is.

Example 2: The procedure of Example 1 was repeated but the product after being pasteurized was spray dried to give a powder having a long shelf life and a strong pleasant cream cheese flavor.

Genus Bacillus and Genus Streptococcus

R.L. Kasik and A.J. Luksas; U.S. Patent 3,674,508; July 4, 1972; assigned to Beatrice Foods Co. produced basic cheese flavor or a Cheddar cheese flavor in a short time by growing a combination of one member of the genus Bacillus and one member of the genus Streptococcus in an aqueous medium containing a protein and a carbohydrate.

It has been found that when a member from the genus Bacillus and the genus Streptococcus are contained in a medium which will support their growth at a reasonable rate, a cheese flavor will rapidly develop. The organisms must be contained in a medium which will support their rapid growth and this medium must contain a protein and a carbohydrate. The proteins which may be utilized by the organisms are not critical and may be animal proteins, e.g., milk, or vegetable proteins, e.g., soy bean protein, and like vegetable proteins. The organisms are nonsensitive to the particular protein and carbohydrate. Likewise, the particular percentage of protein and carbohydrate in the aqueous medium is not critical, but from 0.5 g. of protein per 100 g. of aqueous medium and from 0.5 g. of carbohydrate per 100 g. of aqueous medium to 35 g. of protein and 35 g. of carbohydrate per 100 g. of aqueous medium are suitable, especially 1 g. of protein and 1 g. of carbohydrate to 7.5 g. of protein and 5.5 g. of carbohydrate. Milk itself provides quite suitable proteins and carbohydrates in the above ranges and is, therefore, a suitable medium for growing the organisms. However, it is not necessary to use only milk, since other milk fractions or products may be used.

While any member of the genus Bacillus and the genus Streptococcus may be used to produce a cheese flavor, some of the organisms are toxic or produce toxins which, of course, would be unsuitable for consumption; accordingly, the toxic organisms and the organisms which produce toxins are excluded from use here. Also, as is known, Group 1 and Group 3a of the genus Streptococcus are pathogenic. Accordingly, the genus Streptococcus is limited to Group 2, the Enterococcus Group of Group 3, and Group 4.

While the combination of at least one member of the genus Bacillus and the genus Streptococcus will produce a cheese flavor, the degree and distinctiveness of the flavor developed varies with the particular members chosen. In other words, by choosing a particular member or members from each of the two groups, various specific cheese flavors and varying degrees and distinctiveness of a cheese flavor can be obtained. Hence, certain combinations of the organisms from the genus Bacillus and genus Streptococcus will produce a Cheddar cheese flavor of varying distinctiveness, in addition to the basic cheese flavor.

For example, if it is desired to produce a high degree and distinctiveness of a Cheddar cheese flavor, the organisms which have been found best for this purpose, from the genus Bacillus are *alvei*, *cereus*, *cereus var. mycoides*, *megaterium*, *megatherium*, *subtilis*, *subtilis var. aterrimus*, *subtilis var. niger*, and from the genus Streptococcus are *cremoris*, *durans*, *faecalis*, *lactis*. Also within this group, a certain combination produces quite superior results in terms of the degree and distinctiveness of the Cheddar flavor developed. Accordingly, this combination is considered to be best for these purposes, especially since a specific cheese flavor as well as a basic cheese flavor is produced. The combination is the organisms *megaterium* and *megatherium* from the genus Bacillus and the organisms *S. lactis variation diacetilactis* (citrate fermenter) and *S. lactis-diacetilactis* (citrate nonfermenter) from the genus Streptococcus.

However, when any members of the genus Bacillus and the genus Streptococcus are used to produce a cheese flavor, as described above, the resulting product is a basic cheese flavor and not necessarily a Cheddar cheese flavor, unless the particular organisms noted in connection with Cheddar cheese are particularly used. The basic cheese flavor produced according to the broader aspects of this process is characterized by the cheesy flavor common to all cheeses. Additional flavors are superimposed upon this basic cheesy flavor and these flavors give rise to the particular cheeses such as blue, Cheddar cheese, Swiss, Romano, etc.

As noted above, the protein and carbohydrates may be chosen as desired, but it is most convenient to use milk, a milk fraction or a milk product. In this regard, skim milk is quite suitable as a medium for the process. The milk medium must not have less than 4% and not greater than 35% total solids, the remainder being essentially water and the dissolved natural constituents of milk. This medium is heated from at least 145°F. for 30 minutes., i.e., pasteurization, to 295°F. for 15 seconds, or to sterility at any temperature and time, which processes are well-known in the art.

When the medium is heated to within such temperature and time ranges, the organisms occurring in the medium will be substantially killed or retarded to a point that they are prevented from competing with the organisms of this process. Under the circumstances, the heating step is quite critical. The medium is then cooled to 130°F. or below, and preferably below 100°F. The medium, however, must be maintained above 60°F, and preferably above 85°F. While the medium is maintained within the above-noted temperature range, it is inoculated with at least 10 organisms of the genus Bacillus and at least 10 organisms of the genus Streptococcus per gram of medium. Preferably, for the best results, at least 1,000, e.g., 10,000 or more organisms of each genus are placed in the medium.

While the ratio of the organisms from the genus Bacillus and the genus Streptococcus is not narrowly critical, it is preferred that the ratio be between 25:75 to 75:25 (Bacillus to Streptococcus). Preferably the media with the inoculant therein are stirred sufficiently to distribute the organisms throughout the media. The stirring is, however, not critical since the organisms will naturally distribute throughout the media within a reasonable amount of time. The inoculated medium is then sealed in an airtight manner, e.g., by capping or the like and maintained within the above-noted temperature range for 3 to 5 days during which a fermentation process takes place. Longer times may be used, but it has been found that the growth of the organisms tapers off considerably after 5 days and very little advantage can be obtained in allowing the organisms to grow beyond a period of 5 days.

After the fermentation process, the composition has the basic cheesy flavor described above, except when the organisms mentioned above in connection with a Cheddar cheese flavor are used as the inoculant. In this latter case, the composition obtained has a definite Cheddar cheese flavor as well as the basic cheesy flavor.

After the organisms have been allowed to grow, the product may be used in any desired way for providing the cheese flavor. For example, the product may be concentrated to produce a high degree and distinctiveness of Cheddar flavor or it may be pasteurized and dried as a cheese flavored powder or it may be used to mix with an unaged cheese base or aged cheese to make a flavored cheese or cheese product of excellent quality and taste. As an alternate procedure, instead of sealing the medium in an airtight condition, the fermentation may take place in a container in which the head space is continually flushed with an inert gas such as nitrogen. The flush of nitrogen or the use of a sealed container is necessary since the process must be carried out under microaerophilic conditions.

Example: (A) Production of the Inoculant — Skim milk having 22% total solids was heated to 145°F. for 30 minutes to substantially destroy any pathogens and to reduce the number of competitive organisms. The skim milk was then cooled to 100°F. and placed in a sterile container have 1/4 of its volume as a head space. The container was placed in an isothermal bath maintained at 100°F. The cooled skim milk was inoculated with 3,000 organisms per gram of the medium of the organism S. lactis-diacetilactis (citrate fermenter) and 3,000 organisms per gram of the medium of the organism S. lactis-diacetilactis (citrate non-fermenter). The skim milk was also inoculated with 2,000 organisms per gram of aqueous medium of B. megaterium and 2,000 organisms per gram of the medium of B. megatherium. The inoculated skim milk was maintained at 100°F. for 30 days during which a symbiotic relationship of the organisms therein was established. The product was then suitable as an inoculant for producing a cheese flavor or a cheese flavored product.

(B) Production of the Cheese Flavor — Skim milk having 15% total solids was heated to 145°F. for 30 minutes to substantially destroy pathogens and to reduce the number of competitive organisms. The skim milk was cooled to 100°F. and placed in an isothermal bath maintained at 100°F. The skim milk was then inoculated with the product of Example (A) wherein the resulting inoculated skim milk had a total of 10,000 organisms per gram of skim milk and a ratio of the Bacillus organism to the Streptococcus organisms was approximately 60/40.

The inoculated skim milk was stirred for 35 minutes to disperse the organisms therein. The inoculated skim milk was then sealed in an airtight container having 1/3 of its volume as a head space and allowed to remain at 100°F. for 5 days during which the skim milk was fermented and gases were contained in the head space. Thereafter, the container was opened and the pressurized gases in the head space and the liquid medium was allowed to escape. This product was then suitable for further processing into a finished consumable cheese-flavored product.

A mixture containing 50% of fermented solids on a dry basis of (B), 31.3% acid whey powder, 4% salt, 5% disodium phosphate (a dispersing agent), 0.5% USDA Yellow No. 5, the remainder being dry milk solids was prepared. After thorough mixing, a portion thereof was dispersed with vigorous stirring in water and spray dried to form a cheese powder. The cheese powder was a fine textured powder having a very distinctive and medium degree of cheese flavor.

On a dry weight basis, there was mixed 50% of the fermented solids of (B), 6% aged Cheddar cheese, 16.3% acid whey powder, 15% butterfat, 0.5% Atmos 150 (monoglyceride emulsifier), the remainder being salt, disodium phosphate and USDA Yellow No. 5. This product after being thoroughly mixed was cut into wedges and produced an excellent Cheddar cheese.

In an extension of the work cited above, R.L. Kasik and A.J. Luksas; U.S. Patent 3,667,968; June 6, 1972; assigned to Beatrice Foods Co. describe the use of the process to produce other specific cheese flavors.

According to this improvement, instead of using the ferment alone or mixing with a base such as naturally produced cheese, the ferment is inoculated with specific organisms to also produce a specific cheese flavor. The inoculant may be the specific organisms responsible for producing the specific cheese flavor, e.g., Penicillium roqueforti (blue cheese) and Propionibacter (Swiss cheese), or it may be simply a small amount of the desired natural cheese which will, of course, have the natural flavor producing organisms therein. Only enough inoculant to start the further fermentation is required. This amount is, of course, quite small, e.g., 10 to 100 or more organisms per gram of ferment (either isolated organisms or those contained in a natural cheese).

This inoculated ferment is then further fermented for at least 2 days and will be essentially complete within 20 days, more likely 12 or 14 days.

The further fermenting is carried out at about 40° to 130°F. depending on the particular natural cheese or cheese producing organism used in the inoculation step. The conditions of the further ferment should follow that known in the art for fermenting the natural cheese, e.g., use aeration when fermenting with Penicillium roquefortii, anaerobic to microaerophilic conditions for Italian cheese (reduced oxygen tension), all of which is well-known in the art.

This suspension may be used directly as a liquid flavor composition without further processing, other than pasteurization, to prevent any more fermentation or the composition may be spray dried with or without pasteurization to produce a powdered flavor composition. Alternately, the solids may be separated from the unpasteurized ferment (optionally pressed to remove excess whey) and allowed to age at about 40° to 100°F. for at least 5 days and up to a year or more, if desired. Good flavor will be produced, however, generally within 30 days. This will produce a full bodied cheese of excellent flavor.

Example: The ferment produced in parts (A) and (B) of the example given to illustrate the process described in U.S. Patent 3,674,508 is maintained at 77°F. and inoculated with Penicillium roquefortii. The ferment was further fermented for 5 days with aeration at that temperature. The resulting ferment was divided into 3 parts, one part was pasteurized at 145°F. for 30 minutes and cooled. One part was not pasteurized but immediately spray dried to produce a blue cheese flavored powder. The whey of the third unpasteurized part was separated from the solids; the solids were pressed and cubed to produce an unaged blue cheese. The unaged cheese was ripened at 70°F. for 30 days and produced an excellent blue cheese.

The whey was separated from the solids and a portion of the solids were spray dried to produce a blue cheese powder of excellent taste. The other portion of the solids was pressed and portions of the pressed solids were mixed, serially, with partially hydrogenated corn oil, sour cream, soya protein, and an unaged cheese base to produce respectively a dip, salad dressing, cheese flavor, and blended cheese, all of which had an excellent blue cheese flavor.

Genus Acetobacter and Genus Streptococcus

In further studies designed to find suitable cheese flavors, R.L. Kasik; U.S. Patent 3,729,326; April 24, 1973; assigned to Beatrice Foods Co. describes the use of organisms from the genus Acetobacter and from the genus Streptococcus. This process may be carried out under conditions from microaerophilic to aerobic. The basic cheesy background flavor is more intense than that described in U.S. Patent 3,674,508 and can, therefore, provide the required background flavor in smaller concentrations.

While any non-toxic member of the genus Acetobacter may be used, the process must be carried out in the presence of a nontoxic lactic acid producing organism from the genus Streptococcus. Hence, the process requires that at least one nontoxic member of the genus Acetobacter be grown in the presence of at least one nontoxic lactic acid producing Streptococcus organism. While the ratio of the organisms from the genus Acetobacter to that of the organism from the genus Streptococcus may vary widely, it is preferred that ratios between 20:80 to 80:20 respectively, be used for the best results.

The organisms are grown in a fluid medium with agitation and aeration for a time sufficient to develop the desired intensity of the cheesy background flavor. The specific time required to accomplish reasonable degrees of flavor development will vary somewhat depending upon the amount of agitation and aeration, as well as the temperature under which growth takes place. Additionally, the required growth period will vary somewhat depending upon the specific concentrations of growth substances in the growth medium. Nevertheless, for most applications of the process, growth times as short as 5 hours will produce a significant amount of flavor, although growth times of at least 15 hours are preferred.

There is essentially no upper limit on the amount of time that the organism may be allowed to grow, but the rate of growth, under most conditions, drops off rapidly after about 2 to 5 days, and little growth of the organisms takes place thereafter. Nevertheless, if one desires, the growth could be continued for a month or more, e.g., 2 to 5 months, although no advantage is obtained with these longer growth periods.

The medium for the growing of the organisms is a fluid medium containing a protein and a carbohydrate. The proteins which may be utilized by the organisms are not critical and may be animal or vegetable proteins. The organisms are nonsensitive to the particular protein and carbohydrates. The particular percentages of protein and carbohydrate in the fluid medium are not critical, but from 0.5 g. of protein per 100 g. of fluid medium and from 0.5 g. of carbohydrates per 100 g. of fluid medium to 35 g. of protein and 35 g. of carbohydrate per 100 g. of fluid medium are suitable, especially 1 g. of protein and 1 g. of carbohydrate to 7.5 g. of protein and 5.5 g. of carbohydrate.

In addition, the product produced by this process may be combined with the products produced according to U.S. Patent 3,674,508. These products may simply be combined after their respective growth processes. These growth products may be combined in the liquid state. Thereafter, they may be spray dried to produce a powdered cheese flavor. Alternately, the respective growth products in the liquid state may be simply combined and pasteurized to produce a liquid cheese flavor. Further, the respective growth products may be combined, and after being pasteurized, they may be concentrated to produce a concentrated liquid cheese flavor.

The combination of the growth products of these two processes provides a total cheese background flavor which is superior to the flavors of the products produced by the respective processes. This, in effect, provides a synergistic result between the growth products and is an important feature of this process.

PRESERVATION AND PACKAGING

PRESERVATIVES

Electrostatic Deposition of Sorbic Acid

H.G. Foster and R.H. Mroch; U.S. Patent 2,965,495; December 20, 1960; assigned to Swift & Company describe a method for producing a substantially mold-free cheese by depositing an antimycotic agent on the cheese in sufficient amount to render the cheese substantially mold-proof.

It is possible to spray an antimycotic agent, such as crystalline sorbic acid, through an electrostatic field and deposit that agent on the surface of the cheese. Although sorbic acid is more soluble in oil than in water, if the cheese is heated sufficiently to bring the oil to the surface the subsequent electrostatic precipitation of the sorbic acid is ineffective to produce a mold-free product. Therefore, in treating the cheese it is important to maintain water as the continuous phase on the surface of the cheese, i.e., to treat cheese having substantially no free fat on its surface, an "unoiled" cheese.

What particular phenomena occur between the surface of the cheese and the sorbic acid to bind them is not completely understood inasmuch as the surface of the cheese is a complicated system and not predictable in its characteristics as would be a material with a regular structure. In any event, it is possible to deposit a mold-inhibiting amount of the antimycotic agent through electrostatic deposition provided there is no substantial oil surface on the cheese.

The crystalline sorbic acid is sprayed by means of a spray gun operated at about 30 lb. air pressure. A spray gun of the type used for applying flock finishes is suitable for these purposes. An electrostatic field is then created by highly charging a ring having a series of points inwardly projecting at an angle from its periphery.

Preferably the spraying apparatus is isolated from the electrical apparatus and does not impart any charge to the spray particles nor tend to attract any stray particles of the mold inhibiting material to be deposited. The cheese is placed in front of the ring and is kept at ground potential. The apparatus for dispersing the sorbic acid crystals is set in operation and the spray of crystalline sorbic acid particles is projected through the electrostatic field. The particles become ionized and ultimately are uniformly precipitated, or deposited, on the cheese. The entire apparatus is preferably placed in a metal cabinet or insulated cabinet.

The exact positioning of the elements is not critical. As long as the source of crystalline sorbic acid and the spraying equipment is maintained at one end of the electrostatic field and the material to be coated with the sorbic acid is maintained at the other end of the electrostatic field, the distances between the spraying equipment and the highly charged field and between the charging field and the material to be coated is dictated by practical considerations such as the speed of the particles ejected by the spraying apparatus, the amount of acid flow, the amount of coating material necessary to be deposited and the voltage at which it is desired to operate.

It is apparent that the operation could be made continuous by placing the cheese on a suitable moving conveyor, such as a belt conveyor, which conveyor is kept at ground potential, and depositing the sorbic acid on the cheese while it is being moved through the plane of the sorbic acid dispersion.

The length of time necessary to continue the spraying to precipitate the amount of sorbic acid inhibitor required to produce a substantially mold-proof cheese may be adjusted through the length of time that the spraying apparatus is operative or, in the case of the continuous operations, by the speed of the belt conveyor and/or length of time of the spray. Suitable adjustments in the air flow through the spraying apparatus and the amount of acid flow may be used in conjunction with the above to arrive at a time of exposure suitable for depositing sufficient sorbic acid on the cheese to effectively inhibit mold growth.

Cheese, 99% of which will remain free of mold when stored for 90 days at 45°F. (roughly the same as storage for one month at 60°F.) is considered to be substantially mold-proof. To insure keeping qualities of this type at least 0.02% by weight of the cheese (200 ppm) of sorbic acid is necessary, with 0.1% by weight (1,000 ppm) being optimum. Amounts of sorbic acid up to 3,000 ppm have been deposited by this process with good results.

Example: Slices of Cheddar cheese (both mild and aged) and Swiss cheese were placed on a belt conveyor moving at 18.9 feet per second, maintained at electrical ground, and positioned 15 inches from a charged ring maintained at a voltage of 25,000 volts. A spray gun, positioned 6 inches from the charged ring (and thus approximately 21 inches from the product on the conveyor) was operated at an air flow of 6,000 feet per minute (on the basis of air flow into free space). Sorbic acid crystals were sprayed in an amount of 9.72 grams per minute and directed toward the charged ring.

It was found that under these conditions, as the cheese moved through the area in which the charged particles of sorbic acid were being dispersed, 960 ppm of sorbic acid (0.096% by weight) were uniformly deposited on the cheese slices. The cheese was stored at 45°F. for 90 days and tested at that time for mold growth. Each of the slices was found to be free from mold growth.

Both surfaces of the cheese may be coated to insure that the product will be for the most part mold-proof by running the cheese through the enclosure containing the electrostatic depositing mechanism twice, turning the cheese over for the second run.

Lower Alkyl Esters of 2-Haloacetoacetic Acids

W.E. Rader and M. Legator; U.S. Patent 3,044,844; July 17, 1962; assigned to Shell Oil Company describe an effective preservative based on the finding that the lower alkyl esters of 2-haloacetoacetic acids effectively inhibit and prevent growth and development of microorganisms which cause deterioration of foods. These esters have the formula:

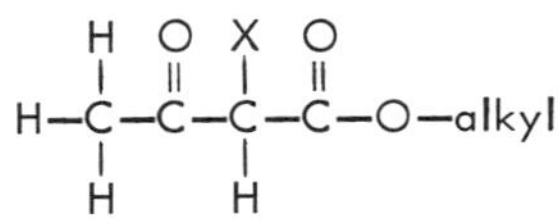

in which alkyl represents the lower alkyl group and X represents halogen. Because they appear to be the most effective biocides, the esters in which the alkyl group contains up to four carbon atoms, and X represents a middle halogen — that is to say, bromine or chlorine — are preferred.

The method of this process thus comprises bringing into close association with a food subject to attack by and deterioration due to microorganisms a minor but effective amount of at least one lower alkyl ester of a 2-haloacetoacetic acid. This has been found to be of value in the preservation of both the cured or ripened cheeses such as Cheddar cheeses and the fresh cheeses such as cottage cheeses and the process cheeses such as cheese spreads and cream cheeses.

These 2-haloacetoacetic acid esters are highly active against a broad spectrum of microorganisms, including bacteria, yeasts, fungi and actinomycetes, and retain this activity in the presence of organic materials. These esters are particularly active against fungi, concentrations of as little as 5 parts per million by weight of the portion of the food providing protection against attack by fungi in many cases.

In the case of bacteria, actinomycetes and yeasts, somewhat higher concentrations of the microbiocidal ester is usually required; even here, though, in some cases a concentration of the microbiocide of as little as 10 parts per million has provided protection against bacteria and actinomycetes. In the majority of cases, a concentration of the microbiocide in excess of about one percent will not be required. Ordinarily, a concentration of at least 50 parts per million is employed to insure adequate protection, but it will not often be found necessary to employ more than about one-half percent of the microbiocide to attain the protection desired.

These esters of haloacetoacetic acids are soluble to some extent in water, and can in many cases be most conveniently and effectively applied in the form of solutions in water. These esters readily disperse in water, so that a solution-dispersion of the esters also can be used. If desired, suitable edible emulsifying agents can be added.

Any technique can be used to attain and insure the necessary intimate association of the esters and the food products to be protected. Where the food product is formed, and is essentially sterile, attack by microorganisms can be inhibited by coating the surface of the food product. To effect this coating, the food product can be dipped in a liquid or semi-liquid composition of the esters, or washed or sprayed or painted with such a liquid or semi-liquid composition, or it can be dusted with a solid composition containing the esters.

The esters can be incorporated in or on materials used for wrapping the food products. Paper, parchment, cloth or other porous materials used for wrapping or gelatinous materials used for coating the food products can be impregnated with the esters, and foil or plastic films and other nonporous wrapping materials can be coated with the esters. Since the esters are volatile, they may be introduced into sealed packages to provide the necessary ester concentration in the gas surrounding the food product. Either the liquid or the gaseous esters can be so used.

Example: Methyl 2-chloroacetoacetate was tested to determine its ability to preserve cheese. Fresh Cheddar cheese was sliced and the slices inoculated with a pore suspension of penicillium obtained from molded Cheddar cheese. The slices were incubated at 15°C. for 24 hours to initiate mold growth. Following incubation, measured amounts of dilute solutions of the test chemicals were spread uniformly over the slices. The treated slices were then placed in Petri dishes alongside similar untreated but inoculated slices of the cheese and the slices incubated for 96 hours at 15°C.

The effectiveness of the test compound for preserving the cheese was indicated by the percent of the surface of the cheese slice remaining free of mold. It was found that at a concentration of 100 parts per million by weight, the test chemical reduced mold by about 50% as compared to the control.

This test is considered to be a very severe test, with all factors favoring growth of the mold. All of the control slices were overgrown with mold. It is evident that methyl 2-chloroacetoacetate is effective in controlling mold growth even under these very severe conditions For comparison: the concentration of the ester was less than one-tenth that of the recommended standard, sorbic acid.

The effect of methyl 2-chloroacetoacetate in controlling the spoilage of cheese was also determined. Fresh small curd cottage cheese was purchased in 1/2 pint cartons. The cheese was removed from the carton, mixed thoroughly with a concentration of the test chemical and replaced in the carton. The cartons were incubated at 15°C.

At 24-hour intervals the cheese was examined for organoleptic changes, primarily sour odor. In addition to this determination, the resazurin dye test was run. This is a dye reduction test which is a measure of the bacteria in dairy products. It is carried out by adding 1/10 ml. of a 0.05 percent sterile aqueous solution of resazurin to 5 ml. of cottage cheese.

The samples are then incubated in the dark at 37°C. for one hour. The dye changes from the original blue to violet, red, pink, and finally white, depending on the number of bacteria present. A blue color after one hour incubation usually indicates under 10,000 bacteria per gram, while the white color usually signifies 100,000,000 or more organisms per gram sample.

The results obtained are shown on the following page. The efficacy of the ester in extending the storage life of the cottage cheese is evident from the results of these tests.

Compound	Concentration in parts per million by weight	Time in Hours after incubation: Slight Souring	Time in Hours after incubation: Sour	Resazurin Test
None (control)	0	48	72	Pink 48 hrs. White 72 hrs.
Methyl 2-chloroacetoacetate	10	96	120	Pink 96 hrs. Pink 120 hrs.
	25	....	120	Pink 120 hrs.
	50	....	120	Red 120 hrs.

Beta-Propiolactone

S.H. Hopper; U.S. Patent 3,050,405; August 21, 1962; assigned to Economics Laboratory, Inc. describes the use of beta-propiolactone (BPL), having the chemical formula

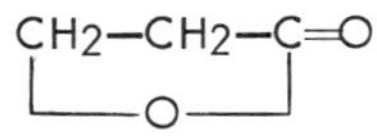

as an additive in the form of a germicidal BPL solution to effectively kill the microorganisms and stabilize food products against deterioration caused by the development of harmful bacteria.

A 10% solution of BPL was used. Of this solution, 2.5 ml. was added to creamed cheese which had previously been inoculated with *Pseudomonas fragi*. This mixture resulted in a final concentration of 0.1% BPL in the inoculated cream cheese. The table shows the results where inoculated and control samples of creamed cheese were utilized together with the results obtained when identical samples of the inoculated cream cheese were treated with a 0.1% aqueous solution of BPL.

	Creamed Cheese		Inoculated and Treated with Chemical*
Time, days	Inoculated	Control	Creamed Cheese
Start	190,000	190,000	210,000
5	210,000,000	180,000	0
10	**	18,000,000	0
15	**	290,000,000	0

*Chemical used was BPL in a final concentration of 0.1%.
**Too many to count.
Note — All samples stored at approximately 45°F.

Further experiments were conducted on samples of Cheddar cheese that were inoculated with a species of Penicillium obtained from a contaminated agar plate. Both controls and experimental samples of cheese were used. The experimental samples were sprayed with aqueous solutions of BPL ranging in strength from 0.1 to 1% and all samples were wrapped separately in Saran wrap, sealed with a hot iron and stored at 50°F. After a lapse of approximately four months, the untreated inoculated samples were moldy and rancid. The samples which had been inoculated and sprayed with the various percentages of concentration of the BPL solution had varying degrees of mold and showed only a slight change in color.

Application of Powdered Antispoilage Agent

H.H. Beck; U.S. Patent 3,390,001; June 25, 1968; assigned to Armour and Company provides a method and apparatus for treating cheese pieces with an anti-spoilage powder to obtain uniform application of the powder, while at the same time protecting the processing personnel from contact with the chemical dust.

In essence, the apparatus has a casing providing a treating chamber having an inlet and an outlet, an open mesh conveyor for advancing the cheese pieces through the inlet and outlet of the casing, a means for supplying powder in the lower portion of the casing, means for introducing air jets into the lower portion of the casing for agitating and suspending the powder, a baffle supported above the conveyor within the casing to confine powder-bearing air currents about the conveyor, and a suction withdrawal conduit communicating with the upper portion of the casing above the baffle.

Additional work designed to allow the application of a powdered antispoilage agent to cheese pieces is discussed by L.J. Hansen and M.B. Westover; U.S. Patent 3,328,176; June 27, 1967; assigned to Armour and Company. Apparatus is described for impinging on all sides of the cheese air streams having a pressure between 10 and 35 pounds per square inch which carry a dry antimicrobial food spoilage powder to embed particles of the powder in the cheese. The antimicrobial powder may comprise crystalline sorbic acid, salts of sorbic acid, metal salts of propionic acid, such as sodium and potassium propionates, hydroxyl esters of benzoic acid, and other known food spoilage inhibitors.

Dipping in Calcium Sorbate Suspension

E. Lück; U.S. Patent 3,391,008; July 2, 1968; assigned to Farbwerke Hoechst AG, Germany describes a preservation process which involves dipping hard cheeses into an aqueous suspension of calcium sorbate and a thickening agent and subsequently drying them. The behavior of calcium sorbate is fundamentally different from that of other sorbates, because after having been applied to the surface of cheese, it guarantees a much longer preservation from mold fungi than sorbic acid per se and the water-soluble salts of sorbic acid.

The calcium sorbate is used in the form of a suspension in an aqueous solution or suspension of a thickening agent, preferably cellulose ethers or cellulose esters such as carboxymethylcellulose and methylcellulose. The thickening agent only prevents the calcium sorbate from running off too rapidly and, furthermore, fixes the individual calcium sorbate particles on the surface of the food. The quantitative proportion of calcium sorbate to thickening agent, calculated, in each case, as weight of solid dry substance, in general is 50:1. In most cases, especially good results are obtained with a quantitative proportion of 20:1 to 5:1. The suspension contains per liter at least 10 grams of calcium sorbate. The content of thickening agent is determined by the increase in viscosity in the aqueous phase. The viscosity must be so high that the suspension only drops slowly off the surface.

Example: After having been compacted in forms and salted, provolone cheese was dipped for 10 to 15 seconds into a homogeneous suspension of the following composition, the suspension was allowed to drop off, the cheese was dried and further treated as usual. After the treatment the cheese was allowed to ripen for three months in the usual ripening rooms.

Composition of Suspension	Percent
Calcium sorbate	10.0
Carboxymethylcellulose (viscosity of a 2% aqueous solution at 20°C. about 2,000 centipoises in a Höppler viscometer)	1.5
Water	88.5

As control samples, cheese of the same production was used which had not been treated or had been treated with immersion baths in which the calcium sorbate was replaced by potassium sorbate or finely powdered sorbic acid. For each test 10 whole cheeses each having a weight of about 2 kilograms were used. The cheese was stored for a prolonged period of time.

	Number of Moldy Whole Cheeses, weeks held					
Treatment	1	3	5	8	10	12
Untreated samples	2	5	9	10	10	10
Sorbic acid	-	-	-	2	3	4
Potassium sorbate	-	-	-	3	3	4
Calcium sorbate	-	-	-	-	-	1

Since it is sometimes desirable to treat the surface of cheese with a preserving agent which does not contain a thickening agent because it dries in a very short period of time, E. Lück; U.S. Patent 3,486,909; December 30, 1969; assigned to Farbwerke Hoechst AG, Germany describes preserving cheese with a suspension of calcium sorbate which is prepared by mixing, while stirring continuously, an aqueous solution of potassium sorbate with an aqueous solution of calcium chloride. The cheese is coated with the suspension of calcium sorbate by immersion.

The aqueous suspension containing calcium sorbate is obtained by double reaction of aqueous solutions of metal sorbates with aqueous solutions of calcium salts. In the preferred reaction of potassium sorbate with calcium chloride potassium chloride is formed besides calcium sorbate. When sodium sorbate is used instead of potassium sorbate common salt is formed in addition to calcium sorbate. Either chloride is physiologically harmless and can, therefore, remain in the immersion suspension. Instead of calcium chloride physiologically unobjectionable calcium salts of other acids may be used, especially of organic acids, for example, calcium acetate.

The salts obtained in addition to calcium sorbate must also be physiologically unobjectionable. The sorbates and calcium salts used for the double reaction are suitably soluble in water to an extent such that at room temperature their aqueous solutions, prior to mixing them, contain sorbate and calcium ions in an amount corresponding to a content of 4 to 15% by weight of calcium sorbate in the suspension obtained after mixing the solutions.

The by-products potassium chloride and sodium chloride, which alone do not have an antimicrobe action or compared with sorbic acid and its salts have a minor action only, obviously intensify the effect of calcium sorbate as in practical tests the preserving agents proved to be more effective than those containing only calcium sorbate as preserving agent. Moreover, with the same content of calcium sorbate the suspensions have a lower tendency to sedimentation than suspensions of dry calcium sorbate in solutions of thickening agents. This effect appears unusual because it should be expected that thickening agents maintain substances in suspension for a longer period of time than salt solutions.

A further advantage of these preserving agents without thickening agent resides in their improved rheological properties. They dry very rapidly so that the treated cheese can be immediately transported to the ripening rooms. Compared with the use of compositions containing thickening agents and having a higher viscosity, these agents permit a considerable time savings.

It has been found that only calcium sorbate suspensions having the specified concentration of effective ingredient, preferably about 10% by weight, are useful. Compositions having a lower concentration of calcium sorbate do not offer any noticeable advantages over potassium sorbate solution and tend to sedimentation. Compositions having a higher content of calcium sorbate are unsuitable for use as immersion baths for mechanical and rheological reasons. The foodstuffs which have been preserved with the suspensions are quite stable. Hard cheese, for example, is stable for many months.

Example: An immersion bath was made by dissolving 11.5 kilograms of potassium sorbate in 64 liters of warm water and, separately, dissolving 4.3 liters of anhydrous calcium chloride in 20 liters of water. The two solutions were combined while continuously stirring. The suspension obtained contained about 10% of calcium sorbate and some potassium chloride.

Provolone cheese was compacted and salted and was then dipped for 10 to 15 seconds in the suspension, dried and further treated in usual manner. After the treatment the cheese was allowed to ripen for 3 months in the usual ripening rooms. There was used as control samples cheese of the same production which had not been treated, or which had been treated with immersion baths containing as active ingredient the same amount of potassium sorbate or finely powdered sorbic acid. In each test 10 whole cheese having a weight of about 2 kilograms were used. The cheese was stored for a prolonged period of time. The results are shown below.

	Number of mouldy whole cheese after (weeks)—					
	1	3	5	8	10	12
Untreated control samples	1	6	10	10	10	10
Cheese treated with sorbic acid				1	3	4
Cheese treated with potassium sorbate				3	3	3
Cheese treated with calcium sorbate						1

COATINGS

Gelatin and Metaphosphate Polymer

H.L. Keil, R.F. Hagen and R.W. Flaws; U.S. Patent 2,953,462; September 20, 1960; assigned to Armour and Company describe a coating composition incorporating gelatin and a metaphosphate polymer. This complex can be formed by dispersing the ingredients in an acidic aqueous solution which then can be used directly for coating the particular item desired. For example, the item to be coated can be dipped into the complex-containing solution to enrobe the item with a transparent, water-resistant coating which permits dehydration of the enrobed item if desired.

To prepare the solution gelatin and a metaphosphate polymer are mixed in an acidic aqueous solution. The acidity can be obtained through the use of a nontoxic acid, such as acetic, gluconic, lactic, hydrochloric, etc. The gelatin may be any gelatin in the range of 80 to 275 Bloom. For the metaphosphate polymer which, when added to an acidic aqueous solution becomes a metaphosphoric acid polymer, a wide variety of salts of metaphosphate polymers can be used. The sodium, calcium, potassium, or aluminum salts of a metaphosphate acid polymer, or any other suitable edible salt, should be employed.

The solution should be heated slightly to elevate the temperature above the melting-point of gelatin, about 42°C. Raising the temperature decreases the viscosity of the coating solution so that upon applying the coating, a thinner coat is achieved. A solution temperature of 42° to 60°C. is preferred. A quantity of acid is added to the aqueous solution so as to depress the pH to below 3.5, optimally, about 2.0.

Gelatin and a metaphosphate polymer, when present in acidic aqueous solution in a ratio of about 4 to 1, react to form a stable complex. Preferred, however, is a concentration of metaphosphate polymer which is about 5 to 50% of the amount of total solids, i.e., the ratio of gelatin to metaphosphate polymer may vary in the range of 19 to 1 to about 1 to 1. Excess of one or the other ingredient varies the characteristics of the coating somewhat, but in any case an important feature of the coating is the above-described complex. The composition of the complex is about 20 to 21% metaphosphate polymer after dialyzing coatings for 72 hours against distilled water.

One coating used contained 1,600 g. gelatin (265 Bloom), 400 g. sodium hexametaphosphate, 120 cc hydrochloric acid and 4,800 cc water. The total solids amounted to 28.9% and the pH was 2.15. In making the coating, the gelatin and the metaphosphate polymer in the form of sodium hexametaphosphate are placed in a glass-lined tank equipped with an anchor agitator. In a separate glass-lined vessel, the acid is mixed with water at about 50°C. The aqueous acid is added with stirring to the dry, mixed ingredients. The stirring produces, after a short period, a homogeneous product with no heat being required during this step. At this point, the pH of the solution is adjusted to about 2.0.

The item to be coated,can be mounted on any convenient hanging means which are then positioned over the dipping tank to immerse the product desired to be coated. It is only necessary that the item remain in the solution for a few seconds. Upon removel and standing in room temperature air, the coating is found to gel in about 1/2 minute. Permitting the item to stand for an additional period of an hour or two completes the drying of the coating. Another formulation of the coating was made up of 200 g. gelatin (265 Bloom), 200 g. sodium hexametaphosphate, 40 cc hydrochloric acid and 1,400 cc water. The total solids in the foregoing formulation were 21.7% and the pH was 1.9.

Example: Six 1-1/2 pound samples of cheese were provided, two being used for controls, two coated with the first formulation, and two with the second formulation, both formulations being applied by the dipping procedure outlined above. After three weeks' residence time in a refrigerator, the two sets of coated samples showed no surface mold, while substantial mold development was found on the untreated cheese samples.

Permeable Resin and Wax

A. Winterberg; U.S. Patent 2,961,322; November 22, 1960 describes the addition to artificial resins of an emulsion of a wax or waxes capable of absorbing water to provide films which dry quickly and are not affected by moisture. A film of polyvinyl acetate is affected by water to a high degree, which becomes particularly inconvenient in cases where the goods themselves contain a high degree of moisture and have also to be stored in moist rooms. Freshly manufactured cheese has to be packed, but the loaves are stored for ripening for a certain period. The loaves of cheese are very moist and they are stored in rooms where the relative humidity is very high, i.e., above 90%. In such cases polyvinyl acetate films do not dry easily.

An emulsion capable of being applied by dipping, spraying or brushing on is prepared from one of the known plastics and an appropriate wax such as lanette wax or beeswax. This emulsion is applied in a thin layer. While drying the thin layer, due to the hydrophilic wax it contains, it absorbs moisture from the cheese (to an extent as such moisture cannot escape through the pores of the film) while at the same time the water content of the plastic emulsion is absorbed. The absorption of moisture by and into the wax components of the films does not impair in any way the pleasing appearance of the wrapping. The cheese is not affected in any way.

By using the combination of plastics with waxes, the disadvantages of an emulsion of polyvinyl acetate are fully removed, namely: the film is not affected by moisture and does not change its properties under continued influence of moisture — color, appearance and surface conditions of the film remain constant; drying proceeds quickly and takes from 3 minutes to half an hour and the films dry in very humid surroundings up to relative humidity 90%. It is necessary to pack the cheese in such a manner that breathing should continue for a certain period and should then be reduced to a certain extent or should be stopped altogether so that the cheese will continue to mature.

The film is applied, the loaves of cheese are stored for ripening and after the conclusion of this period — between 6 weeks and 3 months — the packed loaves are exposed to an increased temperature — up to 60°C. — or up to the melting point of the wax used. In consequence of this heat treatment the wax particles melt and envelop the plastic particles, thereby entirely closing the pores of the covering layer. Thus this layer of film becomes impermeable to air, vapor or gas and further breathing of the cheese is prevented.

The polyvinyl acetate emulsion (a) is made up of 88 parts by weight of polyvinyl acetate of 60% dry substance and 12 parts by weight of a nontoxic plasticizer. The wax emulsion (b) contains 33 parts by weight of a hydrophilic wax and 67 parts by weight of water. The covering layer which allows breathing contains either: (1) 60 parts by weight of emulsion (a), 25 parts by weight of emulsion (b), 14 parts by weight of water and 1 part silicone antifoaming substance; or (2) 50 parts by weight of emulsion (a), 30 parts by weight of emulsion (b), 19 parts by weight of water and 1 part silicone antifoaming substance. These are well suited, owing to their viscosity, for cases where the covering layer is to be applied by being brushed onto the cheese.

Liquid Gelatinate of a Metal

H.L. Keil; U.S. Patent 2,971,849; February 14, 1961; assigned to Armour and Company provides a coating which will act as a grease barrier which may be used as the complete packaging material and will inhibit molding and oxidation. In the process the food is dipped, cascaded or brushed with a liquid gelatinate of a metal selected from the group consisting of aluminum and nontoxic heavy metals. The coating is dried after the food is enrobed. This coating will dry rapidly at room temperature to leave a nontacky surface. The finished coating is transparent, relatively tough and seals in any fat or oil in or on the foodstuff.

The metal gelatinate is prepared by introducing a salt of the metal into an aqueous gelatin solution. The metal reacts with the protein of the gelatin forform a complex termed a gelatinate. Any acid salt of the metal may be employed. A high grade gelatin, e.g., 275 Bloom, is preferably used because the larger molecules give a tougher coating. No heat is required for the reaction, but the temperature should be above that at which the gelatin will gel.

After the reaction is complete, which is substantially instantaneous, the food may be dipped in the resultant liquid gelatinate. The amount of the metal gelatinate in the dip will be determined primarily by the thickness of the coating to be applied. An excellent coating is obtained by using a solution of 20% solids. The preferable range of solids in the solution is between 5 and 30% solids by weight.

After the metal gelatinate coating has been applied, the coated food may be encased in a water and/or water vapor impermeable film. For example, a sheet of polyvinylidene chloride plastic (Saran) or rubber hydrochloride (Pliofilm) may be wrapped and sealed about the coated food. The coated food may be placed in a bag formed of one of the same films with the bag subsequently being sealed, with or without first vacuumizing the bag.

Example: Ten and six-tenths (10.6) pounds of 275 Bloom gelatin was mixed into 46.6 pounds of cold water and the lot heated in a water bath at about 180°F. until the solution reached 142°F. After skimming the foam from the solution, 2.12 pounds of aluminum sulfate (N.F. grade) dissolved in 4.24 pounds of hot water was stirred into the gelatin solution. Before using the resultant product as a dip it was cooled to about 118° to 120°F. This dip is used to coat cheese. The coating obtained is entirely free from grease and will remain that way even when not under refrigeration. It is substantially colorless and transparent.

N-n-Butylbenzamide in Aqueous Vinyl Acetate Coating

P.A. Tillet; U.S. Patent 3,090,689; May 21, 1963; assigned to Societe des Usines Chimiques Rhone-Poulenc, France provides an aqueous polymer emulsion containing N-n-butylbenzamide for application to cheese to form a film which serves as protection against fungal attack.

Suitable aqueous emulsions of polymers may be obtained by known emulsion polymerization methods, which may be carried out in the presence of protective colloids which are physiologically innocuous, such as gelatin or polyvinyl alcohol. Aqueous emulsions of polymer may also be prepared from polymers of another physical form by known means, such as crushing and dispersion of the crushed material in water or by precipitation in the presence of protective colloids, from organic solutions by the addition of water.

All homopolymers and copolymers known for use in the food industry for the coating of food products may be employed. The N-n-butylbenzamide has a plasticizing action on such polymers but if desired the compositions may contain other plasticizers.

To form satisfactory coatings on food products, the emulsions must have an appropriate consistency, which can readily be determined by preliminary test. The desired consistency can be obtained either by diluting commercial aqueous emulsions of polymers with water or by thickening them with noninjurious products such as sodium cellulose glycolate, sodium polyacrylate or methyl cellulose. The quantity of N-n-butylbenzamide in the aqueous emulsions of polymers will vary with the precise use to which the composition is to be put. A proportion of from 10 to 20% calculated on the weight of dry polymer is generally suitable.

The aqueous emulsion compositions may be used by any known method of applying a film coating to a product, for example, by dipping the product in the emulsion or by spraying the product with the emulsion, followed by drying the coating. Clear, transparent, glossy film coatings are obtained which are sufficiently adhesive to the products but can be peeled off when required. Such film coatings remain flexible and have remarkably good resistance to the action of molds, even after prolonged storage. They are particularly suitable for use in applying a protective coating to cheese of the "melted" or "pressed" type.

Example: Cheeses of the pressed type were dipped into an aqueous emulsion containing 85% of a 50% aqueous polyvinyl acetate emulsion and 15% N-n-butylbenzamide, drained and allowed to dry. There was thus obtained on the cheeses a flexible, clear coating. The coated cheeses were found to keep for several weeks without any attack by fungi under conditions in which uncoated cheeses were rapidly covered with mold.

Propylene Glycol, Gelatin and Water

G.R. Whitman and H. Rosenthal; U.S. Patent 3,556,814; January 19, 1971; assigned to General Foods Corporation describes a composition for use in the process of coating food products which is a solution consisting of propylene glycol, gelatin and water, the propylene glycol being present in about 60 to 90% by weight of the solution, the gelatin about 10 to 40% by weight, and the water less than about 5% by weight. The gelatin has a Bloom of at least 25 and is preferably in the range of 150 to 250 Bloom. The use of higher Bloom gelatin permits some water to be added later and included in the final composition without apparently affecting the properties of the coating or film.

To form the preferred composition the propylene glycol is heated to 75° to 100°C., and the gelatin added rapidly and continuously until there is no grittiness to the samples withdrawn from the resulting solution. The solution is then cooled to form a nontacky, warm water-soluble composition that has a melting point of about 50° to 75°C. Its melting point is always below 100°C. The solid composition is either reheated to a temperature about 80° to 140°C., or utilized directly in hot melt form after it has been prepared.

Example: Hot melts and solid compositions of matter were formed by adding 65 grams of 180 Bloom gelatin to 260 grams of ethylene glycol at 90°C. without the addition of water. The same hot melt was formed substituting 1,2,6-hexanetriol for the ethylene glycol, and here also a hot melt and solid composition were obtained in which total solution of the ingredients was indicated. The hot melts were applied to soft cheeses and Cheddar cheese blocks by dipping. Good protection against surface mold growth was exhibited and the films were easily removed by peeling them from the coated cheese.

PACKAGING

Gas- and Moisture-Proof Wrapper

F.O. Coffey, L.J. Hayhurst and G.H. Kraft; U.S. Patent 2,955,045; October 4, 1960; assigned to National Dairy Products Corporation describe a method of packaging pieces or units of cheese in substantially gas-proof and moisture-proof wrappers. The process is applicable to cheeses which do not develop a substantial amount of gas in their shelf life.

The process involves wrapping the cheese in a substantially gas-proof and moisture-proof wrapper, introducing an innocuous, nonmold-growth sustaining, gaseous media which includes a gas which is readily soluble in or absorbable by the cheese and a gas which is substantially insoluble and will not be absorbed by the cheese, sealing the package to provide an air tight package, and holding the sealed package under conditions which are not deleterious to cheese.

One innocuous gas which is soluble in the cheese and which may be used is carbon dioxide. Carbon dioxide is soluble in both the milk fat and the moisture at cheese storage temperatures, is relatively inexpensive, and does not support mold growth. The carbon dioxide is soluble in the cheese at varying rates depending on the shape and surface area of the cheese block being treated. Since the carbon dioxide when dissolved will be dispersed throughout the cheese, a unit which presents a thin section will absorb the gas more rapidly than that which presents a thick section. For example, a 1/2 pound unit measuring 1 inch by 3-5/8 inches by 3-5/8 inches will absorb the gas much slower than the same weight of unit which measures 1/2 inch by 3 5/8 inches by 7-1/4 inches. Temperature does not appear to have a significant effect.

The rate and degree of absorption of the soluble gas into the variety or type of cheese is, in part, related to the amount of soluble gas present in the cheese. Natural cheeses contain a certain amount of carbon dioxide and the amount of contained carbon dioxide is a function of the variety or type of cheese and the time elapsed between cutting and packaging of the cheese, as well as the physical shape.

Freshly cut cheese contains more carbon dioxide than does cheese which has been standing at room conditions and a thin cut will lose its gas more rapidly than a thick cut. Furthermore, Swiss cheese tends to develop a substantially greater amount of carbon dioxide during curing than do other natural cheeses, as for example, Cheddar cheese, and consequently, Swiss cheese normally has more solubilized CO_2. Process cheese has a small amount of CO_2 due to the conditions it is subjected to in manufacturing where it is heated above 140°F. to melt the cheese. It is then blended with emulsifier and packaged. Such conditions result in the volatilization of CO_2.

The age of the cheese before cutting is also a factor in the degree of carbon dioxide which is in the cheese. Temperature is a further factor in determining the amount of CO_2 which is in the cheese at any given time after cutting.

In order to provide a package of cheese in which the wrapper bears the desired relation to the cheese, a particular ratio between the soluble and insoluble gas at the time of wrapping must be established in the package. This ratio is a function of the abovementioned factors.

If puffing of the wrapper is to be avoided and the wrapper is not to be tightly pulled around the cheese, the relative amount of CO_2 in the gas in the wrapper surrounding the cheese should be higher with cheeses having higher amounts of carbon dioxide in the cheese.

Thus, with Swiss cheese, the surrounding gas should contain a substantial amount of carbon dioxide but with Cheddar cheese, the surrounding gas may contain lesser amounts of carbon dioxide. With normal cutting and packaging procedures, the gas used for packaging Swiss cheese should comprise between 50 and 80 percent carbon dioxide gas, the remainder nitrogen. In the case of Cheddar cheese, the packaging gas should comprise between 25 and 70 percent carbon dioxide gas, the remainder nitrogen. Experience with brick cheese indicates a gas mixture from 30 to 70 percent carbon dioxide with the remainder being nitrogen whereas in the case of washed curd cheese the amount of carbon dioxide in the packaging gas should be between 40 and 70 percent. In the case of process cheese, any amount of carbon dioxide in the packaging gas will cause the wrapper to draw around the cheese, i.e., the presence of any carbon dioxide in the surrounding gas results in absorption of the carbon dioxide.

Various films may be employed in wrapping the cheese, but the film should be impermeable to carbon dioxide, nitrogen and oxygen. Such films as Saran (a polyvinylidinechloride), Vinylite (polyvinylacetate, polyvinylchloride, or copolymer thereof), or various laminated products such as Saran laminated to a film of polyethylene, etc. may be used. The packages may be sealed by heat sealing or by electronic means, depending upon the properties of the wrapper being employed.

The air in the package may be removed after the package is partially sealed by vacuum, and the soluble gas or the mixture of soluble gas with insoluble gas may be injected into the package as the vacuum is broken, after which the package may be sealed. This may be accomplished on various types of known apparatus.

Example: A block of sliced (1/8 inch slices) pasteurized process cheese in the form of a rectangular unit measuring approximately 1 inch by 3 5/8 inches by 3 5/8 inches and weighing approximately 1/2 pound is wrapped in a layer of K_2O_2 wrapping material which is a laminated wrapping material. The wrapping material comprises one layer of Saran to which is combined polyethylene which is heat sealable. One end seal and a longitudinal seal are effected by heat sealing bars which soften the polyethylene thereby partially sealing the package.

The partially sealed package is then subjected to a vacuum of about 27 inches of mercury in a vacuum chamber to remove air from the package, the volume of free or excess space in the package being such that after the second end seal is effected, the difference between the volume of the cheese and the capacity of the package is 55 cc (this amount of cheese can absorb about 70 cc of carbon dioxide). The vacuum is broken with a mixture of carbon dioxide gas (70 percent) and nitrogen (30 percent) and the second end seal is effected by heat sealing bars.

The package, after sealing, is loose in appearance and does not cling to the cheese. This package is placed in a cooler at 45°F. for 24 hours and upon examination at that time the wrapper is found to be drawn around the cheese. In the event that an improper seal is made air is drawn in through the fault in the seal and the wrapper will not be so drawn about the cheese.

Incorporation of Aerobic Organisms

J.C. Tupper; U.S. Patent 2,987,403; June 6, 1961 describes a method and package for protecting food from mold and oxidation which involves the incorporation in the package of any appropriate aerobic organisms such as bacteria or yeasts. Yeasts themselves are highly aerobic, as are the bacteria incidentally found therein. Brewers' yeast, therefore, is a desirable inoculant not only in itself, but because of the aerobic bacteria normally found in commercial brewers' yeast. The organisms consume the available oxygen, and any additional oxygen released from the product, and then their growth ceases. Meantime, in the oxygen-free package, mold cannot grow and the oxidation of the product cannot occur.

The character of the container or wrap is immaterial as long as it is substantially impervious. Any of the conventional wrapping materials such as Saran, polyethylene, Saran-coated cellophane, or cellophane coated with synthetic rubber and amorphous wax may be used. It is not absolutely necessary that the wrapper be completely impervious as long as it does not admit oxygen more rapidly than it can be consumed by the aerobic organism content of the package. Any aerobic organism will suffice if it is capable of consuming the oxygen from a food package which has been substantially sealed.

The organism inoculation is prepared by adding one pound of brewers' yeast to thirty pounds of water of acceptable standards of purity. Thorough mixing suspends the yeast in the water. The cheese is prepared in the usual way by cleaning and removing the rind. It may be in blocks or slices of any size or shape. The prepared cheese is dipped briefly into, or lightly sprayed with, the suspension of yeast in water and upon removal, the excess liquid is drained off. The cheese is then placed in a pouch or other wrapper and sealed.

After the packages are completed and either before or after they are placed in the shipping container, they are held at a temperature favorable to growth of the inoculating organism until the oxygen is consumed and replaced by carbon dioxide. For example, in the case of brewers' yeast inoculation of cheese, the product may be kept at about 65°F. for about twenty-four hours. If the product is one which is ordinarily refrigerated, as is the case with cheese, the product is then placed in ordinary refrigeration, usually 33° and 40°F.

Crustless Cheese Pieces

The object of the process described by A. Bertozzi; U.S. Patent 3,060,034; Oct. 23, 1962; assigned to SpA Abele Bertozzi, Italy is to provide a method of preserving hard or semi-hard cheese, especially Parmesan cheese in crustless fragments of a predetermined size, which maintain the characteristic properties of the just cut cheese such as results from splitting cheese wheels having a crust.

These cheese wheels, of an irregular cylindrical form, are taken after an adequate conditioning period of not less than one year, cleaned, deprived of their crust, subdivided to polyhedral fragments of a regular form having smooth surfaces of the desired size. The fragments are confined within adherent protective wrappers, and are subsequently sterilized together with their wrappers by submitting them to the action of a source of heat to seal the surfaces of the cheese fragment which also causes the wrapper to thoroughly adhere to the fragment surfaces. Simultaneously, any air residue between the protective wrapper and fragment surfaces is driven off. The method can be carried out either in an aseptic or a nonaseptic but hygienically satisfactory environment.

Both comprise the preparatory step of removing the crust from the cheese wheel on a lathe. Of course turning is effected to remove the crust only throughout its thickness avoiding any waste of valuable paste which would necessarily arise by cutting by conventional methods. This operation is carried out in the nonaseptic but hygienically satisfactory environment. The cheese wheel deprived of its crust turned to a cylindrical shape is subsequently conveyed through a conduit having a sterilizing action through admission of bactericidal gases, such as formaldehyde or ozone into the conduit from which it is brought to an environment of an aseptic nature afforded by the action of ultraviolet rays.

The cheese cylinder treated as above is subdivided into fragments of a predetermined shape and weight by means of strong wire, the surface of the resulting fragments being almost perfectly smooth, free from discontinuities, any cheese waste being avoided. The cutting wire is made sterile by the same means as employed for sterilizing the environment.

The last step of the method if also carried out in an aseptic environment and consists of packaging the cheese fragments by means of protective sheets which are caused to adhere to the cheese surfaces by the action of heat in order to avoid leaving any air pocket between the cheese and protective wrapper. The resulting protection replaces the natural function of the previously removed crust in all aspects, namely preventing any contamination from the outside and any decay due to oxidation (staling).

Alternatively all the above described operations are carried out in a nonaseptic but hygienically satisfactory environment, preservation of the packaged fragment being obtained through simultaneous sterilization of the fragment and its wrapper by employing a source of heat which raises the surface temperature to not less than 80°C. during 10 seconds, care being taken that a further rise in temperature necessitates a gradual reduction of the period of treatment. An apparatus suitable for carrying out this operation is fully illustrated and described by the author in the patent.

Sorbic Acid and Foil

A. Winterberg; U.S. Patent 3,275,454; September 27, 1966 finds that a coating having practically unlimited protective properties against molds can be prepared by coating the cheese with a coating in which is incorporated the antimycotic agent, preferably crystalline sorbic acid. The coating, preferably a synthetic resin, serves as an adhesive for a further gas-impermeable coating or sheet of plastic material or metal foil. A whole small cheese can be coated, giving a protective coating of adequate duration for all practical purposes.

The process allows whole items to be treated in one operation, combining the coating with the adhesive-sorbic acid mixture (i.e., the synthetic resin containing the sorbic acid) and applying the impermeable cover onto this. The adhesive, i.e., the synthetic resin film, may be applied by dipping if the item is small, or by spraying or brushing if the item is large. The impermeable coat is applied immediately; its ends are folded over, then secured to each other by means of the composite adhesive. The protection due to the sorbic acid lasts practically indefinitely as it is prevented from evaporating through the impermeable outer cover.

Example: 80 parts by weight of a polyvinyl acetate emulsion of a solid content of 50% of dry polyvinyl acetate, 6 to 8 parts by weight of a nontoxic plasticizer, 5 to 12 parts by weight of finely granulated crystalline sorbic acid, and 5 to 10 parts by weight of water are thoroughly mixed. The quantity of the water depends on the temperature of application; during the hotter season less water is required as the emulsion will be sufficiently fluid for the intended application.

A whole cheese was dipped into the emulsion and immediately removed. A thin aluminum foil was wrapped on the still wet cheese; its ends were folded over, covered with the emulsion and attached to each other. During a period of storage of over eighteen months, no molds developed and the cheese retained its appearance and taste.

PROCESS CHEESE AND CHEESE PRODUCTS

PROCESS CHEESE ADJUVANTS

Sodium Aluminum Phosphate Emulsifier

R.M. Lauck, J.W. Tucker and R.E. Vanstrom; U.S. Patent 3,097,949; July 16, 1963 and U.S. Patent 3,244,535; April 5, 1966; assigned to Stauffer Chemical Company describe the process of making an edible pasteurized process cheese composition containing sodium aluminum phosphate. Among the sodium aluminum phosphates which may be used are the known acidic crystalline and amorphous modifications as well as reaction products of the empirical composition: $xNa_2O \cdot yAl_2O_3 \cdot 8P_2O_5 \cdot zH_2O$ where x is a number from 6 to 15, y is a number from 1.5 to 4.5 and z is a number from 4 to 40, and their water-insoluble fractions.

The reaction products and their water-insoluble fractions are initially prepared by reacting aluminum oxide, aluminum hydroxide (hydrated alumina) or sodium aluminate, and sodium hydroxide or sodium carbonate with concentrated phosphoric acid. This reaction is highly exothermic and proceeds to completion in a matter of minutes. The resulting reaction product is a mixture consisting of a water-soluble fraction and a water-insoluble fraction; the relative proportions being dependent upon at least several factors, including mainly, the aluminum content of the reaction mixture, the speed at which the reaction is accomplished, and the degree of agitation during reaction.

Reaction products with higher aluminum content will normally have higher ratios of insolubles to solubles (up to about 2:1) while those with lower aluminum have proportionally lower ratios (down to about 1:2). The soluble fraction is composed of an intimate mixture of two or more sodium orthophosphates. From the hydrolysis characteristics of the insoluble fraction, it appears to be composed of two or more water-insoluble sodium aluminum phosphates.

These compositions provide different pH levels furnishing flexibility in changing the characteristics of the cheese. An immediate development of melt is produced when cheese is emulsified. Crystal formation is completely eliminated at the 3% maximum allowable limit when using these emulsifying or melting salts. Further, the emulsification produced is superior since there is less tendency for leakage of fats during heating of a cheese containing the compositions than with an equivalent amount of disodium orthophosphate.

Example: Three different reaction product compositions were prepared. These compositions were then used in process cheese in their original form, that is, without leaching off the soluble fraction. The analyses of these three compositions are as follows:

Composition I — 14.4% Na; 6.5% Al; 20.3% P; and 22.5% loss on ignition

Composition II — 21.9% Na; 6.2% Al; 18.6% P; and 19.6% loss on ignition

Composition III — 23.3% Na; 5.4% Al; 17.3% P; and 20.1% loss on ignition

A blend of natural Wisconsin Cheddar cheeses was first formulated using 15% mild cheese, 70% medium cheese , and 15% aged cheese. The cheese was ground through a meat grinder with a 1/8 inch plate. Forty pounds of cheese were used for each test. One pound of each of the various phosphate emulsifying agents was added just before addition of the last of the cheese to the cooker. The cooker was a forty pound pilot plant unit with a screw for mixing and means for injection of steam. The screw was driven through an essentially closed cylinder and therefore the cheese could not easily recirculate or escape its action.

Each cheese batch was normally cooked for a period of between three to seven minutes (after phosphate addition) to 160°F. The cheese was then mixed one minute at 160°F., poured into five pound cheese cartons, and heat-sealed by inverting the box with liner. The final cheese was pre-cooled at room temperature and held overnight at 45°F. in a cooler. The samples were then sliced on a rotating circular blade type slicer, packaged in moisture-impermeable film, and stored for a period of one month under temperatures of 90°F., room temperature, 55°F., and 35°F. The resulting properties of the test cheeses are summarized in the following table.

Emulsifying Agent(s) (2.5% of Final Cheese)	Cheese, pH	Melt, percent	Bloom (Hardness) Reading Initial	Bloom (Hardness) Reading Final (35°F.)	Break Characteristics	Crystals One Month
DSP	5.57	21	336	328	Straight	Moderate
Composition I	5.20	10	518	408	Mod. jagged	None
Composition I plus 20% TSP	5.43	11	351	346	V. sl. jagged	None
Composition I plus 40% STPP	5.29	4	500+	500+	Trace jagged	None
Composition I plus 20% SHMP	5.09	7	500+	450	Jagged	None
Composition II	5.53	26	284	276	Sl. jagged	None
Composition II plus 20% TSP	5.72	8	224	248	Sl. jagged	None
Composition II plus 20% SHMP	5.32	1	500+	348	Straight	None
Composition III	5.72	18	257	267	V. sl. jagged	None
Composition III plus 20% SHMP	5.51	8	335	317	Straight	None
DSP	5.59	7	334	312	Trace jagged	Slight

DSP = disodium orthophosphate. TSP = trisodium phosphate STPP = sodium tripolyphosphate. SHMP = sodium hexametaphosphate. Mod. = moderate. Sl. = slight. V. sl. = very slight.

High Sodium Content Sodium Aluminum Phosphate Emulsifier

R.N. Bell; U.S. Patent 3,726,960; April 10, 1973; assigned to Stauffer Chemical Company describes a sodium aluminum phosphate composition of relatively high sodium content which has as a cheese emulsifier, the recognized advantages of sodium aluminum phosphates and in addition imparts to cheeses the quality of maintaining their melt properties over a long period of time.

This sodium aluminum phosphate has the empirical formula $Na_8Al_2(OH)_2(PO_4)_4$ which will be referred to as an (8:2:4) phosphate composition. The sodium aluminum phosphate can be prepared either by crystallization from solution or by a total evaporation process.

The product is generally prepared by reacting aluminum phosphate with an alkaline solution of sodium phosphate. The aluminum phosphate may be either an acidic solution of monoaluminum, $Al(H_2PO_4)_3$, or a thick slurry of freshly precipitated aluminum phosphate, $AlPO_4$.

The (8:2:4) phosphate composition, is formed when the crystallization temperature is high, preferably near 100°C. If the temperature is considerably lower, such as about 30°C., an entirely different phosphate composition is formed. Such a phosphate composition would be a (4:3:3) phosphate composition as described by R.H. Bell; U.S. Patent 3,729,546; April 24, 1973; assigned to Stauffer Chemical Company.

The (4:3:3) phosphate composition, is formed when the crystallization temperature is low, preferably at room temperature, or about 30°C. Both the (8:2:4) and the (4:3:3) phosphate compositions exhibited good qualities.

As cheese emulsifying agents, they have provided improved properties in process cheese such as to stabilize cheese compositions to which they are added. Also, these phosphate compositions have shown some melt increase after 9 days' storage and were approximately equivalent to disodium phosphate after 42 days' storage either at 50°C. or at refrigerated temperatures. The following examples illustrate the process and the aluminum phosphate reaction products.

Example 1: Preparation of (8:2:4) Sodium Aluminum Phosphate by Crystallization — A first solution (A) was prepared by stirring 243 g. of CP 50% sodium hydroxide into 1,400 milliliters of distilled water in a 2,000 milliliter Pyrex beaker. 400 g. of anhydrous disodium phosphate was then added and dissolved. A second solution (B) had previously been prepared by adding slowly while stirring, 18.2 g. of 200 mesh aluminum hydrate (C-31) to 143 g. of hot (100°C.) 75% phosphoric acid. Solution (A) already warm, was heated to 90°C. Solution (B) was diluted with 199 milliliters of distilled water and added slowly to solution (A) while stirring. Two milliliters of the resulting solution diluted to 150 milliliters had a pH of about 10.8.

The hot solution (mixture of A and B solutions) was filtered through a coarse fritted glass filter and transferred to a 5,000 milliliter 3-neck flask fitted with a heating mantle, reflux condenser, and mechanical stirrer. It was held at 95° to 100°C. and stirred for three days while crystallizing. The resulting slurry was cooled to about 50°C., and filtered with vacuum on a coarse fritted glass filter. Two milliliters of the filtrate diluted to 150 milliliters had a pH of about 11.2. The crystals were washed quickly several times with cool distilled water, then with a 10% acetone solution. They were then reslurried in 95% acetone. The slightly damp product was finally dried overnight under vacuum at about 55°C. The dried product weighed 111 g. This represents a 99% yield based on the aluminum oxide.

Example 2: Preparation of 8:2:4 Sodium Aluminum Phosphate by Total Evaporation — 162 g. of anhydrous disodium phosphate was dissolved in a solution of 72 g. of 85% phosphoric acid in 450 ml. of distilled water. This solution was heated to approximately 80°C. and added slowly with good agitation to 102 g. of reagent grade sodium aluminate. The slurry thus formed was dried by pouring it onto a stainless steel surface preheated to approximately 150°C. The dried product was removed from the stainless steel surface and dried overnight at 102°C.

Example 3: Preparation of (4:3:3) Sodium Aluminum Phosphate by Crystallization — A first solution (A) was prepared by stirring 466 g. of sodium hydroxide (50%) into 1,700 milliliters of distilled water in a 2,000 milliliter Pyrex beaker. 340 g. of anhydrous monosodium phosphate (NaH_2PO_4) was then added and dissolved. The resulting solution (A) was filtered through a coarse fritted glass filter and reheated to 80°C.

A second solution (B) was prepared by adding slowly, while stirring, 18.2 g. of 200 mesh aluminum hydrate to 143 g. of hot (100°C.) phosphoric acid (75%). Then, 218 g. of this 50% monoaluminum phosphate solution was diluted with 100 milliliters of distilled water. This was added slowly to solution (A) while stirring. The final solution was filitered through a coarse fritted glass filter and cooled to room temperature. Two milliliters of the solution diluted to 150 milliliters had pH of about 10.9. The liquor was transferred to a 2,000 milliliter beaker and held at 30°C. while stirring with a magnetic stirrer for 6 days.

The resulting slurry was filtered with vacuum on a coarse fritted glass filter, washed quickly several times with cool distilled water then with 10% acetone and finally with 95% acetone. The slightly damp product was dried overnight under vacuum at 50°C. The dried product which passed through a 325 mesh sieve without milling weighed 70 g. This represented an 86% yield on the basis of the Al_2O_3.

Example 4: Preparation of (4:3:3) Sodium Aluminum Phosphate by Total Evaporation — A phosphoric acid solution was prepared by the addition of 65 g. of 85% H_3PO_4 to 100 ml. of distilled water. This solution was added slowly, with agitation, to a solution of 65 g. of reagent grade sodium aluminate in 100 ml. of distilled water. The rate of addition was such that with some external cooling the slurry temperature was maintained below 60°C. The final slurry was dried under vacuum at 40°C. When apparently dry, it was removed, crushed and redried for four hours at 40°C. under vacuum.

Crystalline Canthaxanthin Coloring Agent

The process describes by B. Borenstein; U.S. Patent 3,528,822; September 15, 1970; assigned to Hoffmann-La Roche Inc., is concerned with a composition containing canthaxanthin which is used in coloring process cheese.

The process is predicated on the finding that crystalline canthaxanthin, a compound which is insoluble or substantially insoluble in most nontoxic edible solvents, is soluble in diacetylated monoglycerides. These compositions are prepared by adding crystalline canthaxanthin to a diacetylated monoglyceride and dissolving them therein. When added to process cheese, such solutions impart an attractive color to the cheese which withstands fading when the cheese is maintained at normal refrigerator temperatures, even for extended periods of time.

Example 1: 0.40 g. of canthaxanthin was added to 100 g. of diacetylated monoglyceride containing about 0.02% by weight of butylated hydroxy anisole, about 0.006% by weight of proply gallate and about 0.004% by weight of citric acid. Myvacet 9-40 was the diacetylated monoglyceride employed. The mixture was heated to a temperature of 75°C. and it was maintained at that temperature until dissolution of the canthaxanthin in the solvent was completed.

The solution was evaluated as a colorant for process cheese. One and one-half pounds of uncolored process cheese were broken into small pieces and melted in a double boiler. To the melted cheese, 1.25 g. of the canthaxanthin solution produced was added. The amount of canthaxanthin solution was sufficient to provide 3.3 mg. of canthaxanthin per pound of process cheese. The cheese was poured into molds where it was allowed to cool to room temperature. The process cheese obtained, had an attractive cheese color resembling the hue of cheese colored with a high annatto-content, annatto-paprika mixture.

To determine color stability of the process cheese treated with the canthaxanthin solution, the cheese was maintained at refrigerator temperature for a period of nine months. After about three months at refrigerator temperature, the process cheese was assayed and it was found to have retained about 79% of its initially assayed (3.1 mg./pound) canthaxanthin potency. At the end of nine months, at refrigerator temperature, it was found that the process cheese had retained about 75% of its initially assayed canthaxanthin potency.

Example 2: In this example, a ratio of 1.88 g. of the canthaxanthin-containing solution was added to one and one-half pounds of process cheese to provide a product comprising 4.95 mg. of canthaxanthin per pound of process cheese. This imparted an attractive color to the process cheese. The cheese resembled that colored with a high annatto-content annatto-paprika. The cheese was stored at refrigerator temperature for a period of up to nine months. It was found that at the end of three months at refrigerator temperature, the cheese had retained about 69% of its initially assayed (4.87 mg./pound) canthaxanthin content. At the end of the nine month period, the cheese had retained about 65% of its initially assayed canthaxanthin potency.

Phosphate Modified Starch Thickener

In the process described by H.-A. Rohlfs, W. Koch and G. Scheurer; U.S. Patent 3,615,586; October 26, 1971; assigned to Benckiser-Knapsack GmbH, Germany process cheese and pasteurized cheese of superior texture, aroma, and keeping qualities are prepared by melting the starting cheese with a suitable melting salt and a phosphate-modified starch as thickening agent. The amounts of phosphate-modified starch in such process cheese products are between 1.0 and 2.5% and the total amount of melting salt and phosphate-modified starch between 2.5 and 4.0%.

The process uses a conventionally employed melting or emulsifying salt together with phosphate-modified starch to replace part of the melting salt. The addition of phosphate-modified starch as thickener avoids all the disadvantages encountered with other thickening agents.

Any phosphate-modified starch may be used. Especially suitable is one obtained, by mixing 100 kg. of corn starch in a kneading device, with a solution of 5.56 kg. of sodium tripolyphosphate and 4.17 kg. of disodium diphosphate in 58 l. of water. The amount of water used for dissolving the phosphates is preferably so low that the starch is merely moistened and not converted into a paste, and is dependent upon the solubility of the phosphates used. Usually amounts between 50 and 60 l. of water for 100 kg. of starch are sufficient. The resulting moist mixture is dried at a temperature below 60°C. so that it does not glutinize.

Preferably the moisture content of such dried phosphate-modified starch does not substantially exceed about 10%. The predried starch is then heated to 150° to 170°C. Heating is discontinued as soon as the viscosity of the phosphate-modified starch stirred in an amount of 2% into distilled water has attained a maximum, i.e., does not increase further. This can readily be determined by taking samples from time to time during heating.

The resultant process cheese products have a paste-like consistency, are readily spreadable, have retained the natural cheese aroma, and do not have any off-flavor or off-taste. The quality of the processed cheese is not affected by the manner in which the phosphate-modified starch is added, whether in the dry state or in preswollen form. The amount of phosphate-modified starch is between 1 and 2.5%, calculated for the body of cheese to be processed. The total amount of phosphate-modified starch and melting salt is between about 2.5 and about 4%, calculated for the body of cheese to be processed. Excellent results are achieved with about 2% of phosphate-modified starch and with a total amount of phosphate-modified starch and melting salt of about 3.5%.

An advantage achieved by the process is the improved keeping quality of the resulting process cheese. Process cheese products which contain the usual known thickening agents, for instance, carob bean flour, show strong microbial growth when stored for 20 days at 37°C. In contrast, process cheese products prepared with the addition of phosphate-modified starch were free of microbial growth even when stored for 30 days under the same temperature conditions.

Example 1: To a mixture of 50 kg. of Cheddar cheese and 50 kg. of Emmenthaler cheese there are mixed 1.5 kg. of the melting salt which is a sodium polyphosphate with 59.5% P_2O_5 and of a pH value of 9.1, and 2 kg. of phosphate-modified starch, which has been dispersed in 50 kg. of water and heated to 85°C. for 5 minutes before the addition. The resulting cheese mixture with the melting salt and phosphate-modified starch is melted and processed at 85°C. for 7 minutes while stirring. A smooth, nonropy cheese product which can readily be packaged, is obtained. Half of the resulting process cheese product is homogenized. Both halves are then packaged in aluminum containers which are hermetically sealed.

Example 2: The amounts and composition of cheese-starting material and melting salt as well as the procedure are the same as given in example 1. The phosphate-modified starch, however, is replaced by 1 kg. of carob bean flour. A slightly ropy melt is obtained, half of which is homogenized. Both halves are then packaged in aluminum containers which are hermetically sealed.

The process cheese products obtained were organoleptically tested 3 days after their preparation. The results are given in the following table.

Process cheese prepared according to Example—	Consistency	Taste (Aroma)
1a (not homogenized)	Well spreadable, smooth.	Corresponding to its type, fully aromatic.
1b (homogenized)	do	Corresponding to its type, fully aromatic, slightly inferior to cheese product of Example 1a.
2a (not homogenized)	Pudding-like, elastic, not as well spreadable as the cheese product of Example 1a.	Blank, only slightly aromatic and not corresponding to its type, sandy.
2b (homogenized)	About like the cheese product of Example 2a.	About like the cheese product of Example 2a but somewhat less sandy.

Proteolytic Enzyme Emulsifier

In the process described by T.P. Kichline and L.G. Scharpf; U.S. Patent 3,635,733; January 18, 1972; assigned to Monsanto Company pasteurized process cheese is prepared by forming an intimate mixture containing natural cheese, water, an emulsifying agent and a proteolytic enzyme. The mixture is then pasteurized and cooled. Reworked pasteurized process cheese may comprise a portion of the mixture. The process conditions are fully described in the following example.

Example: Sixteen batch runs were carried out with each batch containing approximately 800 g. of ground cheese having the composition shown in the table. The separate batches were introduced individually into a preheated (150°F.) one-quart steam-jacketed Baker Perkins mixer. After mixing with a single-blade mechanical agitator was begun, appropriate quantities of distilled water (to yield the end moisture content) were added followed by the addition of an enzyme-emulsifying agent mixture containing approximately (a) 5 g. of disodium phosphate duohydrate, (b) 5 g. of trisodium phosphate dodecahydrate, and (c) appropriate quantities (to yield the enzyme level specified; Runs 1-5 contained no enzyme) of the proteolytic enzyme Rhozyme P-11, of fungal origin with a specific acitivity of 110,000 protease units per gram of enzyme. (One protease unit equal to 0.5 milligrams of trichloroacetic acid soluble tyrosine liberated from casein on digestion for 10 minutes at 37°C.)

The mixing was continued for 5 minutes to achieve an end (or "cooker out") temperature of 165° to 168°F. (It took approximately 8 and 10 minutes, respectively, to achieve an end temperature of 175°F. and 186°F. Runs 13-16). The cheese temperature was monitored during mixing by means of a thermocouple which was inserted into the side of the mixing chamber and was connected to a temperature recorder. At the end of this pasteurization or heating and mixing period, which was conducted at atmospheric pressure, the separate samples were poured into individual Teflon-lined trays, covered with Teflon sheets, rolled out to approximately one-eighth-inch thickness and then cooled down to refrigerator temperature of approximately 10°F. After the pasteurized process cheese melt had sufficiently solidified, it was cut into standard size slices, that is, a 3.5 inch x 3.5 inch square. The results of the evaluation of the cheeses are also shown in the table.

As can be seen from the data, the results of the use of a proteolytic enzyme in the preparation of pasteurized process cheese from natural cheese, either with or without the use of rework pasteurized process cheese, yields a process cheese which is an improvement over the process cheese produced by methods which did not use a proteoylic enzyme. Specifically, it will be noted that in Run Nos. 1-5 no enzyme was added. In Run Nos. 6-8, the use of an enzyme at different levels significantly improved the overall characteristics of the end product. For example, the penetration, which is an indication of softness, increased from a low of 10 mm. (Run 1 — no enzyme) to 14, 18 and 21 mm., respectively (Runs 6, 7 and 8 — enzyme present) which results in respectively, an improvement of 40%, 80%, and 110% in softness. Similar improvements are also vividly apparent from the other data given.

	Enzyme based on percent level in emulsifier	Cheese components, percent			pH of cheese slurry	Cooker out temperature, ° F
		Natural[1] barrel	Natural[1] cheddar	Rework[2]		
Batch Run Number:						
1	0	50	50	0	5.6	167
2	0	50	40	10	5.7	167
3	0	40	35	25	5.5	167
4	0	30	20	50	5.8	168
5	0	20	10	70	6.0	168
6	0.1	50	50	0	5.7	167
7	0.5	50	50	0	5.8	168
8	1.0	50	50	0	5.4	168
9	1.0	50	40	10	5.8	165
10	1.0	40	35	25	5.9	168
11	1.0	30	20	50	5.6	167
12	1.0	20	10	70	6.0	167
13	0.5	50	50	0	5.8	175
14	0.5	40	35	25	5.7	175
15	0.5	50	50	0	5.7	186
16	0.5	40	35	25	5.7	187

	Moisture in final product, percent	Flowability of melt	Penetration Millimeters, mm.	Melt spread	Oil off	Melt text	Break	Grain	Slice separation
Batch Run Number:									
1	40.0	1.75	10	24	2.0	2.0	2.25	2.0	2.0
2	40.0	1.50	9	21	1.25	1.50	1.50	1.75	1.0
3	39.5	1.50	7	16	1.0	1.25	1.0	1.0	.75
4	39.8	1.00	5	13	.75	.75	.50	.75	.50
5	39.1	.50	4	11	.50	.25	.25	25	.25
6	39.7	2.5	14	28	2.5	2.5	2.75	2.5	2.75
7	40.1	3.0	18	36	3.0	3.0	3.0	3.0	3.0
8	40.0	3.0	21	41	3.0	3.0	3.0	3.0	3.0
9	40.1	3.0	20	38	3.0	3.0	3.0	3.0	3.0
10	40.1	3.0	18	35	3.0	3.0	3.0	3.0	3.0
11	39.5	2.75	16	25	2.75	2.75	2.75	2.75	2.75
12	39.9	2.50	13	20	2.75	2.75	2.75	2.50	2.50
13	44	3.0	23	42	3.0	3.0	(3)	(3)	(3)
14	44	2.75	18	34	3.0	2.75	(3)	(3)	(3)
15	60	3.0	26	43	3.0	3.0	(3)	(3)	(3)
16	61	3.0	18	30	2.75	2.50	(3)	(3)	(3)

[1] Both barrel and cheddar cheese prepared by the use of rennet as the enzyme-milk curdling agent.
[2] The rework consisted of 50% by weight process barrel cheese and 50% by weight process cheddar cheese.
[3] These tests are not applicable to the process cheese spread and food products.

In further work on the process just described T.P. Kichline and L.G. Scharpf; U.S. Patent 3,692,630; September 19, 1972; assigned to Monsanto Company specify the necessary amount of enzyme for their process cheese. In this process they describe a composition for preparing process cheese which comprises an intimate mixture of a process cheese emulsifying agent in an amount of at least 90% of the composition and a proteolytic enzyme from bacterial or fungal source in an amount to provide 1 casein unit protease activity to about 8,000 casein units of protease activity per gram of emulsifying agent.

In previous work on suitable cheese emulsifiers for process cheese T.P. Kichline, N.E. Stahlheber and J.L. Vetter; U.S. Patent 3,337,347; August 22, 1967; assigned to Monsanto Company describes the use of alkali metal aluminum phosphate compositions of relatively high aluminum content.

These compositions although relatively high in aluminum content, have a relatively high ratio of soluble fractions to insoluble fractions. In fact, the ratio of soluble fractions to insoluble fractions is at least equal to and in most cases higher than ratio of soluble fractions to insoluble fractions of the sodium aluminum phosphate materials of relatively low aluminum content.

The alkali metal aluminum phosphate compositions are characterized by the formula $xM_2O \cdot yAl_2O_3 \cdot 8P_2O_5 \cdot zH_2O$ in which M represents alkali metal, x is a number between 5 and 30 inclusive, y is a number between 4.7 and 15.0 inclusive, and z is a number between 0 and 50 inclusive.

The preferred compositions are substantially amorphous, i.e., an X-ray pattern exhibits no characteristic lines and if lines are present they are usually those chacteristic for dialkali metal orthophosphate, and/or trialkali metal orthophosphate. It should be noted, however, that these compositions can be formed in the crystalline state and although the crystalline forms are useful as cheese emulsifiers they do not appear, in general, to exhibit the advantageous properties as cheese emulsifiers that the amorphous forms of the compositions exhibit.

In addition, the water as indicated in the formula should not necessarily be taken as representing water of crystallization since thermal gravimetric analysis indicates that the water is lost gradually when the compositions are heated to temperatures between 80° and 500°C. and there is usually no clearly defined break such as would be present in the case of free water or a crystalline hydrate. Further, these alkali metal aluminum phosphate compositions are alkaline in nature in that the pH of a 2% by weight aqueous slurry at about 25°C. is between 7 and about 10.

These compositions can be prepared by reacting an alkali metal material, an aluminum containing material, and a phosphorus containing material. The reactants are mixed in any order and, if necessary, water is added in sufficient amounts to provide a suitable reaction medium with the reaction preferably conducted under agitation. The reaction is usually exothermic and the temperature, depending upon the amount of water and reactants used can reach as high as about 70° to 120°C., although it is generally preferred to keep the temperature below about 100°C. Again, depending upon the reactants and the reacting conditions used, an amount of water can be added which will give, because the reaction is exothermic, a relatively dry, pulverulent reaction product.

The process cheese formulations can be prepared by mixing, preferably with the aid of heat, one or more cheese stocks and, as an emulsifier, these alkali metal aluminum phosphate compositions.

The alkali metal aluminum phosphate compositions and especially the preferred sodium aluminum phosphate compositions impart to the cheese stocks not only emulsifying properties, i.e., preventing or minimizing oil separation from the cheese stocks especially when heated, but also such advantageous properties as exhibiting no tendency to crystallize, improving the development of melt, i.e., the ability of the cheese to melt without oil separation, so that little or no curing period is required, and exhibiting the ability to raise the pH of a cheese to the desired level for process cheese formulations which is usually between about 5.5 to about 6.

Example: Into a suitable reaction vessel are charged about 63.9 parts of monosodium orthophosphate and about 36.1 parts of commercial sodium aluminate of approximately 46% Al_2O_3 by weight and a weight ratio of Na_2O/Al_2O_3 of 0.65. Enough water is added to give a slurry concentration of about 50% solids. The reaction mixture is agitated and an exothermic reaction takes place, usually in a short period of time of about 15 minutes, and the reaction product is drum dried to a relatively dry particulate composition analyzing on an anhydrous molar ratio basis as 4.84 Al_2O_3, 13.2 Na_2O and 8.0 P_2O_5. This sodium aluminum phosphate composition is incorporated as a cheese emulsifier on a weight basis in cheese stocks containing (1) slightly aged cheese stock (about 90 days) having a pH of 5.4; (2) medium aged cheese stock (about 8 months) having a pH of 5.5; and (3) aged cheese stock (about 12 months) having a pH of 5.3. The following table gives the results of tests conducted on these cheese stocks.

Cheese Stock	SAP (percent)	DSP (percent)	TSP (percent)	Mixing	Separation		Brittleness	Break	Grain	Melting		pH	Crystals
					Cold	R.T.				Texture	Oil Off		
	3.0			Good				2.5	2.0	3.0	3.0	5.30	1
2	3.0			do	3.0	3.0		2.5	2.0	3.0	3.0		1
3	3.0			do	2.5	2.5		2.5	2.5	3.0	3.0		1

Surface Active Agent

P.M. Lauck; U.S. Patent 3,697,292; October 10, 1972; assigned to Stauffer Chemical Company has found that extrudable pasteurized process cheese utilizing large quantities of salvage cheese can be prepared by incorporating in the blend of the raw cheese, salvage cheese and usual process cheese emulsifiers, from 0.00015 to 2% by weight, of a surface-active agent.

Problems encountered using salvage process cheese in raw cheese blends containing the normal process cheese emulsifiers may be obviated by mixing a small proportion of surface-active agent with the raw cheese and salvage process cheese blend. Up to 100% salvage process cheese can be reprocessed to provide an acceptable process cheese by the incorporation of the surface-active agents in the rework cheese. It had also been observed that the salvage process cheese effect can be reversed. Thus, additions of salvage process cheese can, under certain conditions be used to increase melt rather than decrease melt.

The surface-active agents which are used are intermixed with the usual ingredients found in cheese, including the usual emulsifiers. These emulsifiers, which may be used alone or in mixtures, include monosodium phosphate, disodium phosphate, dipotassium phosphate, trisodium phosphate, sodium methaphosphate (sodium hexametaphosphate), etc.

The amounts utilized are generally based on the raw cheese present in the cheese which is being processed. Any salvage cheese used already contains emulsifiers from its original processing. However, the total amount of emulsifier present in the final product must not exceed the specifications set forth in the Federal Regulations for process cheese which are set at 3%. The total concentration of emulsifier can vary as low as 0.5% and be as high as the legal limit of 3%.

A suitable range for these surface-active agents is between about 0.00015 and 2%, by weight, of the raw cheese including the salvage process cheese. The amount preferred in any given case will, of course, depend upon the particular surface-active agent, the raw cheese selected and the quality of the salvage process cheese, but will generally lie between about 0.0003 and 1%. The range of surface-active agent of from about 0.0003 to about 1% is particularly effective in cheese mixtures containing from about 0.5 to about 50% salvage process cheese. The surface-active agents which can be utilized are:

(a) phosphorylated substituted and unsubstituted sorbitan fatty acid monoesters and their alkali metal salts wherein the fatty acid contains from 12 to 18 carbon atoms;
(b) phosphorylated substituted and unsubstituted polyoxyethylene sorbitan fatty acid esters and their alkali metal salts wherein the average monomer chain length is from 4 to 50 and the fatty acid contains from 12 to 18 carbon atoms;
(c) polyoxyethylene fatty acid esters wherein the average monomer chain length is from 4 to 50 and the fatty acid contains from 12 to 18 carbon atoms;
(d) alkyl and alkoxy substituted phosphates wherein the substituent contains from 12 to 18 carbon atoms;
(e) substituted and unsubstituted monoglycerides wherein the fatty acid contains from 12 to 18 carbon atoms and the substituent is selected from the group consisting of acetylated tartaryl and phosphoryl;
(f) lecithin;
(g) the alkali metal salts of C_4 to C_{20} aliphatic esters of sulfosuccinic acid; and
(h) mixtures of these.

Example 1: 500 g. of a raw cheese blend containing 34% water and comprising 40% of an American cheese having a pH = 5.25, 40% of an American cheese having a pH = 5.30, and 20% of an American cheese having a pH = 5.02 was selected. The cheese was ground, frozen to inhibit bacterial action, thawed at room temperature and equilibrated by storing in a refrigerator prior to the following experiment.

The bowl of a Brabender Plastograph was adjusted to a temperature of 80°C. and the sigma blades were rotated at 150 envelopes per minute. Steam was then passed into the bowl at a rate sufficient to supply approximately 25 ml. of water to the mixture described below. About one quarter of the raw cheese was added, followed by 17.5 g. of sodium aluminum phosphate, 2.5 g. of salt, the remainder of the raw cheese and 37 ml. of water. To this mixture was then added 100 g. (20% based on weight of raw cheese) of good melt salvage cheese having a pH = 5.87, Bloom Gelometer Reading of 220 and a melt percent of 49.

The batch was cooked for three minutes and then removed from the bowl and placed in 150 ml. glass beakers. A paraffin lid was placed on the beakers and the beakers were placed in a refrigerator. After cooling overnight and equilibrating at room temperature, a sample was subjected to the Bloom Gelometer and Melt Tests. The Bloom Gelometer reading (the Bloom) is a measure of the hardness of cheese. The test comprises measuring the amount of lead shot required to force a 13/16" diameter disc a standard distance into a plug of cheese 3/4" thick and having a diameter of 13/16". Higher numbers consequently indicate harder cheese. The example cheese had a Bloom of 200.

The Melt Test is conducted by placing a 1/4" plug of cheese having a diameter of

13/16" in a 250 ml. stainless steel beaker and placing the beaker in a boiling water bath for 4 minutes. The diameter of the cheese plug is then measured and the percent increase in diameter reported as the melt. A melt of about 15 to 20 is acceptable but higher melts are desired. No maximum limit seems to exist. The example cheese had a melt of 12.

Examples 2 through 18: The examples shown in Table 1 were conducted according to the procedure of Example 1 with the exception that a surface-active agent was added to the Plastograph bowl immediately following the addition of the salvage cheese. For comparison purposes in the table, the data of Example 1 is presented. It will be seen from Table 1 that all the compounds are effective to a greater extent in increasing melt.

Moreover, most of the examples show an improvement in blend, permitting the cheese to be folded without breaking. A slight increase in fat loss on paper is also observed which permits easier separation from the paper as well as from other sliced pieces.

TABLE 1: THE EFFECT OF A SURFACE-ACTIVE AGENT ON CHEESE CONTAINING 20% GOOD SALVAGE CHEESE

Ex. No.	Surface-active agent: Type	Surface-active agent: Amt. (gm.)	Melt	Bloom	Fat loss: Melt	Fat loss: Paper	Slice	Break	Strength
1	None	0	12	200	None	Trace	Moderately smooth	Trace, jagged	Very slight.
2	Polyoxyethylene (40) monostearate	1	60	180	Trace	Very slight	Slightly smooth	Trace, jagged, good bend.	Do.
3	Lactylic stearate	5	36	220	None	do	do	do	Do.
4	Distilled (90%) monoglycerides made from hydrogenated lard.	5	48	225	do	do	do	Trace, jagged	Do.
5	Commercial lecithin	5	52	185	do	do	do	do	Do.
6	Acetylated tartrated monoglyceride of vegetable fat.	5	56	170	Trace	Moderate	do	do	Do.
7	Sorbitan monostearate	5	71	193	do	Very slight	do	Trace, jagged, good bend.	Do.
8	Stearoyl monoglyceride	5	48	225	None	do		Trace, jagged	Do.
9	Tris (phosphoryl) polyoxyethylene (20) sorbitan monolaurate.	5	44	210	do	do	Slightly smooth	do	Do.
10	Sodium palmitoyl sorbitan orthophosphate.	5	54	198	do	do	do	Straight	Do.
11	Phosphorylated stearoyl monoglyceride	5	35	226	do	do	do	Trace, jagged	Do.
12	Propoxylated lauryl polyphosphate	5	65	164	do	Moderate	do	Very slightly jagged.	Do.

TABLE 2: THE EFFECT OF A SURFACE-ACTIVE AGENT ON POOR QUALITY SALVAGE CHEESE

Ex. No.	Surface-active agent: Type	Surface-active agent: Amt. (gm.)	Melt	Bloom	Fat loss: Melt	Fat loss: Paper	Slice	Break	Strength
13	None	0	23	225	None	Very slight	Slightly smooth	Straight	Very slight.
14	Polyoxyethylene (40) monostearate	1	61	185	Trace	do	do	Very slight	Do

TABLE 3: THE EFFECT OF A SURFACE-ACTIVE AGENT WITH VARIOUS COMMONLY USED EMULSIFIERS

Ex. No.	Surface-active agent: Type	Surface-active agent: Amt. (gm).	Emulsifier: Type	Emulsifier: Amt. (gm.)	Melt	Bloom	Fat loss: Melt	Fat loss: Paper	Slice	Break	Strength
15	None	0	DSP	11.7	14	220	None	Very slight	Slightly smooth	Trace, jagged	Very slight.
16	Polyoxyethylene (40) monostearate.	1	DSP	11.7	25	170	do	do	do	do	Do.
17	None	0	SC	11.5	16	345	do	do	Slightly rough	Very slightly jagged.	Slightly brittle.
18	Polyoxyethylene (20) sorbitan monoleate.	(1)	SC	11.5	31	290	Trace	do	Slightly smooth	Trace jagged	Very slightly brittle.

[1] 1 milliliter.

The examples reported in Table 2 were conducted in the same manner as the previous examples with the exception that salvage cheese was used which was adjusted to be of poor quality by an experienced blender. This salvage cheese had a melt of 6, a Bloom of 500+ and a pH = 6.01. Because of its poor quality, the quantity added was reduced from 20 to 2.5%, the latter being the commercial upper limit for a satisfactory cheese product. Example 13 is a control containing no surface-active agent. The data in Table 3 again reveals a manifest increase in melt and a desirable reduction in Bloom as well.

In the examples shown in Table 3, procedure of the preceding examples was followed. These examples were conducted to demonstrate the effect of a more active surface-active agent with two of the commonly used emulsifiers, disodium phosphate (DSP) and sodium citrate (SC) on process cheese containing 20% good salvage cheese. Examples 15 and 17 represent controls containing no surface-active agent.

CHEESE-LIKE PRODUCTS

Polyunsaturated Fat "Cream Cheese" Spread

G.D. Elenbogen and M. Baron; U.S. Patent 3,397,994; August 20, 1968; assigned to Vitamins, Inc. describe the manufacture of a dietary spread resembling cream cheese which contains 15 to 40% fat high in polyunsaturation, 5 to 13% phosphoprotein solids, and water. The spread is made by homogenizing the fats, the phosphoprotein, and the water at 160°F. A lactic acid producing culture is added and allowed to incubate until about pH 4.6 is achieved. The mixture is heated to about 165°F., whereupon a vegetable gum is added and the mixture again is homogenized.

The salient feature of this spread is its ratio of polyunsaturated fat to saturated fat which falls in the range of 3:1 to 9:1 with the total fat content in the spread amounting to 15 to 40% by weight. The taste of the spread is appealing and pleasing to the average consumer, since no trace of vegetable oil can be observed. Also, the fats have not been chemically or artifically hydrogenated.

The spread also contains, based on the ultimate weight of the spread, about 0.1 to 2.0% by weight of a stabilizing vegetable gum, about 7 to 10% by weight, of phosphoprotein solids from the group consisting of casein, alkali caseinate and nonfat milk solids, about 0.5 to 5% by weight of lactic acid producing culture, with the rest of the ultimate weight substantially comprising water in the range of about 40 to 70%. The spread resembles cream cheese and is uniformly spreadable having a smooth taste and texture. It is substantially homogeneous at room and refrigerator temperatures and does not separate into phases.

To make the spread, the fat content is thoroughly intermixed with the phosphoprotein solids, the mixing being carried out at about 160°F., after which it is homogenized in a conventional dairy homogenizer at about 2,500 psi. Lactic acid producing culture is added, after which the composition is incubated at about 70° to 73°F. for about 14 hours or until an acidity of 0.70 or a pH range of 4.4 to 4.8 is reached.

Following incubation, the product is heated to about 120°F. for about 20 minutes and then unwanted liquids are drained. A control is run on the product and if the fat content is too low, fat can be added or if the fat content is too high, it can be lowered by the addition of water. The product is then heated to make it fluid and to pasteurize it, the heating being done by raising the temperature to about 145°F. at which point a 30 minute timed pasteurization period is initiated. Continued heating of the product in this period brings the temperature to about 170°F. A gum stabilizer, mold inhibitor preservative, antioxidant, emulsifier, vitamin, or the like can be added during this period.

Following pasteurization, the product is again homogenized at a pressure in the range of about 1,000 psi to 2,500 psi and advisedly about 2,500 psi after which the finished product can be placed in containers.

Example: 200 pounds of cream cheese spread having a total fat content of about 34% is prepared from 66 pounds corn oil, 60 gallons of milk products (skim milk, condensed milk or whole milk) having 2 pounds of butter fat and 10% of the milk products being nonfat milk solids, 36 x 10^6 units vitamin A (U.S.P.), 4 x 10^6 units vitamin D (U.S.P.), 11,000 mg. tocopherol (vitamin E), 1 gallon lactic acid producing starter culture (Streptococcus lactis), 45 g. of mono- and diglyceride emulsifier, 45 g. locust bean gum, 9 g. NDGA (nordihydroguiaretic acid), 16 oz. of sorbic acid, 24 oz. salt, 5 cc diacetyl (dimethyl ketone), and 0.75 cc rennet.

A solution containing 60 gallons of milk products is prepared so that 2.0 pounds of butter fat and 10% nonfat milk solids are present. The corn oil and vitamins A, D and E are added with the mixture being stirred and heated to about 160°F. where it is homogenized in a conventional dairy homogenizer at 2,500 psi. Subsequently, the lactic acid producing culture is added along with about 0.75 cc rennet which serves to facilitate coagulation. The composition is then incubated at about 72°F. for about 14 hours or until acidity is 0.70 or pH is 4.6. Following incubation the composition is heated to about 120°F., for about 20 minutes to destroy the culture after which the composition is placed in cheese bags and allowed to drain unwanted liquids, i.e. whey.

A control is run on the product at this stage of the process to ascertain if the proper fat content is present at which time fat can be added to the product if necessary or the fat content lowered by addition of water.

The product, which is in a semi-solid state, is then heated to place it in a fluid condition and to pasteurize it, this being accomplished by heating the product to about 145°F. at which point a 30 minute pasteurization period is measured. Continued heating of the product brings the temperature in the range of about 170°F. During this time period the locust bean gum, NDGA, salt, sorbic acid, diacetyl and emulsifier can be added. At the end of the pasteurization period, the composition is again homogenized in a dairy homogenizer at a pressure of about 2,500 psi. The finished product can then be placed in appropriate containers by conventional container-filling equipment.

Elastic Cheese with *Streptococcus lactis* on Aqueous Sodium Caseinate

R.L. Kasik and A.J. Luksas; U.S. Patent 3,579,354; May 18, 1971; assigned to Beatrice Foods Co. describe a process in which an elastic cheese product is made by inoculating sodium caseinate containing butterfat with lactic acid organisms in aqueous medium, then inoculating with a micrococcus, fermenting aerobically until the system becomes elastic and plastic, then acidifying to pH 4.5 and drying.

A bland product can be prepared by growing trained Streptococcus lactis and/or micrococci on aqueous sodium caseinate, acidifying, removing the liquid, resuspending with aid of alkali to pH 6-7 and drying.

Example 1: A 10% aqueous sodium caseinate mixture was sterilized and inoculated with Streptococcus lactis (previously trained to grow aerobically on aqueous sodium caseinate by 5 consecutive transfers from one 10% aqueous sodium caseinate mixture to another after 24 hour intervals). The Streptococcus lactis was allowed to grow aerobically for 24 hours at 32°C. until the caseinate had precipitated. The caseinate was resuspended as a 10% solid aqueous mixture by raising the pH to 6.5 with sodium hydroxide. The product was then heated to sterilize it and spray dried to give a sodium caseinate powder which was bland in character and was devoid of the usual undersirable sodium caseinate flavor and odor.

Example 2: The procedure of Example 1 was repeated except that a mixture of Streptococcus lactis and Micrococcus caseolyticus was employed and the growth was continued for 48 hours. The spray dried product was bland like that of Example 1. The procedure set forth above is the preferred way of obtaining a bland sodium caseinate since it can be done readily and does not require washings since the objectionable character is eliminated by the microbiological activity.

Example 3: An aqueous dispersion of 10% sodium caseinate and 5% butterfat in water is sterilized. Then the mixture is inoculated either statically or with stirring with Streptococcus lactis but no aeration is provided and the bacteria are allowed to grow 12 to 48 hours. Then the system is inoculated with micrococci, e.g., Micrococcus caseolyticus or Micrococcus freudenreichii and fermented aerobically with agitation in the liquid state for 2 to 3 days, until the system becomes elastic and plastic in consistency. It is then acidified with any nontoxic acid.

The acidification should be sufficient to coagulate the product. The acidification is preferably 4.5 but can be as high as pH 5.2 if heating, e.g., to 40°C. is used to aid in coagulation. The acidification is desirably to the isoelectric point of the casein. The acidification should be done before adverse flavor develops, as the flavor increases the elastic properties drop off. After acidification the liquid is expelled by drying. The product in this form is elastic and plastic and can be sold as a replacement for Mozzarella or scamorzo cheese. There should be retained enough moisture in the product, (5 to 10%) to insure a plastic consistency. After acidification, the precipitated sodium caseinate can be resuspended in water at 10% concentration, broken down into fine particles by homogenizing at 500 to 1,500 psi followed by spray drying or roller drying to give a powder product. For roller drying it is essential that the body of the product be broken down. The bacteria are allowed to grow at 32°C. Of the fats tried only butterfat was effective in giving the desired Mozzarella cheese type product.

Example 4: A sterile aqueous dispersion containing 10% sodium caseinate and 5% butterfat was inoculated statically with Streptococcus lactis and the microorganism allowed to grow at 32°C. for 24 hour without aeration. The mixture was then inoculated with Micrococcus caseolyticus which was allowed to stand at 32°C. for 3 days while the liquid mixture was aerated and agitated. The system became elastic and plastic. The mixture was then acidified with lactic acid to a pH of 4.5 and the liquid expelled to give an elastic and plastic Mozzarella cheese-like product containing 8% moisture.

Example 5: Example 4 was repeated. The product was suspended at 10% solids in water having a pH of 6.5, homogenized a 1,000 psi and spray dried to give a powder. This product can be sold as such even though it is no longer elastic. By adding 10% water to the product an elastic and plastic product was obtained suitable as a substitute for Mozzarella or scamorzo cheese.

C_4 to C_8 Monocarboxylic Aliphatic Acid in Dry Casein Coagulate

A. Baron; U.S. Patent 3,635,737; January 18, 1972 has found that casein coagulate can be converted into a palatable cheese-like composition. In the process a minor amount of one or more of the C_4 to C_8 aliphatic monocarboxylic acids is mixed with casein having particle sizes in the range 0.5 to 400 mesh. Provided that the water content of the mixture, based upon casein, is below 30 to 40 weight percent, the mix has a good shelf life, i.e., no mold formation occurs in the absence of preservatives under ordinary ambient temperatures. The product is useful per se as a relatively neutral cheese-like base material, and by the addition of suitable flavorants and flavor enhancing agent, it is useful as a substitute for grated cheese.

A granulated cheese-like composition is produced from a casein coagulate having a 30 to 90 mesh particle size and a water content less than about 5 weight percent. In the process the casein is mixed at a temperature of 20° to 50°C. with an aqueous mixture containing isovaleric, butyric, caproic, and succinic acids and sodium chloride. The casein is added portionwise to the aqueous medium and the latter is absorbed into the casein. Efficient stirring of the resulting mixture is continued until a substantially homogeneous mixture results, i.e., a total mixing time of 5 to 20 minutes yields a mix which for all practical purposes is homogeneous. For 100 parts (weight) of the casein, 9.6 parts of water containing 0.7 parts of succinic acid, 1.0×10^{-3} parts of caproic acid and 4.3 parts of sodium chloride is employed.

Preferably the succinic acid is in solution. The sodium chloride in powdered form is added to the succinic solution prior to the mixing with the casein. If desired, a trace of an antioxidant such as butylated hydroxyanisole plus butylated hydroxytoluene or the like may be added to the aqueous solution. Similarly, trace amounts of flavor enhancing agents such as disodium 5'-inosinate and 5'-guanylate or the like may also be added.

Example 1: Into a mechanical mixing unit were charged 40 g. of water at a temperature of about 25°C. containing 3.0 g. of succinic acid solute, 0.13 g. of disodium 5'-inosinate, 0.13 g. of disodium 5'-guanylate and in powdered (~200 mesh) 18 g. of sodium chloride and 25 g. of monosodium glutamate. To this mixture was added with mixing 1.0 g. of a propylene glycol solution containing about 0.6×10^{-2} g. of isovaleric acid and about 1.5×10^{-2} g. of butyric acid. While the above charge was efficiently stirred, 415 g. of commercial casein coagulate (HCl precipitated from skim milk, having a water content of about 5 weight percent, a 30 to 90 mesh particle size, and a natural fat content below about 0.2 weight percent) were charged portionwise to the mixing unit.

The initial portions were small, i.e., 1 to 10% of the total casein to be added and succeeding similarly small portions were added until the water phase appeared to be depleted at which time the balance of the casein was added. Stirring was continued and after 5 to 10 minutes thereof, the mixture was, for all practical purposes, homogeneous. The product had the appearance of an exceptionally pure grated cheese.

It had an excellent mouth feel and the flavor of a good Italian Romano cheese. After more than one month's storage at ambient temperatures (~20° to 30°C.), there was no evidence of mold or bacterial action. The flavor remained excellent and, if anything, the mouth feel was better than initially.

Blend of Fat, Carrier and Flavoring

H.A. Blackstock and N.D. Ross; U.S. Patent 3,689,290; September 5, 1972; assigned to Carnation Company describe a dry particulate food product provided by blending together a vegetable fat: a dry, bland particulate organic carrier, such as food starch: flavoring agent, with or without coloring agent to provide the flavor and appearance of the food, and then coating the resultant mixture with dry powdered coating agent, such as powdered hydrolyzed cereal solids having a smaller particle size than that of the carrier.

As a suitable bland carrier food grade flaked starch particles, gelatinized or ungelatinized, can be used. Ungelatinized flaked food starch is preferred. Other suitable carriers such as flaked corn meal, soy flakes, so-called pearl starch and the like can be employed.

It is important to control the shape and size of the carrier particles, preferably 12 to 100 mesh and generally relatively flat for grated synthetic cheese. The particles must have sufficient strength so as not to produce excessive fines during manufacture, but not so strong as to provide undesirable chewing qualities.

The carrier is present together with fat in the finished product in a combined major proportion. Usually, that combined weight percentage is about 70 to 90, with respect to the total product, the carrier being present in a concentration of about 44 to 74 weight percent and the fat being present in a concentration of about 10 to 40 weight percent.

The fat can be any suitable edible fat, for example, refined cottonseed oil, peanut oil, corn oil, coconut oil, refined soybean oil etc., hydrogenated or unhydrogenated, depending on the oil. The melting point preferably is below 100°F. Edible flavoring agent is employed which includes one or more edible organic dry or liquid ingredients which impart to the product a typical cheese flavor. A small amount of edible, organic coloring is used, for example, 0.05 to 1.0 weight percent of a liquid color simulating the color of cheese.

The coating agent can be any one or more suitable edible powdered coating materials, preferably of vegetable or cereal derivation, and preferably characterized by low hygroscopicity and free flowability. Preferably, the coating agent has an average particle size less than that of the carrier, so that a plurality of the particles of coating agent are required to cover the surface of the fat encapsulated carrier particle. The purpose of the coating agent is to give free flowability to the product.

The coating agents comprise, for example, hydrolyzed cereal solids, such as low DE (Dextrose Equivalent) dry corn syrup solids in powdered form or powdered starch, with or without free-flowing agents. The concentration of coating agent usually is about 10 weight percent.

In preparing the product, the fat, flavoring agent and carrier are blended together, after which the resultant mixture is blended with the coating agent. The blending can be carried out in any suitable manner. Preferably, the dry ingredients (carrier and any dry flavoring, such as salt, etc., are blended together, and a similar procedure is following for blending the liquid ingredients together (fat, liquid flavoring and coloring).

In the latter instance, elevated temperatures (120° to 140°F.) are employed to bring the fat, if necessary, and/or other ingredients into the liquid state and/or facilitate blending. The liquid mixture is then passed into contact with the dry mixture and blended about 10 minutes, after which the coating agent is blended into the resultant mixture for about 5 minutes. The product then is ready for packaging.

Example: A grated Parmesan-type synthetic cheese was produced by first blending in a ribbon-type blender about 65 parts of flaked food starch having an average particle size of about 100 mesh, about 1 part salt (sodium chloride) and 1 part dry particulate Romano cheese flavor. The Romano flavor was disposed on a particulate gum arabic carrier (25% liquid flavor to 75% solid carrier). The Romano cheese flavor was comprised, in parts by weight of butyl butyryl lactate, 5.00; diacetyl, 0.30; isovaleric acid, 1.00; methyl ethyl ketone, 0.20; ethyl butyrate, 1.00; n-butanol, 0.20; n-butyric acid, 2.00; acetoin, 10.00; ethyl propionate, 0.20; isobutyric acid, 1.50; methyl nonyl ketone, 0.20; caproic acid, 2.00; octanoic acid, 2.00; decanoic acid, 4.00; and propylene glycol or vegetable oil, 20.40.

A liquid mixture was then blended in a steam jacket kettle at about 140°F. This mixture comprised 19 parts of a liquid formed of equal parts of hydrogenated coconut oil, soybean oil, and cottonseed oil, to which were added 3 parts of a low temperature (melting point) flavored vegetable fat, 1/2 part of lactic acid, 1/10 part n-butyric acid, 3/100 part annatto food coloring and 1/2 part of imitation Parmesan cheese flavor.

The Parmesan cheese flavor was composed, in parts by weight, of butyl butyryl lactate, 3.00; diacetyl, 0.20; isovaleric acid, 3.00; methyl ethyl ketone, 0.10; ethyl butyrate, 1.00; n-butanol, 0.05; n-butyric acid, 3.00; acetoin, 2.00; ethyl propionate, 0.50; isobutyric acid, 1.00; methyl nonyl ketone, 0.10; caproic acid, 3.00; octanoic acid, 4.00; decanoic acid, 4.00; and propylene glycol or vegetable oil, 25.05.

The resulting liquid was held at 120°F. while being pumped into the dry mixture and is blended for about 10 minutes. To the resultant mixture was then added, with blending, about 10 parts of hydrolyzed cereal solids and blending was continued for about 5 minutes, after which the product was passed to a packaging line, packaged in foil-lined cardboard containers, stored, and later tested.

It was found that the shelf life of the product was more than 12 months when so packaged. Moreover, the product had the color, odor, appearance, texture, taste, and quality of grated Parmesan cheese but was substantially less expensive to produce. The particles of product were closely examined and found to comprise a flaked starch substrate covered with flavored fat, in turn dusted with the powdered cereal solids coating.

Blue Cheese Flavoring

A.J. Luksas; U.S. Patent 3,720,520; March 13, 1973; assigned to Beatrice Foods Co. describes a process in which a blue cheese flavor of accurate blue cheese flavor profile and with substantially no unwanted off-flavors is produced by growing *Penicillium roqueforti* in an aqueous medium containing essentially only sodium caseinate and butterfat. This relatively pure medium provided the growth nutrients required for producing an accurate blue cheese flavor while at the same time eliminating other normal constituents of milk which allow the production of off-flavors.

Further, the combination of sodium caseinate and butterfat forces the *Penicillium roqueforti* to grow in such a manner that both the fat derived ketone flavor profile and the protein derived flavor profile develop in substantially the correct proportions for producing an accurate blue cheese flavor.

Sodium caseinate is not a usual constituent in growth media for producing cheese flavors and it is unusual that the *Penicillium roqueforti* could utilize sodium caseinate as a protein nutrient in producing the blue cheese flavor. The medium must contain sodium caseinate, and casein or alkali caseins other than sodium caseinates are not equivalents in the growth medium of this process.

The fat portion of the medium may be other than butterfat, although butterfat provides superior results to other fats such as animal fat and vegetable fat. When vegetable fat or animal fat is used instead of butterfat, the product which results does not have the clear, complete blue cheese flavor, although some blue cheese flavor is produced.

Accordingly, the process is one for developing blue cheese flavor which comprises aerobically growing *Penicillium roqueforti* with aeration and agitation in an aqueous medium consisting essentially of 5 to 10% sodium caseinate and 3 to 15% butterfat for at least 24 hours at 15° to 40°C. until a sharp, strong, aged, blue cheese odor and flavor develops. The growth period will vary in time depending upon the particular proportions of the constituents of the growth medium and the particular temperature utilized. Optimum growth periods will result with temperatures between 17° and 30°C. and these periods will generally be no longer than 5 days.

It is necessary that the growth take place under aerobic conditions and to this end, aeration and agitation of the fluid aqueous medium should be carried out. It is necessary that the medium remain fluid and be susceptible to aeration and agitation sufficient to insure those aerobic conditions. The medium, therefore, should not have a viscosity greater than 100 cp. during the initial fermentation stage, although viscosities greater than this can be accommodated as the fermentation proceeds.

With this process, any of the mutants of *Penicillium roqueforti* may be used. The growth medium should be first pasteurized or sterilized to prevent the growth of unknown contaminating organisms. This may be accomplished by heating from 160° to 180°F. for 5 to 20 min. or to pasteurization or sterility at any times and temperatures.

After sufficient growth to develop the flavor of blue cheese has been accomplished, the mixture should then again be pasteurized or sterilized to prevent further growth of the organisms. Thereafter, the liquid product can be used directly for flavoring food products or alternately the liquid product can be dried or concentrated in any conventional manner.

Whether the liquid, the concentrated or the dried product is incorporated into a food, the amount used in the food can vary completely at will and consistent with the degree of blue cheese flavor desired. Any food may be flavored including cheeses, such as bland, unaged cheese curds, relatively unaged blue cheese, cottage cheese, salad dressings, dips, crackers, sauces, soups, etc.

LOW FAT CHEESES

Ripened Skim Milk Cheese

R.E. Hargrove and F.E. McDonough; U.S. Patent 3,156,568; November 10, 1964; assigned to the U.S. Secretary of Agriculture produce a cheese product, low in fat content and relatively high in moisture and nonfat milk solids, which resembles Cheddar cheese in texture and flavor.

The process includes adding a small amount of well dispersed cream to skim milk with a resulting improvement in texture of the product; partial development of flavor by a pretreatment of the cream portion; the use of an atypical starter culture; the prevention of rapid development of acid in the cheese milk, almost all of the acid being formed after hooping and pressing; and pressing out in the substantial absence of air.

In general, cream, containing about 20% milk fat is pasteurized at about 175° to 190°F., homogenized, combined with lipase and allowed to stand at about 40° to 50°F. to develop rancidity. About 2% by weight of the enzyme treated cream is combined with about 98% by weight skim milk to provide a cheese milk. The cheese milk is pasteurized at 175°F., and then homogenized at a pressure not appreciably more than 1,000 psig. The temperature of the cheese milk is adjusted to 88° to 90°F.

A starter culture containing *Streptococcus cremoris,* Leuconostoc, and another microorganism such as *Lactobacillus lactis*, *Bacterium linens*, *Micrococcus caseolyticus,* and *Pseudomonas fragi*, is mixed with the cheese milk and, shortly thereafter, rennet is mixed with the inoculated milk to coagulate the curd. After about one hour the curd is cut, stirring is begun, the curd is cooked, whey is drained from the cooked curd, the curd is salted and hooped, the wet curd is pressed out in substantially the absence of air, and the pressed cheese is allowed to cure.

The cream fraction containing about 20% milk fat does, however, have several important functions in this process. The level of milk fat in the cheese milk is very low, about 0.3 to 0.5%. A maximum dispersion of fat globules contributes to the soft and smooth texture of the cheese product. Homogenization of the cheese milk should not be conducted at too high pressure, however, because the protein is altered and results in a hard curd. The 20% cream can tolerate homogenization at much higher pressure, and 3,000 psig was used to disperse the fat in the cream. Then, when the cream and skim milk were combined, homogenization at 1,000 psig was sufficient to provide a uniform dispersion of tiny fat globules in the cheese milk.

The cream containing about 20% fat is also a means of contributing flavor to the cheese. Lipase is allowed to act upon the fat, developing rancidity. While the process was designed to produce a mild, Cheddar-like flavor, more rancidity may be developed to produce a flavor similar to Provolone or Romano.

The pasteurization temperatures used in the process, especially the temperatures of pasteurizing the cheese milk, appears to be a significant factor in determining quality of the curd. To obtain a soft curd, pasteurization is conducted at about 175° to 180°F.

The process is designed to prevent the rapid development of acid in the cheese milk. The inoculated, rennet treated milk is not subjected to the conventional ripening period which is conducive to acid formation and results in a cheese which becomes hard, dry, and fails to show protein break down. In the process the titratable acidity of the milk or whey prior to dipping reaches a value of only about 0.15 as compared with values of 0.35 to 0.4 in a typical Cheddar cheese process.

The starter culture contains either Lactobacillus lactis, Bacterium linens, Micrococcus caseolyticus or Pseudomonas fragi, in addition to the Streptococcus cremoris and Leuconostoc. Faster breakdown is achieved by including one of the additional microorganisms. Although the acid is only about 0.15 at dipping, the organisms continue to function even after salting so that the pressed cheese (usually 21 to 24 hours after dipping) has a pH in the range of about 5.0 to 5.3 and develops a sharp flavor.

The soft curd is conducive to entrapment of air so that the physical and flavor properties are not comparable to that of Cheddar unless the hooped curd is pressed out substantially in the absence of air. Air is removed by pressing out under vacuum, or, alternatively, the curd is hooped and pressed out while submerged to exclude air from the cheese.

The appearance of the product is enhanced by adding cheese color to the cheese milk, after the fashion of a Cheddar cheese process.

Example: Cream was standardized to contain 20% milk fat, pasteurized at 185°F. for 15 seconds, and homogenized at 3,000 psig. A 1% inoculum of a whole milk culture of Pseudomonas fragi was added, and the inoculated cream allowed to stand one week at 40°F. to develop rancidity.

Fully skimmed milk was prepared by separating whole milk. Homogenized, somewhat rancid, 20% cream was combined with skim milk at the rate of about 2 lbs. cream per 100 lbs. skim milk. The cheese milk mixture was pasteurized at 175°F. for 15 seconds, then homogenized at 1,000 psig, and introduced into the vat. Cheese color was added at the rate of 1 ml. per 100 lbs. cheese milk. Starter cultures, one containing Streptococcus cremoris and Leucinostoc and the other containing Lactobacillus lactis, were each added at the rate of 1 lb. starter per 100 lbs. cheese milk.

Rennet, at the rate of 8 ml. per 100 lbs. cheese milk, was mixed as soon as the starter cultures were well distributed in the cheese milk, usually 3 to 5 minutes after adding the starter. After one hour the curd was cut with 1/2" knives and slow agitation was begun. After 15 minutes slow agitation, the contents of the vat were heated to 102°F. over a 30 minute period. After 10 minutes at 100° to 102°F. excess whey was drained, and the curd particles were stirred constantly to prevent matting. After 10 to 30 minutes of draining and stirring, the curd was salted at the rate of 200 g. salt per 100 lbs. cheese milk. The wet, salted curd was placed in cheese hoops, pressed immediately for one hour, and then placed in a vacuum chamber for 45 minutes to remover air from the curd. Following vacuum treatment the hooped curd was pressed again for 24 hours, then removed from the hoops and measured for acidity. As soon as the pH was 5.3 or lower, the cheese was waxed, placed in a 40°F. curing room, and cured 3 to 6 months at 40° to 60°F. Analysis of the product gave 4.26% fat, 54.77% moisture, 1.58% salt, and pH 5.3.

Polyunsaturated Fat Cheddar

K.H. Rhodes; U.S. Patent 3,278,313; Oct. 11, 1966; assigned to Pacific Vegetable Oil Corporation describes a process for producing Cheddar type cheese containing polyunsaturated fat in which a substituted milk is made from skimmed milk solids (or skimmed milk) and polyunsaturated fat. A typical example contains:

	Parts by Weight, %
Milk content:	
Nonfat milk solids	8.45
Milk fat	0.5
Inherent citric acid	0.9
Mono- and diglyceride emulsifier	0.5
Lecithin	0.2
Added citric acid	0.13
Antioxidant (e.g., BHT, BHA, and propyl gallate and citric acids	0.075
Water	86.6
	97.355

These ingredients are added to a pasteurizing vat and heated slowly to 165°F. in about 10 minutes. Then 3.8 parts by weight of polyunsaturated fat, preferably in the form of safflower oil, are added to this heated mixture, and the pasteurizing process is continued for 30 minutes at about 175°F. The mixture is then homogenized at 500 psi, cooled to 36°F., and stored for cure overnight. The cure assures good taste, tempering, and stability.

After curing, the substituted milk is added to a cheese vat and heated to about 90°F., and a Cheddar cheese starter is added at the rate of about 3% of the milk and stirred in thoroughly.

The starter used here is made from the various microorganisms (usually Streptococcus with some Lactobacillus, such as a mixture of Streptococcus lactis, Streptococcus cremoris, Leuconostoc dextranicum, and Leuconostoc citrovorum) and used to introduce the organisms into a very small sample of the safflower milk, and then cultured. The organism grows fairly slowly in this culture, but it does grow and as it grows it becomes somewhat accustomed to its new and different environment. Material from the culture is then taken and injected into another sample of safflower milk to make another culture, and this reculturing is continued until the organism is thououghly accustomed to the environment of the safflower oil filled milk.

It is then suitable for use to make starter that is to be injected into the commercially manufactured filled milk, the starter base also being a safflower oil milk. Cheddar cheese color is also added at this point. Any suitable approved coloring agent can be used. Rennet is then added at the rate of about 3 ounces per 1,000 pounds of the filled milk mixture. The rennet is well mixed in, and then the mixture is left to set. Setting takes about a half hour and is indicated when the curd breaks evenly from the sides of the vat. After setting, the curd is cut by stainless steel wires about 1/4 inch apart, into a series of 1/4 inch cubes. These cubes stand 10 minutes to develop a film on the outside and then are put into the cooking process. Every 5 minutes, the temperature is raised 2° until 100°F. is reached, and the cooking is continued at 100°F. until a handful of curd in a closed fist will not mat together; then it is stopped. A typical cooking time for this product is about 1 3/4 hours. Then the substance is strained of whey, which at this time will have a titratable acidity of about 0.30%.

After draining, the curd stands until it knits together sufficiently so that it can easily be lifted off the bottom of the vat. This happens in about 25 or 30 minutes after draining. The matted curd is then cut into sections of about 1 inch by 2 inches by 1 foot and submitted to the normal cheddaring process, being turned and stacked alternately, until the titratable acidity of the whey reaches a value of about 0.65%. This takes about 2 hours, quite a while longer than it takes on regular Cheddar cheese. Then the curd is milled, and, after milling, is salted at the rate of 2.5% by weight of the original safflower-filled milk.

After salting, the curd is put up in hoops, and is pressed for about one hour in a cheese press. Then heavy linen bandages that retain the moisture and keep the air away from the cheese are put on and the cheese is kept pressed for about 26 hours. After that, the bandages are removed, and the hoop-shaped loaves are cut into bricks of about 2 1/2 and 5 pounds each, and these bricks are dressed with cheese cloth, held to get a slight scab, and then returned to the press for 2 hours. Afterwards, they are stored at 40°F. in a cold room for about two days and are then dipped in paraffin at about 230°F. to put on a wax coating. The cheese is then cured, at least 60 days, the exact number of days depending on how sharp a cheese is desired.

Low Fat Milk Emulsion Cheese Precursor

A. Bratland; U.S. Patent 3,716,377; February 13, 1973 describes an emulsion product suitable for producing cheese which contains up to 85 weight percent fat, of which fat at least 25 weight percent is nonmilk fat, emulsified in a low fat milk fraction having a fat-free milk solids content of from 6 to 12 weight percent. Cheeses are prepared from this emulsion product.

In the preparation of the emulsion product, the fat is emulsified into the low fraction at an elevated temperature of 50° to 60°C. by mechanical means, such as circulation pumping with a centrifugal pump. An important aspect of this is the fact that the taste or flavor of cheeses is governed only to a small extent by the type of fat they contain. What is considerably more significant is the composition of the membrane or film by which the fat globules of the starting material are bounded. Inasmuch as phospholipoids per se are unsatisfactory, a possible explanation is that complex compounds are formed between the former and milk proteins. Thus in manufacturing cheese substitutes, the composition of the membrane or film should be as similar as possible to that forming the boundary layer in natural milk or cream if the flavor of genuine cheese is to be achieved.

The term "low fat milk fraction" used there refers to a variety of extracts, including mixtures of these, which can be separated from milk or from solid or liquid emulsions derived from milk, such as cream and butter, and which contain to varying extents membrane- or film-forming substances. Typical low fat fractions containing the membrane-forming substances are skim milk, buttermilk as well as various serums which can be extracted from milk, cream and butter by removing butter oil. Butter oil serums remaining after refining butter oil to higher degrees of purity provide additional important sources of these substances.

The quantity of membrane-forming substances present in a low fat milk fraction can vary over quite wide limits depending upon what the fraction is and the manner in which it has been prepared. The amount of membrane-forming substances present in the emulsion product is of real significance since too high a quantity prevents the separation of whey during cheese production while too little results in the separation of free fat.

At the same time, the fat-free milk solids content (sometimes referred to as the milk solid non-fat or MSNF content) of the low fat milk fraction is extremely important if a satisfactory cheese is to be obtained, a range of 9 to 10 weight percent being especially convenient. Generally speaking, a low fat milk fraction having a MSNF content exceeding 12 weight percent results in a cheese which is too dry, while with MSNF value below 6 weight percent a fatty product is obtained having a consistency somewhat resembling that of mayonnaise.

An upper limit of 85 weight percent fat is specified for the emulsion product since with fat contents exceeding this figure, a phase-exchange takes place from the desired fat-in-water emulsion to a water-in-fat emulsion. However, it is more usual to prepare the cheese from an emulsion containing 3 to 5 weight percent emulsified fat, although fat contents as low as 0.03 weight percent have been used for certain cheeses. Emulsions of higher fat content can be readily diluted to this level by the addition of skim milk.

Non-milk fats which can be employed in producing the emulsion products include vegetable and animal oils. Preferably, vegetable fat mixtures are used which melt between 25° to 35°C. Such a mixture can consist of liquid and solid soya bean oil and semisolid sunflower oil.

Skim milk produced from natural milk by separating cream having a fat content of 20 to 40% by weight is a source of appreciable quantities of membrane-forming substances. However, skim milk which remains after separating a 35% cream (a cream containing about 35 weight percent milk fat) from natural whole milk, while possessing a satisfactory MSNF content, has per se too low a concentration of membrane-forming substances per unit volume for making most cheeses. For preparing suitable low fat milk fractions in which the fat can be emulsified, it has been found convenient to add predetermined quantities of buttermilk and cream or butter serums. The proportions of the ingredients used in preparing the low fat milk fractions depend on their respective contents of membrane-forming materials and fat-free milk solids.

Suitable low fat milk fractions include those containing (a) skim milk and buttermilk in a ratio of from 2:1 to 1:2 by volume, (b) 10 to 20 volume percent butter serum with the balance being buttermilk, and (c) about 5 volume percent butter serum, 30 to 40 volume percent buttermilk with the balance being skim milk. The skim milk, in this instance, is that obtained by separating 20 to 40% cream from a natural milk containing 5 weight percent milk fat, the buttermilk that obtained by churning the

20 to 40% cream to butter and the butter serum that obtained by separating butter oil from the butter.

It is not customary to use buttermilk alone because it generally has too low a MSNF value. However, this can be overcome by concentration and/or the addition of buttermilk powder or skim milk powder. Moreover it has been found that where a buttermilk/skim mixture contains somewhat less than the desired content of membrane-forming materials, the emulsifying properties of the low fat milk fraction can be improved by adding a emulsifier. Thus where the ratio of skim milk to buttermilk is of the order of 3:1 by volume, the skim milk being obtained from 5% natural milk by separating a 20 to 40% cream and the buttermilk by churning this cream to butter, the addition of 0.2 weight percent of emulsifier results in a satisfactory low fat milk fraction.

Example 1: Natural whole milk having a fat content of 5 weight percent was pasteurized and then treated in an homogenizer of the "Clarifixator" type operated 6,000 to 8,000 revolutions per minute, the milk being fed at about 1,500 liters per hour which was roughly half the normal feeding speed. The temperature of the milk was maintained at 55°C. and butter oil was separated having a degree of purity of 96.5% leaving behind a serum extract containing 0.5 weight percent fat and a milk content solid non-fat (MSNF) content of 9.5 weight percent.

A vegetable fat mixture was prepared having a melting point of about 30°C. and containing 35 weight percent solid soya bean oil (melting point 35°C.) and 30 weight percent semisolid sunflower oil, the balance being liquid soya bean oil. 3 kilograms of this fat mixture were added to 100 liters of the serum extract with stirring accompanied by heating to a temperature of 55°C. whereby an emulsion with a pronounced whole milk taste or flavor was obtained having a fat content of 3.5 weight percent without free fat.

Rennin and a suitable cheese-forming culture were added to the emulsion and the mixture fermented and matured. A cheese was obtained having a fat content of 50 weight percent and a taste which could not be distinguished from cheese produced from conventional whole milk.

Example 2: Buttermilk having a fat content of 0.9 weight percent and a MSNF content of 8.5 weight percent was prepared by churning to butter a cream having a milk fat content of 35 weight percent, while skim milk having a fat content of 0.5 weight percent and a MSNF content of 9.5 weight percent was prepared from whole milk having a fat content of 5 weight percent.

50 liters of the buttermilk, 50 liters of the skim milk and 3.1 kilograms of the vegetable fat mixture of Example1 were heated to 35°C. and an emulsion containing 3.8 weight percent fat, without free fat, was produced by circulation pumping.

The emulsion was separated into a fat-rich fraction containing about 15 weight percent fat and a fat-poor fraction. The fat-rich fraction was homogenized and thereafter the two fractions were mixed together again by circulation pumping to form an emulsion having a similar fat content to the first-formed emulsion and a pronounced milk taste. A cheese was produced as described in Example 1 having a 50 weight percent fat content and a taste which could not be distinguished from that of cheese produced from conventional whole milk.

DRIED CHEESES

Prevention of Curd and Whey Formation

H.E. Bauman, J.L. MacMillan and J.A. Stein; U.S. Patent 2,965,492; Dec. 20, 1960; assigned to The Pillsbury Company describe a process for the rapid manufacture of a dried cheese product of predetermined flavor other than that produced by a coagulant. Briefly, the process consists of fermenting and developing the desired cheese flavor in a casein fluid product such as reconstituted nonfat milk solids, but for such length of time, not exceeding but a few hours, and under such conditions as will not develop a system of curds and whey.

The entire mass is then stirred into a fairly homogeneous fluid mass and rapidly dried down to a pulverulent product having a moisture content of approximately 3% and being the equivalent in eating and flavoring properties to the dried cheese product which normally requires weeks of preparation and reworking to obtain the ultimate dried product.

A characteristic of the process is that flavor development can be easily controlled and terminated at the desired time by drying the product, particularly if the product is spray-dried at a temperature and to a degree of dryness which will prevent further bacterial and enzymatic action.

As a source of casein, whole milk, skim milk, dried milk, or nonfat dry milk solids may be used. A pasteurized source is preferred to prevent the growth of undesirable bacteria which may produce off-flavors during the process. If the initial casein product does not contain fat, an edible fat may be added which improves the physical characteristics of the ultimate product and also furnished a source of material for fat-splitting enzymatic reaction where desired. Also, the presence of fat globules in a dried cheese product tend to retain volatile flavors and, hence, render the product more flavorsome.

To increase the fat surface for hydrolyzing, the fluid casein product is homogenized with its fat content and its fat particles suspended in colloidal form throughout the product. The fluid material containing evenly dispersed fat and casein is then inoculated with flavor-producing substances. Instead of the various types of microorganisms and enzymes which may be used in combination for producing cheese flavors, as a further short-cut in the process, commercial cheese such as Cheddar, Roquefort, Parmesan and Romano can be utilized to furnish microorganisms and enzymes.

After the development period and prior to the formation of a curd system, the entire mass is stirred so as to produce a uniform mass. The fluid product is then rapidly dried down to a moisture level usually in the neighborhood of three percent so as to instantaneously stop further bacterial and enzymatic action. Spray-drying at temperatures above room temperatures are preferred since the product in its natural condition is peculiarly adapted to spray-drying and the dried product produced by such procedure needs no further attention.

Example: A formula using 76.5% skim milk, 7.12% nonfat dry milk solids, 12.05% peanut oil, 2.19% emulsifier, and 2.14% Cheddar cheese was made up. Emulsifiers were added to the oil and the oil homogenized so that a stable emulsion is present during the fermentation. The Cheddar cheese was added to the fermentation mixture to serve as a natural source of bacteria, enzymes and cheese flavor.

The nonfat dry milk solids were dissolved in the skim milk by gentle agitation. The mixture was then heated to 140°F. and the previously melted mixture of peanut oil and emulsifiers were added. This slurry was then homogenized using a first stage pressure of 2,500 pounds per square inch gauge and second stage pressure of 800 pounds per square inch gauge. The emulsion was cooled to 60°F. and the grated cheese was added and thoroughly dispersed. The initial pH of this product was 6.1.

The emulsion was inoculated with 0.5% by weight, of a lactic starter culture. The fermentation was allowed to proceed for eight hours at 75°F. The pH at the end of the fermentation period was 4.9. The fermented material was spray-dried using a pilot plant size spray dryer which operated at an inlet air temperature of 310°F. an upper chamber temperature of 175° to 180°F., and an outlet air temperature of 150° to 155°F.

Drying Process

R.H. Rogers and E.C. Rogers; U.S. Patent 3,056,681; October 2, 1962 describes a process of manufacturing dehydrated powdered cheese in which the cheese is first reduced to a lump form, and is then circulated in water which forms a carrier for the cheese and regulates the cheese temperature. The percentage of solids with respect to the circulating mixture may vary from 35 to 42% by weight. The circulation may be maintained by a pump. It is preferred to maintain the circulation for at least thirty minutes to assure complete pasteurization.

The preferred temperature of the circulating mixture is 160°F., this being at least the approximate melting point of the cheese. The circulatory flow also breaks up the lumps, reducing them to the form of quite small particles.

The mass of cheese and water is now cooled to substantially 140°F. and is then homogenized at a pressure ranging from 2,500 to 5,000 psi. The product is then spray-dried at an air temperature not above 165°F. and a product temperature not exceeding 140°F. As the drying operation proceeds, the dried powder is progressively removed from the dryer, being then cooled to approximately 90°F., by employing mechanical refrigeration. The product is then deposited in suitable containers, where it remains for at least forty-eight hours to allow the fat in the cheese to set up. Then, by use of a suitable screen or screens, the product is sifted to eliminate any lump formations, being then ready to package for the trade.

The delay in screening and packaging the product is of decided importance since in absence of such delay, the product will form at least a few objectionable lumps within two or three weeks after sifting. By minimizing temperatures employed in the process, any material escape of volatile flavors is eliminated, and the resultant cheese is more full flavored.

Application of Edible Absorbent Powder During Thawing

M.B. McCadam; U.S. Patent 3,184,318; May 18, 1965 describes in detail the production of a granular cheese product based on the finding that discrete particles of cheese can be preserved as such and packaged by first freezing the cheese particles, thawing the particles and, during the thawing operation, applying an edible absorbent powder to the particles. During the freezing, the natural moisture in the cheese (about 38% by weight for natural American Cheddar cheese) crystallizes and expands, rupturing the natural cheese fats and solids in the cheese particles. As the temperature of the cheese particles rises above the freezing point, the free moisture and fat in the particles moves to the surface and upon contact with the particles, the powder is quickly absorbed by this surface moisture and fat, the powder penetrating the cheese particles and drying their surface areas, there, preventing adhesion of one particle to another.

Natural American Cheddar or other cheese such as Swiss, Roquefort, and blue is reduced granular or particle form by a grinding machine. The cheese pellets are then placed in a freezing room below 32°F., and maintained at this temperature until the pellets are completely frozen, producing the necessary crystallization and expansion of the moisture in the cheese.

The frozen cheese pellets are then placed in an open rotating drum and allowed to rotate slowly with the drum heated to 85° to 100°F. This specific range is the optimum temperature for the production of granular cheese adapted to be stored at normal room temperatures.

As the temperature of the cheese pellets rises above 32°F., thawing begins at the surface and the free moisture and fat in the cheese pellets moves to the surfaces of the pellets. This point is readily determined by visual observation of the pellets as they are rotated, and immediately at this point, an edible absorbent powder such as low moisture nonfat milk powder is slowly added to the pellets. Upon contact with the pellets, the dry powder is quickly absorbed by the surface moisture and fat on the pellets, penetrating into the pellets and drying their surface areas.

As the temperature of the cheese pellets rises, additional fat and moisture is continuously drawn to the surfaces of the pellets and is further absorbed by the addition of more powder. The addition of powder and rotation of the pellets during the heating operation is continued until all the pellets have reached the desired temperature and further absorption of the powder stops. The absorption of the dry powder in each pellet creates a low moisture protective area surrounding the center core of each pellet, reducing the moisture content of the surface area to a point where mold growth is discouraged and where the pellets have no tendency to stick or agglomerate together, yet not appreciably affecting the natural flavor and melting qualities of the original cheese.

While nonfat milk powder is the preferred material, other edible, low-moisture absorbent powders can be used for example, whey powder, wheat or other grain flour. The amount of absorbent powder used may vary within reasonable limits, depending mainly upon the specific type of cheese, type of powder and processing or heating temperature utilized. Sufficient powder should be utilized to absorb the moisture released during the heating operation and to provide a moisture content in the surface areas of the particles sufficiently low to prevent appreciable adherence of one particle to another during the heating step, as well as during storage at the desired conditions.

Amounts of powder in excess of such sufficient amount are not especially harmful, but add nothing to the quality of the product and are preferably avoided, particularly in the case of white-appearing powders such as milk powder, which would affect the appearance of the finished product unless colored substantially the same shade as the natural cheese. The amount of powder can be varied from 5%, or even less, to as high as 50% or more, based upon the weight of the cheese.

Example: 1,000 pounds of fresh natural American Cheddar cheese was reduced to pellets approximately 1/4 inch in maximum dimension and 1/16 inch in minimum dimension in a grinding machine, and the pellets were then placed in trays and frozen at a temperature of minus 10°F. The frozen cheese pellets were then placed in an open rotating drum and rotated slowly at a temperature of about 90°F. until the temperature of the pellets rose above freezing (32°F.), and immediately at this point, nonfat milk powder containing approximately 3% moisture, was slowly added to the pellets.

The heating and addition of the powder was continued until the entire mass of pellets had reached the temperature of about 90°F., at which point 250 pounds of powder had been added to the pellets. The treated pellets were removed from the drum and packaged in transparent cellophane bags for room temperature storage and use.

Extrusion Method

A process for preparing dried cheese and especially dried cheeses of high fat content which can be stored without refrigeration and added to commercial food products which are subsequently packaged or frozen is described by R.F. Dale; U.S. Patent 3,573,931; April 6, 1971; assigned to Universal Foods Corporation. The product is free flowing when comminuted and is adapted for use in automatic dispensing equipment.

The cheeses include blue or blue vein cheese, such as Gorgonzola or Roquefort. These cheeses are produced by inoculation of a culture of the organism *Penicillium roquefortii*. Other cheeses of the lower fat content varieties, which may be dried by this process include Mozzarella, cottage cheese, creamed cottage cheeses and ricotta.

The process comprises extruding a high moisture content cheese with a ram type extruder into elongated noodle form having a diameter of up to 3/32 inch. The extruded noodle is dried at elevated temperatures up to about 115°F. to a moisture content about 17%. Cheeses which contain up to 53% fat are adapted for processing by this method to produce a storage stable product which does not require refrigeration.

Example 1: A blue cheese having a fat content of 52% was broken up into chunks and fed into the hopper of a piston or ram type extruder (8 inches in diameter) equipped with an orifice plate having 2/32 inch orifices. The temperature of the cheese during the extrusion was 60°F. The noodle formed by the extruder was arranged into a bed about 1 1/3 inches deep, on the perforated trays of a forced air dryer, and dried at a temperature of 110°F. for 3 1/4 hours. During drying the direction of air flow through the bed was recovered at forty minute intervals.

The product had a final moisture content of 5.3%. The product was stored at room temperature (70°F.) for six weeks in polyethylene lined cartons. At the end of this storage time the product showed no signs of appreciable deterioration. When the dried cheese was formulated into a salad dressing, it achieved full flavor and texture.

Example 2: A Mozzarella cheese having a fat content of 41% was extruded as in Example 1, and dried at 100°F. for 4 1/2 hours to a final moisture content of 12%. The dried cheese was stored in polyethylene lined cartons at room temperature (70°F.) for four weeks. No signs of deterioration were observed. When placed on a pizza that was subsequently baked, the Mozzarella cheese rehydrated well and had full flavor and texture.

Example 3: Bakers' cottage cheese containing no fat, at a moisture content of 75% was extruded as in Example 1 and dried 3 1/3 hours at 110°F. to a moisture content of 8.6%. Air flow was reversed as in Example 1. The dried product was ground to a fine powder and stored in a polyethylene lined carton at room temperature for three weeks. At the end of this time rehydration of the dried product with warm water essentially restored the original texture and taste.

R.F. Dale; U.S. Patent 3,573,930; April 6, 1971; assigned to Universal Foods Corporation includes in the process the step of mixing a high fat cheese of normal moisture content with an emulsifier. The emulsifiers are those approved for the addition to foodstuffs and include sorbitan monooleate, lactyl stearate, diglyceryl stearate, etc. The amount of emulsifier ranges from about 1/2 to 1 1/2% based on dry cheese solids. Usually the emulsifier is incorporated into the cheese in a water solution and by mixing at elevated temperatures to melt the cheese. The cheese-emulsifier is then processed exactly as described in U.S. Patent 3,573,931 above.

Freezing and Drying

A process for preparing shelf-stable, free flowing dehydrated cheese products in granulated or powdered form is described by H.J. Izzo and C.O. Burton; U.S. Patent 3,694,231; September 26, 1972; assigned to Gentry Corporation. These products are prepared by stimultaneously introducing chunks of cheese and liquefied cryogenic gas into a comminution chamber so that the cheese is frozen and ground. The resulting comminuted cheese particles are dried from the frozen state in a fluidized bed while maintaining a temperature below about 21°C.

The cheese is simultaneously comminuted and treated with a liquefied cryogenic gas which reduces the temperature of the cheese to below -150°C. The cryogenic gas, such as nitrogen gas, is fed directly into the comminution chamber simutaneously with chunks of untreated cheese. The gas is preferably in the liquid state, but it may be partially in the liquid state and partially in the gaseous state. The cryogen is permitted to contact and commingle with the cheese to reduce its temperature to between -10° to -210°C. or below.

Temperatures above about -10°C. have been found to unfavorably affect the physical structure of the cheese attained by the low temperature grinding operation thereby resulting in a dehydrated product of inferior quality. Futhermore, it is preferred that the grinding operation be carried out in an oxygen-free atmosphere consisting essentially of an inert gas such as nitrogen, thereby diminishing the oxidation of the flavor and aroma characteristics of the cheese. Other inert cryogenic gases, such as carbon dioxide, ammonia, zenon, krypton, argon, neon, helium, and chlorinated and fluorinated hydrocarbons are also suitable. To obtain a satisfactory granulated or powdered product, the cheese should be comminuted to a particle size of about 1/4 inch to 30 microns. The rate at which cheese is fed into the comminution chamber will vary depending upon the type and size of comminuting device employed. Any grinding apparatus capable of being adapted for operation in an atmosphere of cryogenic gas may be utilized such as hammer mills or stone mills. Operating speeds for hammer mill types to provide an adequently ground material are from 1,000 to 14,000 rpm depending upon the apparatus used. The finely divided frozen cheese leaving the comminution chamber should be in free-flowing form.

To produce a satisfactory dehydrated cheese in powdered form for immediate consumption, the grinding operation must be coupled with a carefully controlled drying procedure. It has been found that not only are the precise temperature and moisture condition of the drying operation critical, but also the basic nature of the dehydration operation itself.

Therefore, frozen, comminuted cheese at a temperature of not higher than -10°C. is dried in a fluidized bed under precisely controlled conditions to a maximum moisture of 18%. The initial temperature of the air in the fluid drying bed is 5° to 10°C. and within 5 to 10 minutes it is heated to a temperature of 16° to 24°C. The air flowing through the bed should not have a relative humidity exceeding 60%. The rate of air flow up through the bed should be between 1,000 and 10,000 cubic ft./min., depending on amount of product being dried.

Depending upon the amount of moisture originally present in the cheese, the drying operation is continued for 45 minutes to 2 hours until the water content is reduced to 18% by weight of the total product. In addition, it is essential that the temperature of the cheese never be allowed to rise above 24°C., preferably 21°C., during the drying operation. High temperatures during the drying operation have been found to effect unfavorably the quality of the final product.

Specifically, high temperatures drive off flavor and promote oiling off thereby retarding drying and causing the cheese particles to cake. Oil on the surface of the cheese particles causes development of off flavors and rancidity and retards dehydration.

Example 1: 300 pounds of blue cheese having a moisture content of 45% by weight and liquid nitrogen were simultaneously fed into the grinding chamber of a Fitzpatrick Hammer Mill which had been precooled to a temperature of -250°C. This mill operated at a speed at 7,000 rpm and had screens with openings of 0.0078 inch in diameter. The blue cheese was introduced into the hammer mill at a rate of 300 pounds per hour and the liquid nitrogen was introduced into the grinding chamber at the rate of 600 pounds per hour (2 pounds of liquid nitrogen per pound of blue cheese).

The 300 pounds of blue cheese were ground in the hammer mill until the particle size of the cheese was reduced to less than 0.0078 inch in diameter. Care was taken during this grinding operation to maintain the temperature of the grinding chamber at -225°C. After the grinding step was completed the product was removed from the grinding chamber. The temperature of the product upon leaving the grinding chamber was about -170°C. The ground frozen product was then allowed to stand for a few hours until it reached a temperature of -12°C. After this period, 30 pounds of the ground product at a temperature of -12°C. was introduced into fluidized bed type dryer having an air velocity of 2,500 cubic feet per minute. The air temperature in this dryer was between 10° and 13°C. and the air had a relative humidity of between 15 and 30%. The drying was carried out for a period of one hour and fifteen minutes without raising the temperature of the cheese to greater than 16°C. After this period, the dried, ground cheese was removed from the dryer. The final product was a free-flowing blue cheese powder having a moisture content of 18% by weight.

Example 2: 400 pounds of Mozzarella cheese with a moisture content of 46% by weight was fed into a Fitzpatrick Hammer Mill having the characteristics outlined in Example 1. Simultaneously with the addition of the Mozzarella cheese to the grinding chamber of the Fitzpatrick Hammer Mill, there was also added to the grinding chamber, liquid nitrogen at the rate of 600 pounds per hour (1.5 pounds of nitrogen to 1 pound of cheese). Prior to the addition of Mozzarella cheese and liquid nitrogen to the grinding chamber of the hammer mill, the grinding chamber had been precooled to a temperature of -250°C.

The Mozzarella cheese was ground in the hammer mill until its particle size was reduced to less than 0.0078 inch in diameter. Care was taken so that the temperature of the grinding chamber during the operation was maintained at -225°C. After the grinding was completed, the product was removed from the grinding chamber. The temperature of the product upon being removed from the grinding chamber was -170°C. This ground, frozen product was allowed to stand for a few hours until it reached a temperature of -12°C. However, care was taken that the temperature of the ground frozen product was kept at about -12°C. during this period of standing.

Thirty pounds of this product at a temperature of -12°C. was fed into a fluidized bed type dryer having an air velocity of 2,500 cubic feet per minute, and an air temperature of between 13° to 16°C., said air having a relative humidity of from 15 to about 30%. Care was taken so that the temperature of the cheese during drying never was greater than 10°C. After one and a half hours in the dryer, the cheese was removed. The moisture content of the dried powdered Mozzarella cheese was 18% by weight.

Example 3: Fifty pounds of farmer's cheese (cottage cheese type) having a moisture content of 80% was mixed at about 4°C. with 50% by weight of an anti-caking material consisting of whey, starch, salt, calcium silicate and potassium sorbate. The resulting material resembled a heavy dough which was broken into smaller pieces and fed into the grinding chamber of a hammer mill.

At the same time, liquid nitrogen was metered into the grinding chamber. The speed of the hammer mill was 3,000 rpm and the size of the screen openings of the hammer mill was 1/2 inch, resulting in a particle size of 1/4 inch or less. The material from the grinding chamber was collected in a fiber drum and the temperature of the product was about -200°C. The coarsely ground frozen cheese was then transferred from the fiber drum to the container of a fluid bed dryer without allowing the temperature of the cheese to reach -10°C.

The container was then positioned in the fluidized bed dryer and the dryer started. Air at a temperature of 12°C. was drawn through the dryer at a rate of about 2,500 cfm. The air was heated to about 18°C. as it passed through the plenum of the dryer before coming into contact with the cheese. The cheese was then allowed to dry for approximately 180 minutes. Upon analysis, the moisture content of the cheese was found to be 17%. Soft cheese prepared in this manner must be carefully treated with anti-caking agent so as not to mask the natural flavor of the cheese which is of a very weak character. To improve the flavor of dehydrated soft cheeses, therefore, the original moisture content of the cheese is preferably reduced by physically squeezing it prior to mixing with the anti-caking material. For example, the soft cheese may be placed in cheese cloth and mechanically pressed to separate the excess enclosed water. As a result, the pressed cheese would then be mixed with a relatively small amount of anti-caking material in the range of 15 to 20%.

Thus, free-flowing shelf-stable cheese products are prepared in powdered or granulated form. These powdered cheese products are characterized by the texture, appearance and overall organoleptic properties of the cheese material from which it was made.

CHEESE VARIATIONS

Peanut Cheddar

R. Baumann; U.S. Patent 2,989,400; June 20, 1961 describes a process for making peanut Cheddar cheese in which the structure, texture, firmness and other qualities of the American Cheddar cheese remain unaltered, but which has the flavor of peanuts. Most important, the cheese holds its body and is not compromised in any way by the additional flavor of the peanuts.

The milk is pumped into a pasteurizer. The milk is pasteurized at a temperature not less than 161°F. and is held for 20 seconds in the pasteurizer before being cooled to 87°F. It is then pumped into the vat. After 1,000 pounds of milk is in the vats, a starter is added, for instance, Marshall No. 9 culture. The starter should not be less than 1%, and at this point the milk is agitated at low speed.

1 1/4 ounces of coloring is added to every thousand pounds of milk. Then the milk is agitated at high speed to mix the coloring thoroughly with the milk. The agitator is stopped and a Marshall rennet test is made to determine the time that it will require until the setting of the milk. After the time is determined and expires, the vat of milk is set with 3 1/4 oz. of rennet to every thousand pounds of milk. After a 15 minute wait, the curd is cut into small cubes, for instance of a size of one-quarter of an inch.

After cutting, the curd is started cooking for 45 minutes to a temperature of 101°F. after which the curd is cooked 30 minutes and the whey is drawn off. After there is but very little whey left in the vat to interfere with the operation, the curd is cut into blocks 6 inches wide and 14 inches long. These blocks are turned every 15 minutes. The first time the blocks of curd are turned, the second time the blocks are piled two high for compression, and the third time the blocks are piled three high for further compression, and the fourth time the blocks are again piled three high.

Following this, an acid test is taken to determine if the cheese is matted enough and if so, the acid should be 40 acidity. After matting the blocks of the curd are put through a curd mill and cut to small inch and a half cubes. The curd is agitated and washed with water of a temperature of 110° to 115°F. The water is then drawn off the curd.

The curd is then salted with 2 3/4 pounds of salt to every 1,000 pounds of milk. After the salt has been added to the curd there is another 10 minute wait and it is at this time that peanuts are included. In the batch discussed, 18 ounces of shelled salted peanuts are used. These 18 ounces of salted peanuts are used for every 100 pounds of cheese. Four ounces of the peanuts are partially crushed in order to have the appearance of the peanuts in the cheese and to provide small discrete peanut fragments.

The other 14 ounces are ground very fine. The fine grindings of the peanuts is important because this distributes the flavor throughout the cheese. Salted peanuts are mentioned as a preference over unsalted peanuts since the resulting oil tends to encourage the peanuts to adhere better.

When the peanuts are added to the curd, the curd is agitated. Agitation takes place for 7 minutes at a temperature of 75° to 78°F. Following this, the cheese is hooped and put into the press and after an hour it may be dressed or wrapped. The resulting cheese has a structure, texture and firmness as any other American Cheddar cheese.

Emulsifier Cheese Product Without Milk Solids

P.P. Noznick and E.A. Bernardoni; U.S. Patent 3,097,950; July 16, 1963; assigned to Beatrice Foods Co. describes a blue cheese product free of off flavors and devoid of any objectionable browning action sometimes experienced where milk solids are present since the process eliminates use of milk solids.

Example 1: American blue cheese is ground and added to a vat having an agitator and containing tap water with about 4% disodium phosphate dissolved in it (based on the final dry weight of the cheese powder), the solution having a temperature of about 110°F. and the cheese becoming fluidized under the action of the emulsifier, the temperature, and the agitation to produce a mixture containing about 40% solids for ultimate spray drying. This mixture is further heated at about 145°F. for about 30 minutes so that the phosphate emulsifies the cheese and the emulsion is simultaneously pasteurized.

Thereupon the emulsion is homogenized at about 145°F. under about 1,500 psig and then spray dried. About 3% of the disodium phosphate is useful. The initial temperature is at about 100° to 120°F. and the solids content is about 30 to 45% for ultimate spray drying.

The emulsion heating or second temperature stage is about 145° to 150°F. for 30 minutes to an hour to effect pasteurization and emulsification and produce a spray driable aqueous emulsion containing about 30 to 45% solids. Homogenization is at about 145° to 150°F. for the emulsion using a pressure range of about 1,000 to 2,000 psig followed by spray drying.

Example 2: A blue cheese (Roquefort) wheel was ground, and about 97% added to the tap water containing about 3% disodium phosphate (based on the dry weight of the final cheese powder) in solution. The solution temperature was 130°F. The emulsion was produced simultaneously with pasteurizing at 145°F. for 30 minutes and then spray dried. The initial temperature of the solution was varied between about 120° to 140°F. and was preferably 130°F. The disodium phosphate varied between about 1 and 5% and was preferably about 3%.

Homogenizing was carried out as described and in some examples was omitted. Homogenizing however is preferred. Pasteurizing-emulsifying was at a temperature of about 145° to 150°F. for 30 minutes to an hour.

Example 3: Disodium phosphate was dissolved in hot water at about 120°F. Ground Cheddar cheese was added and the mixture heated to and held at about 150°F. for about 45 minutes with strong stirring. The emulsion was now homogenized at 1,000 pounds psig at 150°F. and thereafter spray dried to provide a Cheddar cheese powder.

For a mild Cheddar about 5% phosphate is generally used and the range is about 3 to 5%. For sharp Cheddar the range is about 1 to 3% and about 3% is usually employed. The hot water solution of disodium phosphate varies between about 100° to 150°F. The heating and simultaneous pasteurizing varies between 140° to 160°F. and the hold time from about 45 minutes to one hour.

Homogenization takes place at from about 500 to 2,000 pounds per square inch gauge and with the emulsion at about 140° to 160°F. These spray dried powders are useful for cheese spreads, macaroni and cheese combinations, and coatings for crackers and baked products among other varied uses.

The Cheddar is added initially as broken cheese, cut or ground. The emulsion contains about 30 to 45% cheese solids in the aqueous suspension, preferably about 35%, before spray drying. Various cheese colors may be added to the emulsion, notably just before pasteurization and emulsifying, such as the water-in-oil type.

Blue Cheese Dressing

B. Heinemann; U.S. Patent 3,108,879; October 29, 1963; assigned to Producers Creamery Company describes a method for the manufacture of salad dressing containing blue cheese which retains its characteristic appearance even after severe agitation. In addition the process provides a method for growth of *Penicillium roqueforti* mycelial pellets.

It has been found that these objectives may be attained by growing mycelial pellets of *Penicillium roqueforti* by submerged culture techniques. The growth of pellets is separated from the nutrient medium and may be dried intact. The dried pellets may be added to a blue cheese salad dressing at the completion of the mixing process, thereby restoring an attractive and identifying appearance.

The cultured, dried mycelial pellets of *Penicillium roqueforti* are added to the finished dressing in an amount from 0.005 to 0.2% by weight. In many instances, the incorporation of 0.1% by weight of the dried pellets will restore the salad dressing to its original appearance. The pellets are added, after complete blending of the dressing, with only sufficient agitation to distribute the pellets uniformly.

It has further been found that mycelial pellets of *Penicillium roqueforti* may be conveniently produced by submerged culture techniques under aerobic conditions. The submerged culture process involves the growth of the Penicillium mold in an aqueous nutrient medium and accompanied customarily by agitation. When the submerged culture process is carried out in small vessels, such as flasks of relatively small capacity, the process is frequently referred to as the shake culture process because the agitation is produced by continuously shaking the culture by suitable mechanical means. When containers of larger capacity are employed, agitation of the liquid nutrient medium is ordinarily accomplished in part by mechanical stirring and in part by aeration, the latter functioning primarily to supply oxygen for growth of the mold.

In the production of mycelial pellets of Penicillium roqueforti, a suitable aqueous nutrient medium is inoculated with spores or viable mycellium and then continuously agitated in the presence of air for a period of more than 12 hours. The mycelial growth is harvested and then subjected to freeze drying, thereby maintaining the mycelium in a viable state.

The Penicillium roqueforti mycelium may be cultured in many different media. Essentially, the medium must contain, in water, a carbohydrate as a source of energy, a source of nitrogen, either organic or inorganic, and a source of essential mineral nutrients.

Example: Three mixes are prepared. Mix A contains 16.5 lb. water, 10.0 lb. vinegar (5%), 3.5 lb. starch and 12.5 lb. blue cheese. Mix B contains 1.5 lb. dried egg yolk, 1.0 lb. salt, 10.0 cottonseed oil, 0.25 lb. Tween 80, 4.0 lb. vinegar, 1.0 lb. water and 1 oz. chondrus extract. Mix C is 0.5 ounce dried mycelial pellets of Penicillium roqueforti.

The ingredients in Mix A are blended together at about 60°F. and pasteurized at 180°F. for a time sufficient to cook the starch and then promptly cooled. The ingredients in Mix B are homogenized to form a stable emulsion and then blended with Mix A. When blending is complete, Mix C is added with only sufficient agitation to distribute the pellets uniformly.

Cheese-Based Survival Bar

L. Jokay; U.S. Patent 3,121,014; February 11, 1964; assigned to the U.S. Secretary of the Army describes the production of a freeze-vacuum dehydrated carbohydrate-enriched cheese-base survival-type food bar which is prepared from a fatty cheese, dehydrated ground potatoes, dehydrated soluble starch and fat. As the cheese, blue cheese is preferred. Other fatty cheeses with well aged flavor, e.g., of the Roquefort, Swiss or aged American type may also be used. They contain more than 25% fat and are thus included as fatty cheese.

Preferably, the dehydrated potatoes used are produced from cooked diced potatoes that were subjected to a sulfiting and calcium chloride pretreatment before dehydration (to prevent discoloration), and ground after dehydration to a fineness of about U.S. Standard No. 8 to No. 30 sieve size. The dehydrated water-soluble starch is preferably of the precooked gelatinized, potato starch type and the fat is preferably hydrogenated shortening.

These ingredients are compounded so that the final composition contains 7-9 parts by weight, proteins, 13-15 parts by weight fats and 70-75 parts by weight carbohydrates. Since cheese is the principal protein source for the food bar, it may be necessary to add a small additional amount of soft cheese, such as uncreamed cottage cheese to the preferred formulation described in the example, to raise the protein content to the foregoing desired food standards.

Example: A food bar containing the above ingredients is compounded using a food mixer. About 85 parts of blue cheese are dispersed in an excess of water (about 400 parts) above room temperature (preferably about 140°F.), and about 200 parts of dehydrated precooked ground potatoes are then mixed into this slurry until a smooth paste is produced.

A second batch is prepared by mixing about 30 parts of hydrogenated shortening with an excess of water (about 200 parts), also above room temperature sufficient to melt the shortening (preferably about 140°F.) and about 112 parts by weight of potato starch are blended with this mixture of water and melted shortening, which causes the starch to rehydrate.

This second batch is then combined with the first batch and blended at about 140°F. until a homogeneous mixture is obtained. This mixture is then chilled to a solid pliable mass, without freezing. This mass is molded into bars of the desired size by pressure molding on a hydraulic press. The size of these bars should be larger than the desired size of the final dehydrated product, as they will shrink during the dehydration step. Thus, if a dehydrated bar of an area of 90 x 46 mm. is desired, the press mold should have an internal dimension of about 95 x 51 mm.

Similarly, the thickness of the molded (undehydrated) bar should be about 5 to 10% in excess of the desired final thickness; thus, if the final thickness is intended to be 14 mm., the bars should be initially molded to a thickness of 15 to 15.5 mm. The molded bars are then freeze-vacuum dehydrated.

The bars are frozen to -15° to -20°F., and vacuum-dehydrated in a chamber equipped for obtaining and holding a very low absolute pressure (high vacuum). The chamber should be equipped with shelves which can be heated or cooled by means of a circulating liquid medium. The drying chamber should be thoroughly cleaned and free of all traces of foreign odors coming from a disinfectant, a washing solution, or from products previously dried.

The shelves should preferably be chilled, using the water or other refrigerant before and during loading and while the vacuum is being drawn on the oven. However, if cooling of the shelves is impossible, the frozen trays should be kept frozen and as cold as possible before loading the oven. To avoid thawing of any portion of the frozen cheese bars while they are being transferred to the dryer, arrangements should be made to accomplish the transfer quickly and to draw a vacuum of 1.5 mm. or lower in not to exceed about 8 minutes after completion of the transfer. About 30 minutes after reaching 1.5 mm. absolute pressure or less (preferably 600 to 800 microns), heat is applied to the circulating medium.

The final temperature of the circulating medium should be about 110°F. This temperature should be reached in about 4 hours or more. The rate of heating should be so regulated that the plate temperature, for the first two hours should be about 50°F., and about 90°F. for the next two-hour period. The absolute pressure in the drying chamber at no time should exceed 1,500 microns (preferably 750 microns).

When these conditions have been established, they are to be maintained throughout the drying process, or until the product contains not more than about 4% moisture, preferably 2 to 2.5%. A typical bar, prepared as above, contains about 7 parts proteins, 14 parts fats, 72 parts carbohydrates, 4 parts ash, 1 part salt, and 2 parts moisture, per 100 parts by weight.

Blue Cheese Dressing with Cottage Cheese

In further work designed to improve blue cheese salad dressings B. Heinemann; U.S. Patent 3,189,464; June 15, 1965; assigned to Producers Creamery Company describes a process in which a quantity of dry curd cottage cheese is added to blue cheese salad dressing. The addition enhances the flavor while providing uniform cubes or particles of soft cheese which are attractive to the eye and impart a pleasing particulate nature to the dressing as it is dipped or poured from its container. Preferably, the cottage cheese is added in an amount from 0.5 to 30% by weight of the dressing.

The salad dressing may be prepared by following conventional procedures with the added step of blending a portion of dry curd cottage cheese into the final mixture. For example, blue cheese is mixed with certain ingredients which are cooked to produce a starch paste. This mixture is then blended with an emulsion base formed of other ingredients. The cottage cheese is finally added to the above blend.

It is preferred, however, to pasteurize a mixture of ingredients other than the cottage cheese. This mixture is then homogenized and cooled, resulting in a prepared salad dressing. A second mix consisting of water, vinegar and cottage cheese is then heated to temperatures below pasteurization and blended with the prepared salad dressing. If desired, a small amount of dried cultured mycelial pellets of *Penicillium roqueforti* may be added to the cottage cheese mix.

Example: The following mixture was heated to 185°F. for five minutes, cooled to 160°F., homogenized at 1,000 pounds pressure and cooled at 45°F.:

Ingredient	Amount
Potato starch	27.0 g.
Blue cheese	230.0 g.
Corn oil	225.0 g.
Water	373.0 g.
Salt	20.0 g.
Egg yolk (powdered)	12.0 g.
Vinegar	13.0 g.
Citric acid	11.0 g.
Potassium sorbate	0.8 g.
Emulsifier (Tween 80)	4.5 ml.

The following mixture was then prepared, heated to 130°F. for five minutes and cooled:

Water	91.0 ml.
Cottage cheese	70.0 g.
Vinegar	3.5 ml.
Dried mycelium of *P. roqueforti*	0.5 g.

Cheese, Phosphate and Polyglycerol Partial Ester Composition

C.W. Tatter and P.P. Noznick; U.S. Patent 3,421,904; January 14, 1969; assigned to Beatrice Foods Co. describe the preparation of emulsified cheese compositions by mixing water, a water-soluble nontoxic phosphate, a fermented type cheese and a polyglycerol partial ester of a higher fatty acid. The product can be spray dried. The phosphate serves as a peptizing agent for the protein in the cheese and reacts with it to break down the particle size. The most suitable polyglycerol esters are partial esters, i.e. they have one or more free hydroxyl groups hence have both hydrophilic and lipophilic characteristics.

The proportion of the solid materials usually is within the range of 85 to 95 parts cheese, 1 to 10 parts phosphate, and 0.5 to 15 parts polyglycerol fatty acid ester. The product after homogenization usually has a pH slightly on the acid side, e.g. between 5.0 and 7.0.

The amount of water is not critical and the mixture prior to spray drying usually has 30 to 40% solids. The mixture is pasteurized at 145° to 160°F. for 20 to 30 minutes. Homogenization is usually at 1,500 psi to 2,500 psi.

The polyglycerol partial higher fatty acid ester should be preliquefied before adding to the aqueous mixture or it can be added as a finely divided solid at 145° to 160°F., for example, it is important that the polyglycerol ester be allowed to be melted prior to homogenization of the mixture to insure uniform dispersion.

Example: A mixture of 94.0 parts Cheddar cheese solids, 3.0 parts disodium hydrogen phosphare, 2.0 parts decaglycerol tristearate (Drewpol 10-3-S), 0.25 parts Hansen's double strength cheese color and water to make 33% solids was prepared.

The disodium phosphate was added to the water, then the Cheddar cheese was slowly added while heating to 145° to 150°F. Next the decaglycerol tristearate was added in a preliquefied condition at 145° to 150°F. The mixture was then pasteurized at 150°F. for 25 minutes and then homogenized at 2,000 psi.

A very smooth fluid product was obtained which had a pH of 6.0 and exhibited excellent stability (e.g. when chilled to 40°F. and allowed to stand overnight). The product was spray dried to give a powder which was readily reconstitutable with water, e.g. to form a 50% solids emulsion that was smooth and had good stability.

Food Spread Containing Curd and Fat

J.A. Schaap and L. Torenvliet; U.S. Patent 3,502,481; March 24, 1970; assigned to Lever Brothers Company found that a cheese-like spread may be obtained by using the curd and other milk proteins obtained by heating milk to a high temperature in the presence of a nontoxic calcium salt. The calcium salt may be added before, during or after heating the milk to the desired temperature, provided that if the calcium salt is added to the milk after heating to that temperature, the milk should be at 70° to 150°C.

The combined coagulum of whey protein and curd is referred to here as milk protein curd to distinguish it from the whey protein-deficient curd which ordinarily forms the raw material of cheese and processed cheese manufacture. Skim milk, whole milk, reconstituted skim milk and reconstituted whole milk are suitable sources of curd.

Calcium chloride in a quantity of 1 to 5 g. per liter of the milk to be treated is preferred. A preferable temperature for treating the milk with the calcium salt to curdle the milk protein curd is from 80° to 90°C.

To achieve the desired texture and spreading qualities, it is necessary to ensure that the finished spread contains from 5 to 25% by weight of fat, from 5 to 20% by weight of protein and from 50 to 70% by weight of water.

The fat may be added to and mixed with the milk protein curd after separation of the residual whey to supplement any residual fat derived from the milk itself. It is, however, preferable to ensure that at least a proportion of the fat is present in the milk when curdling with the calcium salt occurs, and the qualities of the curd are further improved if the fat is first emulsified with the milk before curdling takes place. A particularly suitable fat is a vegetable fat, for example groundnut oil.

Milk protein curd as prepared by calcium precipitation, may also be used in the manufacture of traditional ripened cheese and processed cheese. More particularly, the milk protein curd obtained after separation of the whey may be heated with melting salts, including polyphosphates according to standard procedures.

In the preparation of processed cheese from the milk protein curd, the milk protein curd may first be ripened by starter organisms and then heated with melting salts.

USING WHEY

DRYING WHEY

Removing Whey Ash

M.E. Hull; U.S. Patent 3,166,486; January 19, 1965; assigned to Armour and Company describes a process in which cheese whey is subjected to treatment for the removal of the greater part of the whey ash, but not in excess of the amount which would result in insolubilizing proteins. For example, by removing 60 to 65% of the whey ash, it is found that the whey globulins and lactalbumins remain in solution during the ash removal, and thereafter the remaining whey material may be evaporated to increase the whey solids to about 60%, and the lactose then crystallized and separated from the protein.

The whey ash may be removed by any suitable method. The electrodialysis process for removing salts using ion exchange membranes is found to be highly effective in reducing the salts in whey. Electrolytic desalting or any other suitable method may be employed. After the removal of the salt or minerals which tend to interfere with the crystallization of the lactose, the removal being limited to the extent that the protein constituents remain in solution, the remaining whey material may be evaporated until the whey solids are in the neighborhood of 60 to 65%, or at such concentration that crystallization of the lactose readily occurs. After the lactose has crystallized, this product may be separated from the protein by centrifuging or by any other suitable means.

A pH of 6.0 to 6.8 is preferred in the crystallizing step. In a continuous operation, it is preferable to return the wash water used in the washing of the lactose, etc. to the preceding crystallizing step to increase the final yield of the product. In a typical operation in which cheese whey is subjected to electrodialysis, using ion exchange membranes to reduce the salt content approximately 60% after which the treated whey is concentrated in an evaporator to 60% solids and the lactose allowed to crystallize, followed by removal of the lactose by centrifugation and washing, it is found that recoveries as high as 87.5% are obtained, whereas theoretical yields calculated for such conditions are 88.5%.

In the electrodialysis method, the ion exchange membrane units are made up of many electrolytic cells consisting of pairs of anionic and cationic exchange membranes. By the use of an electric current, the cations and anions are forced through these selective membranes which thereby reduce the salt content of the solution.

Example: Approximately 30 gallons of cheese whey were treated in three lactose runs of 10 gallons each, using the wash water from each preceding run to add to each batch and thereby increasing the lactose yield. Prior to the runs, the 30-gallon bath had been subjected to electrodialysis using exchange membranes to remove 63% of the ash, the temperature being maintained at about 24° to 38°C. In each of the Tests 1, 2 and 3, the treated whey was evaporated to a final total solids concentration of about 60%. Such concentrations, the pH, and other conditions of the lactose runs, together with the results, are set out in the following table.

	1	2	3
Whey, grams	27,000	27,600	28,500
Wash water, grams	1,026	1,522	1,289
pH, after adding NaOH	6.55	6.55	6.80
Curd test, percent	50	45	55
Final TS percent	61.2	60.4	60.0
Condensed whey, grams	2,907	3,075	3,370
Lactose, grams	1,300	1,239	1,364
Lactose, moisture percent	12.8	8.7	10.3
Lactose, protein percent	1.0	0.85	0.74
Lactose, ash percent	0.35	0.35	0.40
Lactose, recovery percent, H_2O free	85.8	85.6	87.5
Lactose, yield 100 whey, H_2O free	4.2	4.2	4.3
Mother liquor, grams	1,337	1,315	1,632
Mother liquor, TS percent	39.3	42.2	41.9
Mother liquor, protein percent	15.7	17.3	16.6
Mother liquor, ash percent	3.5	3.8	3.9
Mother liquor, lactose percent	20.1	21.1	21.4
Mother liquor, viscosity, MacMichael, degrees	--	7	--
Mother liquor, pH	--	6.15	--

Spray Drying Without Nonhygroscopic Additive

The process described by F.P. Hanrahan; U.S. Patent 3,222,193; December 7, 1965; assigned to the U.S. Secretary of Agriculture seeks to make more extensive use of cottage cheese whey with an efficient means of drying without impairment of its nutritive value. Cottage cheese whey contains hygroscopic materials, especially lactic acid, which interfere with removal of water, and spray drying of concentrated cottage cheese whey without the use of a nonhygroscopic additive as a drying aid is usually unsuccessful. The incompletely dried particles settle and adhere to the walls of the spray dryer, giving a sticky, lumpy product. This cannot be prevented by increasing the temperature of the air in the dryer as the cottage cheese whey solids become burnt or scorched and a large part of the nutritive value of the whey is destroyed.

According to the process, a liquid food is pasteurized, concentrated to about 40 to 50% solids to produce a liquid food concentrate, the concentrate is forced under a first pressure into a mixing zone, an inert, relatively insoluble gas at a pressure in excess of the first pressure is injected into the concentrate in the mixing zone to produce in the mixing zone a concentrate uniformly supersaturated with respect to the gas, and the gas-concentrate mixture is conducted to a spray drying zone and spray dried.

In this process there is a marked decrease in the bulk density of the particles to be dried compared to the density of the particles in the normal spray drying process. This lowered density is obtained by spraying a concentrated liquid food into the drying chamber as relatively large droplets of foamy or puffed material. This method of foam preparation injects gas into the liquid to be dried an instant before the newly formed foam is actually sprayed. When gassing or foam production in the concentrate before drying is achieved as in this process, the particles remain in orbit for relatively long periods of time providing unique drying results. Unmodified cottage cheese whey can be obtained as a free-flowing powder.

Example 1: Cottage cheese whey was pasteurized and concentrated in a vacuum pan evaporator to 45% solids. The concentrate had a pH of 4.0. The concentrate was heated at about 130°F. and pumped with a high pressure reciprocating pump operating at 1,800 psig into the gas injector and mixing chamber. Nitrogen gas was injected into the gas injector and mixing chamber using 2,000 psig pressure at a uniform rate. The combined flow of gas and whey concentrate mixture was sprayed into the dryer.

The spray dryer was a commercially available nine-foot model of a cyclone type [Advance in Food Research, vol. II, p. 414, Academic Press Inc. (1949)]. The concentrate flow rate into the mixer was 60 gallons per hour and the gas flow rate was 120 standard cubic feet (at 2,000 psig) per hour. The nozzle opening was 0.040" diameter. The temperature of the dryer air was 270°F. The product was a free-flowing powder with a moisture content of 2.40%, a density of 0.21, and particle size ranging from 200 microns (60%) down to less than 50 microns (10%).

Example 2: The cottage cheese whey concentrate had a solids content of 45% and a pH of 4.5. The drying process was performed as described in Example 1 with the exception that the temperature of the dryer air was 260°F. The dry, free-flowing product had a moisture content of 1.20%, a density of 0.20, and the particle size ranged from 204 microns (60%) down to less than 68 microns (10%).

Recovering Lactose from Whey

D.R. Braatz; U.S. Patent 3,431,113; March 4, 1969; assigned to Peebles Foods, Inc. describes a process for recovering lactose from a rennet type cheese whey in which the whey is concentrated at a temperature in excess of 49°C. and the lactose crystallized at temperatures above 38°C. under quiescent conditions existing for at least six hours. The process employs no added chemicals for the crystallization of lactose.

The process enables the recovery of a relatively pure lactose without prior removal of lactoprotein, and the recovery of a high quality mother liquor having a minimum amount of protein degradation from which a high grade dried food product may be produced. This food product is excellently suited for human consumption and forms a most desirable base for an infant food.

In practice, a quantity of rennet type cheese whey, an American Cheddar type cheese whey being preferred, is condensed to a solids content in the range of 42 to 70% and the condensed whey removed from the condensation step at a temperature in excess of 54°C. The condensed whey is transferred to suitable containers where the lactose is permitted to crystallize from the condensed whey while the whey is maintained at a temperature in excess of 380°C. until at least 75% of the later separated lactose has crystallized. The crystallized lactose is then separated from a mother liquor.

Vacuum or air evaporators may be used for the concentration of the cheese whey in preparation for the crystallization step. Normally, the cheese whey is first pasteurized by being passed through heat exchange equipment prior to condensation and here the whey is subjected to a time-temperature treatment preferably not exceeding 74°C. for approximately 30 seconds or an equivalent heat treatment in terms of whey protein denaturation. During condensation the temperature of the cheese whey is desirably maintained below 74°C. to minimize whey protein denaturation. The process utilizes a relatively high temperature crystallization. The condensed whey is removed from the condensation step at temperatures in the range of 60° to 71°C. The minimum critical temperature for the removal of the condensed whey from the evaporator is governed by the solids content of the whey concentrate.

The whey concentrate is transferred to a crystallization zone and there permitted to crystallize while the whey is held for a prolonged period in excess of 38°C. The whey is held there for at least six hours (desirably at least 12 hours) in excess of 38°C. Preferably, the condensed whey is held above 49°C. for at least six hours of the crystallization period. Whey concentrate placed in containers of 44 gallons or larger and held at ambient room temperatures will require significantly more than six hours for the condensate temperature to drop below 38°C. Quiescent conditions in the crystallization zone favor the formation of large, easily separable crystals and for this reason agitation is desirably avoided during the crystallization of the lactose, at least during the first 12 hours of the crystallization.

It has been generally experienced that where the whey concentrate is allowed to cool at a faster rate, small lactose crystals are likely to form which are not uniform in size and which are generally difficult to separate. The cheese whey is condensed to a solids content from 58 to 62%. Lactose crystallization will commence in this process without "seeding." The carrying-on of the crystallization at the elevated temperatures and the avoidance of agitation of the whey concentration during crystallization favor the formation of large, relatively pure, lactose crystals which are easy to separate from the mother liquor. It is believed that such conditions are conducive to optimum crystallization for several reasons.

First, the lactose in the serum portion of the condensed whey is maintained below its labile area with respect to supersaturation and more near its optimum point for forced crystallization in its metastable area. A supersaturated solution in the labile area will undergo spontaneous crystallization which results in numerous and very small crystals. Another factor that is thought to favor optimum crystallization is that the higher crystallization temperatures increase the rate of conversion of beta lactose to alpha lactose which is the form of lactose that crystallizes from solution. A solution of lactose contains about 1.6 parts of beta lactose to 1 of alpha lactose. It is characteristic of lactose solutions to maintain this equilibrium. As alpha lactose crystallizes from a saturated solution, some of the beta lactose is converted to alpha lactose to maintain the equilibrium. The conversion of beta lactose to alpha lactose increases as the temperature is increased. Since the crystallization of lactose is dependent upon the concentration of alpha lactose, it is essential for the optimum crystallization rate to maintain conditions which favor the conversion to the alpha form. The third factor involved is the relatively low viscosity. The higher temperatures of the crystallization step provide a lower viscosity concentrate. Higher viscosities decrease the rate and size of crystal formation since such inhibit the migration of soluble lactose and lactose nuclei to the crystal centers. Maintaining the temperature of the whey concentrate at an elevated temperature during crystallization, at least during the early hours, greatly increases the rate and size of crystal formation.

Generally, the rate of cooling of the whey concentrate during the crystallization step is so regulated that the temperature of the concentrate does not drop below 38°C. during the first six hours of crystallization. Normally, containers of 44 gallons or more of the whey concentrate held at ambient room temperature and having initial temperatures in excess of 49° to 54°C. will remain well above 38°C., usually above 49°C., during the first six hours of crystallization. After the initial crystallization period which may extend to 20 hours or more above 38°C., the condensed whey is most often held for six hours or more at temperatures below 38°C. The lactose is then centrifuged from the mother liquor. If there is to be prolonged storage of the crystallized condensed whey, it is desirable that it be maintained at a temperature below 10°C.

The centrifuged lactose may be purified by washing, recrystallizing, centrifuging, again washing, and drying, or it may be dried directly from the centrifuge as an edible lactose for use as a food ingredient. The protein of the mother liquor recovered in the process is a high quality material and may be employed in human foods. The mother liquor from the centrifugation may be readily dried in a conventional spray drier to recover an edible, partially delactosed, dried whey. The whey product has a light yellow color and a very slight acidic flavor. The product is readily dispersible in water and other aqueous fluid products.

Example: A batch of American Cheddar type cheese whey is pasteurized by passing through a heat exchanger and being heated to a temperature of approximately 74°C. for 30 seconds. Then the pasteurized whey is condensed in a vacuum evaporator to a solids content of 62% with the whey at no time being heated to a temperature in excess of 74°C. The whey is removed from the evaporator at a temperature of 60°C. and placed in 44 gallon drums. The drums are held in a room having a temperature of 22°C. and at the end of 24 hours have cooled to 37°C.

At six hours, the temperature is 49°C. At the end of 28 hours, substantially all of the later separated lactose has crystallized. The barrels are held without agitation. The lactose crystals are separated by centrifuging from a mother liquor which is later dried and found to have a moisture content of 3.5%, a fat content of 3.0%, a protein content of 22.8%, 53.2% lactose, and 17.5% minerals. The centrifuged lactose crystals are purified by washing, recrystallizing, centrifuging, washing, and thereafter drying.

Recycling Part of Dried Whey

The process described by J.G. Moore and E.B. Pinkel; U.S. Patent 3,537,860;

November 3, 1970; assigned to Blaw-Knox Company is directed to drying acid or cottage cheese whey which constitutes almost half of the whey produced and which poses a special problem to dry, by large scale, continuous operation without additives of pH modification, into a stable, commercially nonhygroscopic, nongritty, granular, and preferably whole whey product suitable for human consumption. The lactose, which in a typical cottage cheese whey composition on a dry basis constitutes 64 to 67%, presents the pricipal difficulty, the balance, on a dry basis, being about 12.5% protein, principally lactalbumin, 0.5% fat, and 11.5% ash, principally inorganic minerals.

To obtain a stable product, the lactalbumins can first be coagulated by a predetermined and controlled temperature and time treatment. Such precoagulation step can be omitted, however, if the subsequent concentration is carried out below the coagulation temperature (140°F.) of the lactalbumins, up to the time they are removed, as by dissolving them. Following this, the whey is preconcentrated at controlled low temperatures to inhibit production of the hygroscopic beta-anhydride crystals, initial concentration of the raw whey being under vacuum and below the critical temperature of 200°F., above which the formation of only beta-anhydride crystals occurs. Thereafter the concentrate is cooled under agitation to the temperature necessary to crystallize the lactose essentially in the form of the alpha-hydrate.

The crystallization step is followed by further concentration, with or without centrifuging out of the coagulated lactalbumins and crystallized lactose, which in one form requires additional cooling and stirring to promote crystal growth for such centrifuging. The last stage of processing involves hot air drying with a concurrently moving stream of heated air, followed by cooling, grinding, screening and bagging. The feature of the process involves withdrawing a part of the whey after being air dried (at least 50% by weight of the feed of fresh material to the air dryer), and recycling it to an earlier stage of the process, such recycling involving the use of a blender to blend the earlier stage components and the air dried whey into a condition suitable for reentry into the air dryer.

Addition of Divalent Metal Ion

In the manufacture of cheese from milk, approximately half of the milk solids are coagulated as cheese, the remaining solids being contained in the residue or whey. Unfortunately, raw cheese whey also contains lactose and lipids, and the latter significantly lowers the efficiency of the recovery of valuable protein from whey and removal of the undesirable whey constituents particularly when using molecular sieves. J.M. Attebery; U.S. Patent 3,560,219; February 2, 1971 seeks to provide a process for the removal of lipid material from whey without denaturation of the protein. In addition the process provides for the removal of lipid material from raw cheese whey and the subsequent separation of protein by contacting the essentially lipid-free whey with suitable molecular sieve materials, membrane separation techniques, etc.

Solubilized lipid (including lipoprotein and phospholipid complexes as well as lipids per se) may be removed from aqueous solutions of them with proteins and/or lactose having a divalent metal ion concentration below 0.075 molal by the addition of a divalent metal ion, such as calcium preferably adding a sufficient quantity to raise the divalent ion concentration to at least 0.0750 molal, at approximately room temperature, and by the adjustment of the pH in the pH range of 7.0 to 7.5, thereby causing precipitation of lipid, separating the precipitate, and recovering the protein and/or lactose from the remaining liquid. The preferred divalent metal ions are selected from Groups I, II, IV, VI and VIII of the Periodic Table, including metals such as copper, magnesium, calcium, zinc, strontium, cadmium, barium, tin, lead, iron, cobalt, and nickel. Group II and Group VIII are preferred. Addition of these divalent metal ions is accomplished by adding those water-soluble organic or inorganic salts which are nontoxic and preferably tasteless and colorless in low concentrations when the products are intended for use in foods.

The calcium ion concentration of raw Cheddar cheese whey is approximately 0.025 molal. After the pH of such Cheddar whey is adjusted to at least 7.0, the addition of calcium ion in increments provides successive yields of precipitate at room temperature. As the level of calcium ion approaches 0.0875 molal, the weight yield of precipitate reaches a maximum after which further addition of calcium ion produces decreasing amounts of precipitate. It has been found that the precipitate at the varying increasing levels of calcium ion concentration has essentially the same ratio of the respective constituents, including lipid, nitrogen calculated as protein, ash, lactose and calcium.

In addition, the weight yields of each of the constituents in the precipitate increases at increasing calcium ion concentration levels, at least until the total level of calcium ion reaches approximately 0.0875 molal. Accordingly, the weight percent yield of protein, lactose, ash and calcium ions as well as the lipid material, increase in the precipitate as the calcium ion level is increased. The turbidity frequently observed with raw Cheddar whey is due largely to the presence of extremely fine, well dispersed particles. These particles can be removed by clarification with a high speed chamber bowl clarifier. Analysis of the removed particulate material indicates about 50 weight percent lipid and 30 weight percent protein.

This preliminary removal of dispersed particles can be used to increase the efficiency of the divalent metal ion precipitation procedures of this process and may permit the elimination of remaining lipid with a minimum addition of divalent metal ion, thus reducing the loss of protein and/or lactose in the precipitate formed during the precipitation process. The supernatant remaining after removal of the precipitate formed in the divalent metal ion precipitation process can be concentrated to at least 58% solids and introduced into the lactose and/or protein recovery steps. When such concentrated supernatant liquid is introduced into a bed of molecular sieve material or other separation equipment (e.g., electrodialysis, ultrafiltration membrane, etc.) to remove the salts and lower molecular weight materials, the protein recovered from the resulting effluent had excellent whippability at a level of 1.5 weight percent in water.

Some temperature dependence is noted in the devalent metal ion precipitation procedures. The yield of precipitate is lower at precipitation temperatures below about 68°F. At these lower temperatures the yield of other components is also depressed, due to the lower precipitate yield, but the ratio of most of the constituents remains constant. The ash and calcium contents are exceptions and appear to remain constant over the various temperatures (52° to 140°F.). The increase in protein yield observed at temperatures of 140°F. and higher is apparently due to thermal decomposition or denaturation of the protein at these higher temperatures.

2,500 ml. of raw Cheddar cheese whey was adjusted to pH 7.0 from pH 5.85 by adding 8.4 ml. of 10% NaOH after raising the temperature to 73°F. Then 10.61 g. of $CaCl_2$ was added, and the pH (5.9) as adjusted to and maintained at pH 7.0 with 5.7 ml. of 10% NaOH. The calcium ion concentration was 0.0375 molal. After 20 minutes the mixture was clarified to separate the heavy flocculent white precipitate from the supernate. Repeating the above procedure without the addition of calcium ion produced essentially no precipitate. The pH of 2,000 g. of the clarified supernate was adjusted to 5.8 with 50% HCl. Upon the addition of 3.83 g. of $CaCl_2$ and pH adjustment to 7.0 with 10% NaOH a small amount of white flocuulent precipitate formed and was separated after 30 minutes.

The calcium ion concentration was 0.0627 molal, assuming no loss in the precipitate. Taking 1,638 g. of the last supernate and adjusting the pH to 5.65 with 50% HCl, 4.24 g. of $CaCl_2$ were added and 10% NaOH was introduced to raise the pH to 7.0 and maintain constant pH for 20 minutes. No precipitate formed. The calcium ion concentration was 0.101 molal. When the above procedures were duplicated with a monovalent metal salt, i.e., NaCl, substituted for the $CaCl_2$, no precipitate was formed.

In still another series of 5 runs, using 2,500 ml. of raw Cheddar cheese whey and the single precipitation procedure as described earlier, the effect of the divalent metal ion concentration (calcium in these data) is evident. The conditions utilized are given in Table 1, and the precipitate recovery and composition are shown in Tables 2 and 3. RCHW indicates raw Cheddar cheese whey.

TABLE 1

	Initial		Reaction temperature, ° F.	Ml. 10% NaOH		$CaCl_2$	
	Temperature, ° F.	pH		Adjust	Hold	Grams used	pH after added
Number:							
1	52	5.60	86	6.83	1.82	----------	----------
2	50	5.55	74	9.31	3.92	3.9379	6.35
3	60	5.75	74	6.35	7.06	7.8583	6.20
4	56	5.65	74	7.34	7.65	11.7744	6.00
5	45	5.70	81	6.94	7.70	15.6536	5.80

TABLE 2

	Final level of Ca^{2+} (molal)	Precipitate recovered from clarifier		Percent solids remaining after subtract, percent moisture
		Wet (gm.)	Dry (gm.)	
Sample:				
1	.0267	----------	1.2088	----------
2	.0407	39.4	8.5500	21.9
3	.0545	73.0	13.4647	18.5
4	.0683	84.8	15.6063	18.4
5	.0822	93.6	16.9455	18.1

TABLE 3

	Final Ca^{2+} level (molal)	Obtained, grams dry precipitate	Protein		Ash (gm.)	Ca^{2+}		Ratio, in precipitate, gm. protein	
			Grams	Percent loss*		Grams	Percent**	Gram, ash	Grams, Ca^{2+}
Number:									
1	.0267	1.21	0.394	1.8	0.498	0.197	7.3	0.79	2.00
2	.0407	8.55	1.71	7.9	3.42	1.44	34.6	.50	1.19
3	.0545	13.5	2.26	9.9	4.98	2.04	36.8	.45	1.11
4	.0683	15.6	2.70	12.5	5.94	2.51	36.0	.45	1.08
5	.0822	16.9	2.76	12.7	6.15	2.60	31.2	.45	1.06

*Percent loss indicates percent of original amount present in 2,500 ml. RCHW and lost to recovery by processing through precipitation.

**Percent of total amount of Ca^{2+} originally present in the reaction system which appears in the precipitate.

Flash Cooling

J.J. Becker; U.S. Patent 3,615,663; October 26, 1971; assigned to The DeLaval Separator Company describes a process which relates to the production of powder from the whey obtained when making cottage, baker's or cream cheese and usually having a pH value of 4.2 to 4.5. The object of the process is to provide a method for treating acid whey so that it can be spray dried without formation of oversized lactose crystals which will not drop from the walls of the spray drying chamber but instead become glazed on the walls and must be ground or chipped off. According to the process, the acid whey is heated to a pasteurizing temperature of 142° to 175°F. and then condensed in an evaporator, after which the condensed whey is flash cooled to a temperature of 32° to 50°F.

The whey is then held in a storage tank of the type having a cold wall exposed to the stored content and provided with an agitator which sweeps the exposed surface of the cold wall, the whey being held at a temperature of 38° to 44°F. by circulating a coolant through the cold wall section of the tank. During this storage, the lactose in the whey forms very small crystals and acquires a smooth texture in which the lactose crystals are undetectable by the naked eye or by feel. The attainment of this condition generally requires at least four hours of storage. When the whey thus condensed is spray dried, the glazing of lactose crystals on the walls of the spray drying chamber is avoided. Moreover, the particle size of the resulting product may be varied by adjusting the pressure at which the whey is delivered to the spray nozzle for atomizing the whey in the drying chamber; and the particles can be readily processed to provide a powder of the desired characteristics.

Figures 9.1a and 9.1b are schematic drawings of a system which can be used to produce a nonhygroscopic acid whey powder as described above. The process is described more fully with reference to them and to the following example.

FIGURE 9.1: NONHYGROSCOPIC WHEY POWDER SYSTEM

a.

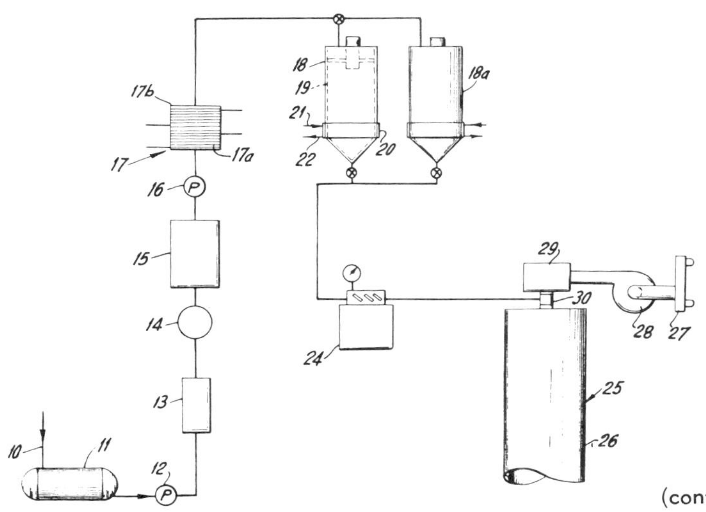

(continued)

FIGURE 9.1 (continued)

b.

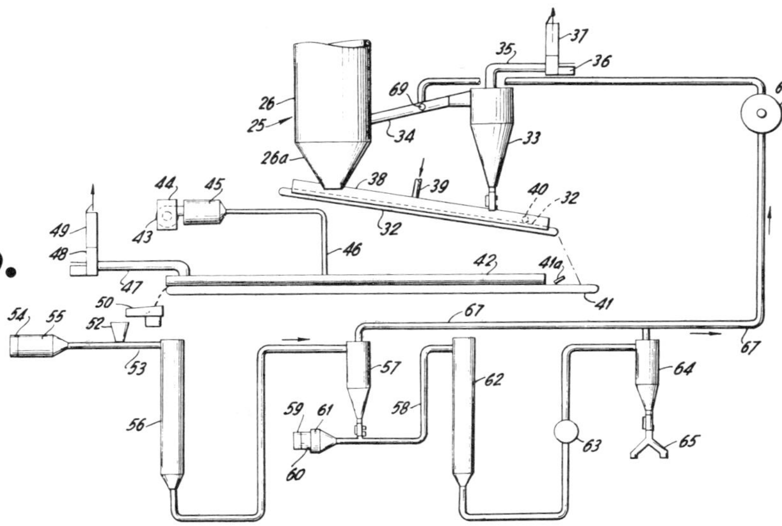

Source: J.J. Becker; U.S. Patent 3,615,663; October 26, 1971

Acid whey having a pH value of 4.2 is pumped through pasteurizer (13), where it is heated to a temperature of 162°F., and then through evaporator (15) where it is brought to a solids content of 50%. The condensed whey leaves evaporator (15) at a temperature of 120°F. and is forced by pump (16) through flash cooler (17) at a pressure of 45 psig, the whey being flash cooled to a temperature of 45°F. In storage tank (18), the flash cooled whey is held at a temperature of 45°F. for a period of 24 hours while operating the agitator (19) at a speed of 8 rpm and during this period a solids content of 35 to 52% and pH value of 4.0 to 4.4 are maintained. The whey from storage tank (18) is delivered to the nozzle of spray dryer (25) at a temperature of 45°F. and a pressure of 3,000 psig.

The air from heater (29) is forced through neck (30) at a temperature of 310°F. and at a rate of 10,000 cubic feet per minute, the whey being delivered to the spray nozzle at a rate of 130 gallons per hour. Of the total amount of dried material obtained from drying chamber (26), 20% by weight is delivered to primary cyclone (33). The powder discharged from this cyclone has a moisture content of 5.5%, and the powder dropping through the bottom of drying chamber (26) has a moisture content of 12.5%. The powder discharged from belt (41) is at a temperature of 100°F. and has a moisture content of 5.5%. The air from heater (55) is at a temperature of 300°F.; the temperature of the air from cooler (60) is 60°F.; and the temperature of the air delivered at (69) to duct (34) from blower (68) is 85°F.

Concentration, Crystallization and Removal of Lactose Crystals

L.H. Francis; U.S. Patent 3,615,664; October 26, 1971; assigned to Foremost-McKesson, Inc. describes a process which will enable the use of whey by-products in the manufacture of valuable food products. The following description of the process is given in conjunction with Figure 9.2.

FIGURE 9.2: FLOW SHEET FOR LACTOSE CRYSTAL REMOVAL

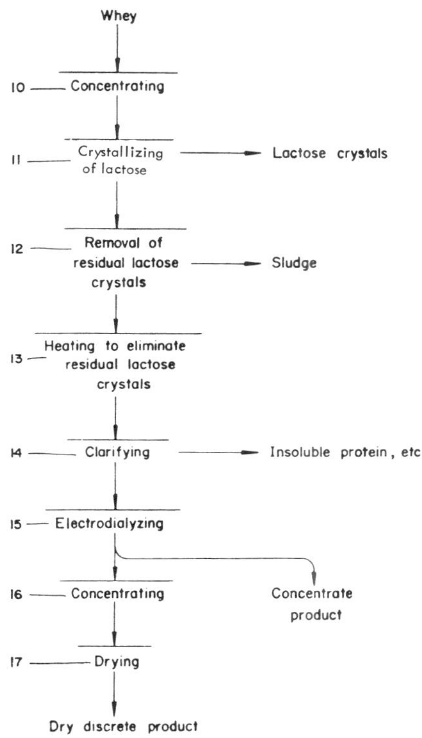

Source: L.H. Francis; U.S. Patent 3,615,664; October 26, 1971

The whey used as source material for the process is most commonly derived from the manufacture of Cheddar cheese or cottage cheese. Cottage cheese whey normally has a titratable acidity of 0.5 to 0.6% (expressed as lactic acid). The acidity of Cheddar cheese whey may be from 0.12 to 0.30%. The ash content of cottage cheese whey may be from 10.0 to 11.5% (dry solids basis), and that of Cheddar cheese whey from 8.5 to 9.0%. It may be either high heat or low heat whey, depending upon the character of heat treatment to which it may have been subjected.

Step (10) of the process concentrates the raw liquid whey preparatory to removing a substantial part of the lactose content. Concentration may be carried out by vacuum evaporation to from 50 to 60% solids, after which lactose is permitted to crystallize in step (11), and the lactose crystals removed by hydraulic or centrifugal separation or by employing a filter or a screen reel. The amount of lactose removed in step (11) may amount to 40 to 60% of lactose present. However, in some instances and with special procedures to be described, from 79 to 85% of the original lactose content may be removed. The partially delactosed whey retains some lactose in solution and some lactose in the form of relatively small crystals. In step (12) the whey is subject to centrifuging or screening for removal of the major portion of such small crystals. In step (13) the partially delactosed whey is subjected to heat treatment to a temperature level of from 110° to 120°F. to eliminate any residual lactose crystals. In step (14) the liquid material is subjected to clarification by treating it in a centrifugal clarifier. Clarification is carried out at this point to remove insoluble material present, particularly insoluble protein or curd fragments.

Assuming the use of high heat cottage cheese whey, the liquid material may in a typical instance have a pH of 3.8 to 4.2. Substantially the same pH limits are applicable to low heat cottage cheese whey. Cheddar cheese whey at this point may range to pH 4.5 for sour or acid whey and to 6.0 for sweet whey. The next step (15) is to subject the liquid material to electrodialysis. The equipment used for this purpose may be a plurality of membrane stacks connected in series with the whey flowing through ducts concurrently to a brine stream. The electrodialyzing operation should be controlled with respect to pH, percentage of solids, and temperature, to avoid precipitation of protein, gelation or lactose crystallization throughout electrodialysis.

Thus when the source material is high or low heat cottage cheese whey, the liquid at the beginning of the electrodialyzing treatment should be maintained between pH 3.9 to 4.2, the solids concentration should not exceed about 33%, and the temperature should not exceed about 110°F. Since the pH of sour Cheddar cheese whey at this point may range to 4.5, and sweet wheys to 6.5, the overall range for cheese wheys is from 3.9 to 6.5. The treatment of a batch should be continued to produce the desired reduction in ash content. To produce a product of the type desired, it is preferable to carry out electrodialysis to reduce the ash content from 20 to 55%. Some wheys are more critical than others with respect to precipitation of protein and gelation. Thus the protein of cottage cheese whey tends to be critical at pH values at or near the isoelectric point (e.g., from 4.2 to 4.65), whereas the protein of Cheddar cheese whey is not so critical in this respect.

Therefore, in the processing of cottage cheese wheys the pH is controlled whereby the processing steps, including electrodialysis, are carried out within a pH range that is either below or above 4.2 to 4.65. Generally it is desirable to process sour cottage cheese whey within the previous specified range of pH 3.9 to 4.2 because this minimizes or makes unnecessary the use of neutralizing chemicals. It may be pointed out in this connection that any chemical added for pH adjustment before electrodialysis serves to increase the mineral content. For the same reason, neutralized sweet cottage cheese whey preferably is processed at a pH above the isoelectric point, and within a range of the order of 4.65 to 6.5. Since Cheddar cheese whey tends to be less critical with respect to precipitation and gelation, the processing of such whey may be carried out within the general range of 3.9 to 6.5, but preferably at values which minimize or make unnecessary the use of added chemical.

The electrodialyzed material from step (15) is subjected to treatment to convert it to a discrete dry product. Preferably, this involves concentration in step (16) as by vacuum evaporation, followed by spray drying in step (17). Immediately prior to concentration, the material may be neutralized, as by the introduction of a suitable neutralizing chemical (e.g., lime), to adjust the pH slightly upward toward neutrality.

Concentration may be from 50 to 55% solids. Preferably, the drying conditions maintained are such that the divided product discharged from the drying chamber has a moisture content of 12 to 18%. This moist discrete material is then subjected to secondary drying with tunnel dryers or dryers of the shaker type. The final product preferably has a moisture content of the order of from 6.0 to 8.0%. A desirable feature of the process is that the ratio of combined mass of sodium and potassium to the mass of calcium in the product is reduced markedly when compared with the source material before dialysis. Table 1 gives analysis for a typical high heat cottage cheese whey which has previously been partially delactosed and then dimineralized to the 20 and 50% levels.

TABLE 1

	Partially delactosed cottage cheese whey					
	Approximately 20% demineralized			Approximately 50% demineralized		
	Before demin.	After demin.	After drying	Before demin.	After demin.	After drying
Lot No.			SD527			SB510
Lactose	44.22	46.41	42.57	42.80	52.56	48.10
Protein	21.95	23.07	18.34	17.93	23.02	20.07
Ash	17.08	13.75	16.17	17.09	8.27	10.05
Acidity	17.37	17.41	6.60	18.40	10.80	6.45
pH	3.8	3.8	5.25	3.9	4.1	4.85
Solids	30.62	29.52		29.79	28.54	
Moisture	(1)	(1)	7.30	(1)	(1)	7.80

[1] Moisture free basis.

Table 2 includes the same data as in Table 1 for a typical high heat Cheddar cheese whey.

TABLE 2

	Partially delactosed cheddar cheese whey					
	Approximately 20% demineralized			Approximately 40% demineralized		
	Before demin.	After demin.	After drying	Before demin.	After demin.	After drying
Lot No.			SE506			SE505
Lactose	46.37		44.67	46.37		48.10
Protein	22.52		21.43	22.52		21.84
Ash	17.45	13.73	14.33	17.45	10.52	11.97
Acidity	9.6	9.1	6.5	9.6	9.45	6.3
pH	4.6	4.45	5.35	4.6	4.6	5.32
Solids	32.37	28.0		32.37	27.5	
Moisture	(1)	(1)	8.40	(1)	(1)	7.20

[1] Moisture free basis.

In Tables 1 and 2 the data under the heading "Before Demin." has reference to whey treated in accordance with the flow sheet preparatory to electrodialysis in step (15). The data under the heading "After Drying" is for the dry discrete material after being partially neutralized and then converted to a dry product in the manner previously described, namely by spray drying with a moisture content of 12 to 18%, followed by secondary drying to reduce the moisture content to about 7.0 to 8.0%. Partial neutralization before drying was carried out by introducing calcium hydroxide in the amount necessary to adjust the hydrogen ion concentration to pH 5.0 to pH 5.5.

Example: High heat raw cottage cheese whey at pH 4.2 was concentrated to 54% solids and lactose permitted to crystallize. The larger lactose crystals were removed by the use of a reel (U.S. Patent 2,768,912). The resulting concentrate at about 80°F. was then subjected to centrifuging for removal of a sludge containing smaller lactose crystals. The concentrate after such centrifuging comprises whey from which 55% of the original lactose content had been removed. This delactosed whey was then heated to 120°F. and held at this temperature for a period of 30 minutes. This served to dissolve remaining small lactose crystals. The liquid material was then subjected to clarifying treatment by passing it through a centrifuge (DeLaval BRPX-309 self-opening clarifier). This served to remove suspended protein solids of the type that can be removed by application of centrifugal forces of the order of 4,000 to 5,000 G.

The material at that time had a hydrogen ion concentration of pH 4.0, and a solids concentration of 28.7%. This material was then electrodialyzed in a four stack demineralizing plant (Ionics, Inc.) in a conventional manner. The electrodialyzing operation was controlled with respect to voltage and current to prevent overheating (e.g., above 100°F.) with resulting precipitation of protein or gelation. During this treatment the pH was within the range of 4.0 to 4.5. Electrodialysis was continued to effect substantially 50% demineralization. The electrodialyzed material, at pH 4.5 was then partially neutralized by the addition of small amounts of calcium hydroxide to pH 5.0, concentrated to 50% solids by vacuum evaporation, and then converted to a dry powdered material by use of a conventional hydrate-type spray dryer together with a secondary dryer.

Analyses before and after demineralization were substantially as shown in Tables 1 and 2. The material obtained was readily amenable to spray drying to form a stable, free-flowing hydrate product. In general, this product was resistant to caking and had good keeping properties without developing discoloration. It was a high protein product well suited for human consumption.

Previous work by L.H. Francis; U.S. Patent 3,447,930; June 3, 1969; assigned to Foremost-McKesson, Inc. illustrates a similar process by making use of Cheddar cheese whey.

Example: The source material used was raw fresh fluid Cheddar cheese whey. It was low heat in that there was no significant denaturation of the protein content. Also it was sweet whey in that the hydrogen ion concentration was about pH 5.8. The whey was passed through a preheater where it was rapidly heated to about 155°F. for a holding period of about 30 seconds. It was then concentrated to 28% total solids by passing it through a vacuum evaporator of the collandria type with three effects. During vacuum evaporation the whey was not heated to a temperature in excess of 155°F., and there was no significant denaturation of the protein content in the concentrate.

After a storage period during which the concentrate was permitted to cool, the concentrate was preheated to 120°F., and the hydrogen ion concentration adjusted within the limits of about pH 6.2 to 6.4, by the addition of potassium hydroxide. The concentrate was then clarified by centrifuging in a bowl type intermittent unloading clarifier. This served to remove small quantities of insoluble suspended casein and whey protein substances. The concentrate was then electrodialyzed in a four stack demineralizing plant (Ionics, Inc.) in a conventional manner. The electrodialyzing operation was controlled with respect to voltage and current to prevent heating above about 100°F. During electrodialyzing the pH was within the range of about 6.3 to 4.8. Electrodialyzing was continued to remove about 50% of the ash content. Thereafter the demineralized concentrate was further concentrated to about 61.4% total solids using a preheater and single effect evaporator, with temperatures and time factors controlled to prevent any significant protein denaturation. The concentrate was then heated to 155° to 158°F. and introduced into a water jacketed crystallizing tank. The concentrate was held in the crystallizing tank for two hours without substantial cooling but with agitation, after which cooling water was circulated through the jacket to gradually lower the temperature of the concentrate from 160° to 100°F., at a regular and even rate of decrease over a fourteen hour period.

The concentrate together with lactose crystals of harvestable size was then fed to a basket centrifuge where lactose crystals were removed. The lactose yield was about 50% of the original quantity present. The partially delactosed and demineralized concentrate resulting after lactose removal was again concentrated to about 48% solids, using vacuum evaporating equipment of the Peebles type as in the first concentrating operation. Temperature and time factors were controlled to avoid any significant protein denaturation. This concentrate was then converted to an anhydrous dry powdered material by use of conventional drying equipment. The resulting dry product analyzed as follows:

	Percent
Protein	27.44
Lactose	58.60
Ash	8.42
Moisture	2.00

The crude lactose harvested in the above example was found to contain 97.25% lactose, 1.67% protein, and 0.55% ash. While the concentrate in the example was spray dried to produce an anhydrous powder, for a commercial product it is preferable to employ a conventional hydrate type spray dryer with a secondary dryer to form a hydrate type product.

Drying Acid Whey and Sweet Skim Milk Solids

Acid whey having a pH of 4.6±0.2 is produced in substantial volume as a by-product in the manufacture of cottage, baker's, Neufchatel and cream cheeses. This by-product contains useful nutrients but most of it is discarded as a troublesome waste product with but a minor part being utilized as relatively unprofitable animal feed. Methods devised for drying and using sweet rennet whey and the like have not been adaptable to sour whey due to the high acidity and difficulty of drying it. For example, if acid whey is concentrated and spray dried in unaltered form, the resulting powder is hygroscopic and adsorbs water. This results in substantial caking at the bottom of the spray dryer making handling of the dried product difficult.

E. Lundstedt; U.S. Patent 3,708,307; January 2, 1973; assigned to H.P. Hood Sons, Inc. provides a process for converting acid whey to a useful product suitable for human consumption. The process comprises the steps of heat denaturing unneutralized, acid whey, separately heat denaturing the serum proteins of a liquid milk product containing sweet, skim milk and spray drying the mixture by separately atomizing the whey and the milk products simultaneously through separate nozzles in a single spray dryer. The whey and milk products are fed to the spray dryer in quantities such that the acid whey solids do not substantially exceed above 43% by weight of the combined acid whey and skim milk solids. The process is characterized by an absence of a neutralization step thereby avoiding the presence of neutralization products in the dried product. The product is a free-flowing, nonhygroscopic powder suitable for human consumption such as for use in the baking industry for the making of bread, cake and the like. The process is explained in detail in reference to Figure 9.3 which is a flow diagram of the process.

FIGURE 9.3: FLOW SHEET FOR CONVERTING ACID WHEY

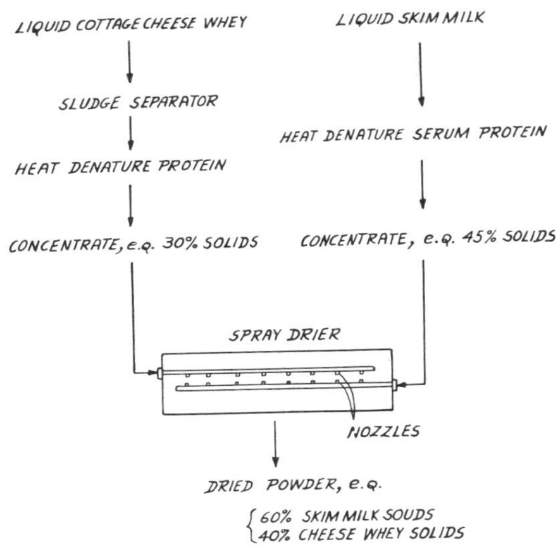

Source: E. Lundstedt; U.S. Patent 3,708,307; January 2, 1973

Cheese whey and skim milk, heat treated to produce at least 80% of the serum proteins in a denatured state can be spray dried simultaneously but separately in the same dryer. Acid whey can be employed in a weight ratio of acid whey solids to skim milk solids of 2 to 3.

Example: Fresh cottage cheese whey at a pH of 4.5 was pumped through stainless steel pipes from cheese vats to a 30,000 pound per hour self-opening sludge separator set for hourly automatic discharge to separate the casein dust from the whey. This separation was found desirable since the presence of heated casein particles will clog the screen to the homogenizer and the spray nozzles. The casein-free whey was then passed through a pressurized hot well in which the whey was heated to 222°F. for 2 minutes or its equivalent in time and temperature to heat denature at least 80% of the albumins and globulins.

The whey was then condensed to 30% solids in a two-stage vacuum evaporator, withdrawn and stored in a separate vat at 115°F. After cleaning the lines with water, skim milk was heated to 222°F. in the above hot well for 2 minutes to heat denature its serum proteins and then condensed in the same evaporator to about 45% solids or approximately 22° Be. at the exit temperature of 115°F. The two condensed products were then connected to separate homogenizers or positive pumps set to deliver a constant volume at 1,500 pounds pressure. The whey was conducted at 115°F. to the top nozzle sections of the Roger's Drier, and the skim milk was delivered at the bottom nozzles being heated to 160°F. The dryer was a five header 600 series dryer with No. 20 spray nozzles with No. 66 core inserts. The inlet temperature in the spray dryer was 330°F. and the outgoing air was 193°F. The yield was 1,700 pounds of whey-skim milk powder per hour, compared to 3,000 pounds per hour yield of skim milk powder by the same apparatus operating at 2,000 pounds pressure.

The operation of the dryer was the same as for drying skim milk excepting for the use of the lower pressure which was considered desirable to provide somewhat increased drying time for the whey. While the two sets of spray nozzles are shown for convenience in the figure as being fed from opposite sides, they are fed from the same side in the actual dryer. The dried powder was removed from the bottom of the dryer by scrapers and passed via conveyors to sifters and baggers. The powder was somewhat lighter in color than dried skim milk, was free-flowing and nonhygroscopic, and was readily dispersible in water. Inspection of the particles indicated that the droplets of the two liquids coalesce in drying.

It is believed that the more rapidly drying skim milk particles absorb the tacky whey droplets and facilitate drying. The resulting unitary character of the particles prevents separation and stratification of the powder during storage and transit. The casein of the skim milk is not coagulated in the dryer since evaporation of water keeps the temperature below the coagulation temperature provided excess whey is not employed. The powder produced as described above contains about 26% protein and 60% lactose and has solids derived from acid whey and skim milk in the ratio of 2 to 3 or 40% and 60%, respectively.

It therefore represents a dried mixture of equal volumes of skim milk with 9% solids and acid whey with 6% solids. It has a pH, on reconstitution, of about 5.5 and a casein coagulation temperature of about 130°F. Products of this process should not have a pH substantially less than 5.5. The powder was found suitable for the making of bakery products and produced bread comparable to products made with plain skim milk solids on a 3% level of the flour. The lower pH permitted its use in the continuous bread making process where skim milk solids (pH 6.5) cannot be used.

PRODUCTS FROM WHEY

Lactalbumin Phosphate in Dough

P.J. Ward, J.D. Johnson and R.G. Robertson; U.S. Patent 3,061,442; Oct. 30, 1962; assigned to National Dairy Products Corporation describe the preparation of a whey composition which finds use in making dough.

Whey solids can be obtained in the manufacture of various cheeses, such as in the manufacture of American type cheeses and in the manufacture of Swiss type, Parmesan type and cottage type cheeses. In the manufacture of such cheeses, a whey is provided which includes a substantial amount of milk sugar, milk salts, including calcium, sodium, phosphate and potassium ions, along with certain milk proteins, primarily lactalbumin. Depending upon the type of cheese which is being made and the manner of carrying out the make procedure, the whey solids comprise somewhat different amounts of salts and protein. In general, with higher acid cheese make procedures, larger amounts of salts are provided in the whey, whereas with lower acid cheese make procedures, lesser amounts of salts are in the whey. The milk will vary from place to place in the United States, during the year, from season to season, and from plant to plant, so that the whey will contain different amounts of salts and milk protein. Furthermore, the whey solids will vary in pH, the higher acid make procedures providing whey solids with lower pH values, and the lower acid make procedures providing higher pH values to the whey solids. The whey from the manufacture of Swiss cheese, which utilizes a lower acid make procedure, is preferred.

To provide the desired calcium ion concentration, various edible, harmless calcium salts such as calcium phosphates, calcium chloride, calcium oxide, calcium carbonate, calcium sulfate and calcium lactate may be used. The desired level of calcium in the product is between 1.1 and 1.9%. The phosphate ion may be provided by means of the calcium compound, i.e., a calcium phosphate compound. The amount of phosphate in the composition, measured as P_2O_5, should range from 1.5 to 2.5%. It is not necessary to add sodium salts to the product though the addition of small amounts of sodium salts are not harmful. The amount of sodium should range from 0.5 to 0.8%.

In the manufacture of the product it is not only important to provide salts in the desired amounts but it is also important to add these salts in such amount as to provide a product having a pH in the range of 6.8 to 7.5. The amount of salts which will be added will, of course, be in proportion to the acidity of the initial whey, more neutralizing salts being required for higher acid wheys and, of course, less neutralizing salts being required for lower acid wheys. For purposes of neutralizing the solids, calcium oxide is a highly desirable material to add as it not only neutralizes but provides a source of calcium ion. However, neutralization may be effected by the use of sodium hydroxide and other well-known harmless neutralizing compounds. Since the end product is desirably spray dried, ammonium hydroxide is not desirable.

It is additionally important that the salts be added in such amounts and proportion that the end product will have a buffering effect at a pH from 5.15 to 5.4. This appears to be highly important in the manufacture of bread and other fermented dough products. A hydrophilic agent to bind moisture in the dough should be used. The agent should have hydrophilic properties in the pH range of 5.0 to 7.5 and should not provide a sticky dough which would interfere with its handleability. A most highly satisfactory hydrophilic agent is sodium caseinate.

Example: 95 pounds of Swiss cheese whey was taken and to this was added 1.386 pounds of dicalcium phosphate; 0.925 pound of $CaCl_2 \cdot 2H_2O$; 0.764 pound of calcium oxide and 2,000 pounds of sodium caseinate. The materials were blended in a ribbon blender for ten minutes to assure uniform mixing of the chemicals throughout the product. The resulting product has a pH of about 7.2 and was buffered at a pH of about 5.23. The resulting product was made into bread in a conventional way.

Dough with Whey Solids

Much effort has been directed at attempts to use whey as a substitute for nonfat milk solids in the manufacture of bread, since great quantities of whey are obtained as a by-product in the manufacture of cheese. It is with this in mind that A.D. Singleton and I.B. Eggen; U.S. Patent 3,525,627; August 25, 1970; assigned to Kraftco Corporation provide a yeast-fermented dough constituent which includes whey solids, buffering and fortifying salts and minerals, a hydrophilic agent comprising corn flour and a water absorbing agent comprising soy flour. The corn flour may be the product commercially available by that name, or it may be a corn flour which has been partially dextrinized and pregelatinized. The soy flour may be any commercially available soy flour prepared from defatted soy beans.

More particularly, the whey solids of the dough constituent comprise the serum solids in the whey obtained from cheese making operations, and may also comprise additional amounts of lactalbumin. Lactalbumin is the proteinaceous material which is recovered from whey by heat treatment of the whey, which denatures a portion of its contained protein, followed by separation of the denatured protein from the remainder of the whey by filtration or centrifugation. The recovered protein is lactalbumin. Additional lactalbumin, when used, is provided in the dough constituent preferably at a level of 2 to 3%.

The whey solids may have been heat treated in accordance with known procedures as to substantially denature the soluble protein. Such heat treatment is usually effected at 190° to 280°F. for from 15 seconds to 30 minutes. The buffering and fortifying minerals and salts of the dough constituents comprise harmless edible compounds which provide a dough constituent having a pH in the range of between about 6.5 and 8.5 when the dough constituent is mixed with water to provide a dispersion having 10% by weight of the dough constituent. The buffering and fortifying minerals and salts provide a buffering effect in the dough system in which the constituent is used at a pH of between 5.1 and 5.4. Such buffering and fortifying salts and minerals are in addition to those occurring naturally in the whey solids of the dough constituent.

Example: Whey obtained from the manufacture of Cheddar cheese is concentrated by vacuum evaporation to a solids content of about 50%. To such whey concentrate is added calcium oxide at a level sufficient to increase the pH of the whey concentrate to about 6.5. Corn flour and lactalbumin are separately dispersed in cold water to provide a first dispersion. In this connection, the water temperature should be maintained below about 140°F. to avoid difficulty in subsequent processing. The dispersion is added to the whey concentrate. Thereafter, calcium chloride, dicalcium phosphate and an additional amount of calcium oxide are also dispersed in water, and this dispersion is added to the mixture of whey concentrate and the first dispersion. The resultant mixture has a pH of 8.0 to 8.4. The resultant mixture is then spray dried, and the dry material provides a mixture comprising:

Ingredient	Percent by Weight
Whey solids	84.7
Lactalbumin	2.0
Calcium chloride	1.1
Dicalcium phosphate	1.3
Calcium oxide	0.9
Corn flour	10.0

Soy flour, partially dextrinized, and gelatinized corn flour are then added to the dried mixture to provide a dough constituent having 20% soy flour and 26% corn flour. The dough constituent has a pH of about 7.2. The dough constituent is utilized to make bread by the conventional sponge method.

Soy Sauce from Whey

A.J. Luksas; U.S. Patent 3,558,328; January 26, 1971; assigned to Beatrice Foods Co. describes a process in which whey is fermented with a combination of a lactase enzyme and *Saccharomyces cerevisiae* under aerobic conditions with agitation to obtain a fermented type soy sauce. This can be mixed with nonfermented soy sauce of hydrolyzed vegetable protein to give a fermented oriental type soy sauce flavor.

In further work in producing a soy sauce from fermented whey A.J. Luksas; U.S. Patent 3,552,981; January 5, 1971; assigned to Beatrice Foods Co. discusses an improvement in the process without the use of *Saccharomyces cerevisiae* since the resulting fermented composition exhibited a very disagreeable odor, which was not acceptable for commercial utilization of the produced flavor. The improvement comprises inoculating whey with a lactase producing organism such as *Saccharomyces fragilis*, *Saccharomyces lactis* and *Saccharomyces unisporum*, fermenting the innoculated whey with aeration and stirring to produce the soy sauce flavor (a very disagreeable odor also develops), mixing the fermented whey with the soy sauce flavor developed with hydrolyzed vegetable protein and salt and heating to at least 150°F., whereby the disagreeable odor is substantially completely removed, while the soy sauce flavor is retained. The whey may be milk whey.

After being heated to at least 150°F. and up to the boiling point of the mixture, the disagreeable odor is removed and the resulting product has an intense flavor. The fluid whey is pasteurized before adding the lactase producing organism. Following pasteurization there is added to the fluid whey a lactase producing organism. The lactase producing organism is introduced in amounts of at least 10 organisms per cc of whey, especially 1,000, e.g., 10,000 organisms per cc of whey. After adding the organism, the system is aerated and agitated throughout the fermentation. Preferably fermentation is carried out at 75°F., but this can be varied, e.g., from 60°F. to 105°F. Stirring and agitation are continued for 12 to 24 hours.

An intense flavor develops as a result of the organism's action on the whey, and the disagreeable odor also develops. After the odor is removed by adding the hydrolyzed vegetable protein and salt followed by heating from 150°F. to the boiling point, e.g., by spray drying, the resulting intense flavored composition can be combined with a nonfermented soy sauce, e.g., imitation soya sauce flavor, and/or additional hydrolyzed vegetable protein with or without flavorings to obtain a fermented oriental type soy sauce flavor.

Example: Whey was pasteurized at 161°F. for 15 seconds. Then there was added 10,000 organisms per cc of *Saccharomyces fragilis*. The mixture was continuously agitated strongly for 18 hours and at the same time air was passed into the mixture constantly. The mixture had an intense flavor but with a disagreeable odor. To 60 parts of the mixture was added 35 parts of hydrolyzed soya protein and 5 parts of salt (all parts by weight). The resulting mixture was heated to 160°F. with vigorous stirring while passing air through. After dwelling at 160°F. for 5 minutes the mixture was cooled. An intense flavor remained without the previous disagreeable odors.

Replacement for Sodium Caseinate

As a by-product of cheese production, whey has long been discarded as waste. In recent times world-wide shortages of protein have directed a considerable effort to the recovery of whey protein as a food source. While a relatively small proportion of whey is utilized in animal feed, and a proportion is dried and utilized for human consumption, a large proportion of whey is still discarded as waste. This has resulted not only in a loss of recoverable protein as a food source but has also contributed to the pollution of many streams, lakes and rivers.

The very composition of whey mitigates against its use as a food source. Whey contains only from about 5 to about 10% solids, the remainder being water. Thus, to even produce dry whey solids, a disproportionately large amount of water must be removed. Large scale drying techniques have in recent years been developed, however, which make the production of dry whey solids both technically possible and economically feasible. A very large proportion of the dried whey produced, however, still is utilized as animal feed. Utilization of dried whey in food compositions for human consumption has not proved feasible because of the functional deficiencies inherent in the dry whey solids themselves.

One particular food area where the use of whey has proven functionally deficient is in nonbutterfat containing dairy product compositions such as the coffee whiteners, whipped toppings, and the like. While such products could theoretically utilize dry whey solids, they have been found functionally deficient. Dry whey used in nonbutterfat coffee whitener compositions, for example, results in products which exhibit oil separation, and feathering in hot coffee. While this is nutritionally unimportant, it is aesthetically unacceptable. Liquid coffee whiteners utilizing dried whey exhibit not only the foregoing deficiencies but additionally are subject to phase separation which again is also unacceptable from a market standpoint.

R.H. Ellinger and M.G. Schwartz; U.S. Patents 3,615,661 and 3,615,662; October 26, 1971; and U.S. Patent 3,620,757; November 16, 1971; all assigned to Stauffer Chemical Company find dry whey solids functionally suitable for use in nonbutterfat dairy products by simple admixture with a particulate alkali metal polyphosphate in which the alkali metal ion has a molecular weight greater than 10 and less than 50, i.e., sodium or potassium. The linear alkali metal polyphosphates suitable for use are those having a P_2O_5 content of 66% or higher such as the sodium hexametaphosphates (Graham's salt) and potassium polymetaphosphates.

Whole liquid whey is the liquid normally obtained from the manufacture of cheese. This liquid is residual from the formation and removal of the curds. The whole liquid whey contains protein, principally lactoglobulin and lactalbumin, soluble salts or minerals of the milk, fat, lactose and a large amount of water. For purposes of this process dry whey solids include dry whey containing up to 100% whey protein. Whole dry whey generally contains at least 5% by weight and usually about 12% by weight whey protein but this amount can vary depending on such factors as the milk source, the cheese process, etc.

The amount of linear alkali metal polyphosphate utilized in the whey composition will vary slightly depending upon the particular dry whey solids employed and the particular recipe in which they are to be used. In general, however, amounts of at least 10% based on the percent whey protein are satisfactory although it is preferred that amounts from about 25.0 to about 35% be used. The whey polyphosphate compositions are formed by the simple admixture of the component ingredients. It is, of course, desired that these mixtures be thoroughly blended to provide a uniform mixture. Any conventional blender or mixers suitable for blending dry powders can be used. It is desirable that the phosphate and whey particles be of relatively similar size to avoid gravity separation in transit. A whey polyphosphate mixture having particle sizes from 80 to 125 mesh has been found to be suitable. These mixtures are useful in nonbutterfat dairy products and it has been found that they are particularly useful in those areas in which sodium caseinate has been used. The mixtures serve effectively as a replacement for sodium caseinate on equal protein basis in the whole or in part.

Egg White Substitute

In the manufacture of various types of cheese, the curds are separated from the whey which is generally disposed of as waste. While relatively small quantities of the waste whey have found utility in animal foods prepared either by spray-drying or used as the whey liquid, the large proportion of this waste has been generally discharged into streams and rivers contributing to the pollution thereof. It is believed, therefore, that the finding of a use for a product from this whey constitutes a major advance.

The soluble protein in milk, contained in the milk whey, may be precipitated with long chain phosphates after removal of colloidally dispersed solids (casein). About one-fourth of all of the proteins contained in milk are solubilized in the whey. The casein may be precipitated from milk by acidification to pH 4.6, by salt precipitation, or by the action of rennet. The lactalbumin phosphate contained in the food compositions of this process has the major proportion of its protein moiety in substantially undenatured water-soluble form. In this regard, it is important to note that a common problem in milk processing is denaturation of the protein. Heat denaturation is most common. The principal results of heat denaturation on whey proteins are a decreased solubility of the proteins in both acid and salt solutions and an increased activity of sulfhydryl groups (Jenness, R., "Effects of Heat Treatment on Serum Proteins," J. Agr. Food, 75, 1954).

Whey proteins are presently thought to be a mixture of α-lactalbumin and β-lactoglobulin (Advances in Food Research, vol. 10, 1960, pp. 45-49). Whole liquid cow milk has been found to contain from 0.4 to 0.5% by weight lactalbumin (mean value) and from 0.11 to 0.26% β-lactoglobulin. In whey these proteins are believed to be present in a ratio of from 2:1 to 10:1 lactalbumin to β-lactoglobulin. The effect of heat on β-lactoglobulin is perhaps most pronounced. Changes in whey protein caused by heat denaturation have usually limited its utility in food applications. The lactalbumin phosphate in this process, which contains its protein in an undenatured form, is not so limited and is useful in many applications where denatured protein is unsuitable.

The recovery of protein from whey as phosphate includes both lactalbumin and lactoglobulin. The term lactalbumin phosphate as used here is intended to include both the lactalbumin and lactoglobulin. These components are generally not separated in their recovery.

S.A. Broadhead; U.S. Patent 3,706,575; December 19, 1972; assigned to Stauffer

Chemical Company has found that undenatured LAP serves effectively as a replacement for egg white. This replacement can be made on a requirements basis broadly from 1:1 to 2:1 lactalbumin phosphate to egg white depending on the formulation and other ingredients utilizing them. Thus, compositions normally containing egg whites can be prepared by admixing the ingredients of the compositions together with an effective amount of a mixture which is functionally equivalent to the egg whites, the mixture comprising undenatured lactalbumin phosphate and/or egg whites. "Functionally equivalent" as used here is intended to include the physical and chemical properties normally attributed to egg whites.

Dried egg white and egg yolk can be purchased commercially as such or as ingredients in prepackaged mixes. Lactalbumin phosphate can effectively replace egg white in all such food compositions. This is an important fact since lactalbumin phosphate is recovered in large measure from whey which has in most instances been discarded as waste, limited amounts only being used in animal feeds. It should also be noted that lactalbumin phosphate can replace egg white in either the whole or dried condition and can even be used with eggs essentially alone in such food applications as omelets to enhance the fluffy light character of the food without adversely affecting the flavor.

In addition to replacing egg white in such food compositions per function, lactalbumin phosphate provides to the food composition equal or superior nutritive values. The lactalbumin phosphate can be defined as the product produced by "reacting" long chain polyphosphates with liquid whey and drying the resulting precipitate without denaturation of the protein. Suitable long chain polyphosphates are metaphosphates of the formulas $(NaPO_3)_x$ and $(KPO_3)_x$, having an average chain length (represented by x) of at least 30. These straight chain inorganic polymers, which are sometimes known as hexametaphosphates, are to be distinguished from the cyclic or ring metaphosphates. Where potassium metaphosphate is used, it is preferable to first react the metaphosphate in solution with a sodium salt such as sodium sulfate, sodium chloride, sodium pyrophosphate, sodium hexametaphosphate (short chain) etc., to improve solubility.

Mixtures of sodium and potassium glassy metaphosphates are also suitable for precipitating lactalbumin phosphate. The preparation of undenatured protein metaphosphates by reaction between an alkali metal metaphosphate and soluble protein is described in an article by Briggs, D.R., The Journal of Biological Chemistry, vol. 134, June 1940, pp. 261-272. After treatment of the whey protein with a glassy metaphosphate by this process, the whey is centrifuged to collect a heavy gumlike coagulum of lactalbumin phosphate which is then dried. To prevent denaturation of the protein, the whey temperature is maintained preferably below 160°F. For the same reason, it is preferable to dry the coagulum by spray-drying at low temperatures.

Bread Flavor Concentrate

R.H. Bundus and A.J. Luksas; U.S. Patent 3,466,177; September 9, 1969; assigned to Beatrice Foods Co. produce a synthetic bread flavor by reducing the lactose content of whey which has from 6 to 40% total solids, adding yeast and growing the yeast for a period of up to 48 hours with intimate mixture of air during the growth period. This bread flavor concentrate can be prepared by using a whey diluted 1:10 with water. This particular dilution is not critical and can be more or less depending on the lactose content of the whey. The lactose concentration should be no more than 1%, ideally 0.5%, with a minimum lactose concentration of 0.1%. The whey, either before or after dilution, should be heated to a high temperature for a short period of time, e.g., 185°F. for 30 minutes, 245°F. for 15 minutes, or 310°F. for 15 seconds. The heat treatment should be after dilution. Preferably, the whey, either before or after dilution, is filtered or centrifuged to remove sediment or suspended matter or sediment precipitated by heat (e.g., due to the heat treatment set forth above).

The clear medium is subjected to fermentation in the presence of lactic acid. Thus there can be added any of the known microorganisms which produce lactic acid such as S. lactis, L. bulgaricus, L. lactis, S. citrovorus, etc. Fermentation is allowed to take place and is normally complete in 12 to 24 hours at a favorable growing temperature, e.g., 70° to 90°C. for all except thermophilic organisms. Growing of the bacteria is preferably initiated under aerobic conditions although anaerobic conditions can be employed. Then fermentation is continued for 24 to 48 hours after inoculation with bread flavor-causing organisms, e.g., yeast of the Saccharomyces group such as Sacchoromyces cerevisiae or unisporum. The yeast must be one that can grow on lactic acid. The preferred whey is cottage cheese whey, however, any cheese whey may be used.

The yeast is grown initially by an aerobic process in which there is preferably 1.5 to 1.6 ppm of oxygen. However, as the assimilation continues the oxygen content goes down, eventually to zero. It is critical that the oxygen content should not be increased or maintained at a high level or the bread flavor either is not produced at all or the process of forming the bread flavor is slowed. A constant oxygen feed is maintained during the fermentation but the growth of the yeast organism reduces the oxygen content of the media. The product obtained is essentially water and bread flavor. It can be concentrated 10 to 100 times, preferably 100 times, and sold as bread flavor or it can be dried to a stable product. The concentrate can be added to bread of various types and can also be used to impart a bread flavor to nonbread foods, e.g., to rice stuffing for fowl, dips made from cheese or sour cream, etc.

Example: 300 liters of cottage cheese whey having 16% total solids and 5% lactose was diluted with water 1:10 to obtain a mixture containing 0.5% lactose. The diluted mixture was heated at 185°F. for 30 minutes, centrifuged to remove suspended matter and the clear medium subjected to fermentation with Streptococcus lactis at 82°F. for 18 hours under aerobic conditions. Then the fermented mixture was inoculated with Saccharomyces cerevisiae and vigorously aerated at 85°F. for 48 hours.

The oxygen content at the time of introduction of the Saccharomyces was 1.6 ppm but this gradually went down to almost zero during the fermentation. The product had a strong bread flavor and was concentrated 100 times by removing the water under a vacuum. The product was then refrigerated and stored for use. It was suitable for adding to bread dough prior to baking. This procedure was repeated but instead of concentrating the bread flavor produced it was spray dried to give a stable synthetic bread flavor.

R.H. Bundus and A.J. Luksas; U.S. Patent 3,466,174; September 9, 1969; assigned to Beatrice Foods Co. describe a variation of this process in which the protein nitrogen content of the whey can be enhanced by the use of nitrogen sources such as ammonium nitrate, potassium nitrate, sodium nitrate or ammonium phosphate. Preferably diammonium phosphate is added to the whey culture because it gives better growth. Usually 0.1 mol of diammonium phosphate is used per liter of whey but this can be increased or decreased if desired, e.g., there can be used 0.2 mol/liter or 0.01 mol/liter.

Example: 300 liters of cottage cheese whey having 16% total solids was pasteurized and then inoculated with a useful combination of yeast and enterococcus serological group D which occur together in Citron and vigorously aerated in a 100 gallon vat with a Cowles dissolver, for 18 hours at 85°F. The resultant product was very strong in bread flavor and the product was spray dried to a powder. When the powder was used as a flavor source in chemically leavened bread it yielded bread possessing a yeast-raised flavor. This procedure was repeated but there was added 30 mols of diammonium phosphate to the whey prior to adding the yeast-enterococcus culture. The yield of the spray dried bread flavor was increased substantially.

Another approach to making a dough flavor concentrate is presented by R.H. Bundus and A.J. Luksas; U.S. Patent 3,485,641; December 23, 1969; assigned to Beatrice Foods Co. The preparation consists of heating whey until the protein is coagulated out and removed. The clear solution is divided into fractions A and B. Fraction A is diluted with water from 1:4 to 1:20. Fraction B is concentrated to 1/5 to 1/12 the volume and the crystallized lactose removed. Fraction B is then diluted with either water or solution A and is inoculated with bread flavor forming organisms to form bread flavor. In the procedure, whey is heated in conventional fashion, e.g., at 185°F. (20 minutes) to 300°F. (15 seconds) until all the protein is coagulated. The whey is then centrifuged or filtered to remove the protein.

The preferred whey is cottage cheese whey, however, whey obtained in the manufacture of Cheddar, Swiss, Neufchatel or cream cheese may also be used. The clear solution is divided into two fractions. Fraction A is diluted with water to give a dilution of 1:20. The resulting product is designated Solution A. Fraction B is concentrated preferably to 1/10 its volume. Fraction B is concentrated by heating in a vacuum at 90° to 140°F., the boiling point of the solution under the vacuum. The concentrate is cooled to room temperature or below (32°F), the lactose is allowed to crytallize and is removed by centrifuging, filtering or other conventional technique. The resulting product is designated Solution B. Solution B can be employed by itself in preparing bread flavor providing it is diluted at least 1:1 with water. Lactic acid can be added to give a pH of 4.5.

Alternatively and preferably Solution B is diluted with Solution A. As little as 2% of Solution B can be used with a 1:20 dilution Solution A, but preferably the ratio of Solution A to Solution B is 1:1 with Solution A being a 1:10 dilution. When Solution A is the source of the lactic acid it is necessary to add an organism known to produce lactic acid from lactose such as *S. lactis*, *L. lactis*, *L. bulgaricus*, *S. citrovorus*, *S. cremoris*, etc. Fermentation is allowed to take place and is normally complete in 24 hours at a favorable growing temperature which is 70° to 90°F. for all except the thermophilic organisms. The fermentation can be carried out either under static conditions or with agitation.

It is also possible to add lactic acid or an acid acting lactate to a mixture of Solution A and Solution B. The lactic acid containing Solution B diluted with either water or Solution A (or a mixture of water and Solution A) is then inoculated with bread flavor causing yeast organisms, e.g., yeast of the Saccharomyces group such as

S. cerevisice or S. unisporum. The yeast must be one that can grow on lactic acid. The yeast is grown in the fluid medium under aerobic conditions with vigorous agitation. However, as the assimilation continues the oxygen content goes down, eventually to zero. A constant oxygen feed is maintained during the yeast fermentation but the growth of the yeast organisms reduces the oxygen content of the media. The product obtained is essentially water and an intense bread flavor together with yeast cells. It can be employed as such, can be further concentrated 10 to 100 times, or it can be spray dried to a stable product. If the product is used as a liquid bread flavor without drying, the yeast cells are desirably removed by centrifuging or filtering with a Berkfeld or Zeiss filter.

If the product is to be dried there should be added a bulking agent since the ferment is merely water, bread flavor and yeast cells. When a bulking agent is added there is no need to remove the yeast cells although they may be removed. Any bulking agent can be used which dries readily and retains the flavor. Fat containing bulking agents are preferred because they retain the flavor better in spray drying. The bulking agent can be added to the liquid bread flavor in an amount of 0.1 to 100% by weight of the liquid bread flavor. Since the liquid bread flavor is mostly water the addition of even a small amount of bulking agent significantly increases the bulk of the dried product. The amount of bulking agent is not critical.

Example 1: 300 liters of cottage cheese whey having 6.5% total solids and 4.5% lactose were heated at 185°F. until the protein had coagulated (about 20 minutes). The coagulated protein was removed by centrifuging. The clear solution was divided into two equal parts, Fraction A and Fraction B. One part of Fraction A was diluted with 9 parts of water to give a 1:10 dilution to give Solution A. Fraction B was concentrated under vacuum at 100° to 110°F. (the boiling point) to 1/10 its original volume. The supernatant liquid was then poured off as Solution B from the residue of crystallized lactose.

Example 2: A mixture of one part of Solution A with one part Solution B was inoculated with Streptococcus lactis and fermented at 32°C. (89.6°F.) for 24 hours with agitation. Then the mixture was inoculated with Saccharomyces cerevisiae and vigorously aerated at 85°F. for 24 hours with constant agitation. The air was introduced into the system at 0.2 cu. ft. per hour, per gallon of media. The oxygen available gradually went down to almost zero during the fermentation. The liquid product had an intense bread flavor even after the yeast was removed therefrom by filtration.

Example 3: To 100 parts of the liquid product of Example 2 after filtration to remove the yeast there were added 20 parts of corn starch and the mixture spray dried to give a powder having the intense bread flavor carried on the starch bulking agent.

Example 4: To 100 parts of the liquid product of Example 2 prior to filtration there were added 15 parts of whole milk and the mixture spray dried to give a powder having the intense bread flavor and yeast carried on the gluten.

Example 5: To one part of Solution B there was added 1 part of water and the pH brought to 4.5 with lactic acid. Then the mixture was inoculated with Saccharomyces cerevisiae and vigorously aerated at 85°F. for 24 hours with constant agitation. The air was introduced into the system at 0.2 cu. ft. per hour per gallon of media but the oxygen gradually went to zero during the fermentation. The liquid product had an intense bread flavor. The bread flavor remained even after the yeast was removed by centrifuging.

A variation on the basic process by which a strong yeasty odor is imparted to the product by either (1) hydrolyzing whey with a lactase, fermenting the product with yeast and adding the product to synthetic bread flavor, (2) growing lactose fermenting yeast on whey and combining the product with the flavor, (3) forming the flavor in the presence of yeast which ferments lactose, or (4) forming the flavor in the presence of lactose fermenting bacteria is described by R.H. Bundus and A.J. Luksas; U.S. Patent 3,466,176; September 9, 1969; assigned to Beatrice Foods Co.

Several different techniques can be employed to increase the desired odor. The lactose in whey can be hydrolyzed with a lactose enzyme to split the lactose to galactose and glucose and then the yeast is introduced. Glucose and/or galactose can be employed as a growth medium for yeast. The product can then be blended with the synthetic bread flavor to intensify the odor.

Example 1: β-Galactocidase was added to cottage cheese whey to split the lactose into glucose and galactose at room temperature. After the glucose and galactose were formed *Saccharomyces cerevisiae* was added and allowed to ferment for 24 hrs. at room temperature. A very intense yeasty odor was produced. This was blended 1:1 with the ferment of bread flavor produced in the previously described processes (prior to spray drying). The product was useful as such and also could be spray dried, roller dried, etc. The yeast ferment and the bread flavor can be spray dried together or spray dried separately and then blended together. Another method of increasing the odor of the synthetic bread flavor is illustrated in Example 2.

Example 2: Lactose fermenting yeast, specifically *Saccharomyces fragilis*, was grown on cottage cheese whey at room temperature to obtain a very intense yeasty odor. The product was spray dried and combined with the spray dried powder as above in a 1:1 ratio to produce a synthetic bread flavor with a strong yeasty odor. If desired, the yeast fermentation product of this example can be combined with the synthetic bread flavor prior to spray drying. *Saccharomyces lactis* and other lactose fermenting yeast can be employed in place of *Saccharomyces fragilis*.

The resultant dried product is then normally hydrated at the time it is to be used as a flavor supplement in yeast raised doughs or as a flavor additive in chemically leavened doughs, e.g., refrigerator doughs or dry biscuit mixes. When chemically leavened doughs are employed the bread flavor is supplied entirely by the bread flavor additive.

Flavor Potentiator

R.M. Hamilton and S.F. Ziccarelli; U.S. Patent 3,720,519; March 13, 1973; assigned to Beatrice Foods Co. describe the production of a flavor potentiator based on the synthetic bread flavor produced in the previously cited patents. It was found that when that growth product is dried to a low moisture content there is a pronounced loss of bread flavor and the low moisture content dried product exhibits relatively little if any bread flavor whatsoever. Apparently, drying the growth product to low moisture contents produces some yet unknown change in the growth product which produces a product of an essentially different nature. While this change is not understood, nevertheless, when the growth product described in those patents is dried to a low moisture content, it functions, not as a bread flavor, but most effectively as a flavor potentiator. The product of the previously cited work can be directly dried to a low moisture content or it may be concentrated before drying. If the product contains over 50% water by weight it has been found desirable to concentrate it to 50% or less by volume, e.g., 30 or 40%, to improve drying efficiency. This concentration may be carried out at 112° to 130°F. in a vacuum. Alternatively, the product may be cooled to 110°F., seeded with lactose in an amount of 0.05% by weight and cooled to 70°F. or lower to force crystallization. The drying step can be accomplished as desired and conventional drying techniques are quite acceptable.

It is preferred that a spray dryer be used, since the dried product is then in a powder form which is a convenient form for application to foods. In any event, the drying step must reduce the moisture content to a low level. While the moisture level is not narrowly critical, the moisture level should be at or below 5% by weight of the dried product and preferably 1 to 0.5%. It is necessary that such low moisture contents be accomplished during the drying step, since as noted above, some change occurs during drying to a low moisture content which converts the growth product from a bread flavor to a flavor potentiator with essentially no flavor itself at all. The drying temperatures are not critical as long as the product is dried to the required low moisture contents. For example, when the growth product is spray dried, inlet temperatures as high as 350°F. may be used. Lower temperature may be used all the way down to room temperature, but these lower temperatures unduly prolong the drying step.

Example 1: Into a 3,000 gallon tank was placed 1,000 gallons of cottage cheese whey which had previously been concentrated to 12% by weight of solids. The whey was pasteurized, i.e., heated to 185°F. for 20 minutes, and then cooled to 79°F. The cooled whey was inoculated with 1% by volume of the whey with an inoculum having 10^7 yeast cells per cc of *Saccharomyces delbrickii*. The yeast and whey mixtures were aerated by stirring and bubbling air therethrough at a rate of 30 to 40 cfm. Growth of the yeast was carried out for about 20 hours at 79°F. and until a steady pH of about 5.6 to 6.3 was reached. The growth product was pasteurized, i.e., at 185°F. for 20 minutes, and the product was concentrated to 45% solids by weight in a vacuum pan.

The concentrated product was spray dried in a conventional manner at 350°F. inlet and 150°F. outlet to form a powder having a moisture content of 2% by weight of the dried product. The remarkable activity of the potentiators is illustrated by the removal of the bitter flavors of brewed coffee. As a further remarkable illustration, spices such as pepper and clove, tend to mask weaker food flavors, but when the potentiators are also present, the action of the spices is retarded sufficiently that the weaker food flavors can be fully sensed.

To illustrate the above, in a laboratory model system, two samples of equal amounts of oil-soluble and water-soluble flavors, blended with an equal weight of water, were tasted by a panel of 12 tasters. The first sample contained 0.5% of the dried yeast fermented whey while the second sample contained no dried yeast fermented whey. All 12 of the panel members picked the sample having the dried yeast fermented whey as a sample having two flavors, while the members picked the sample without the dried yeast fermented whey as having only one flavor. They could taste the oil-soluble flavor but could not taste the water-soluble flavor in the sample without the dried yeast fermented whey. It should be noted that the yeast fermented whey should be used in a hydrated form, i.e., the dried yeast fermented whey must be placed in the presence of water. If the dried yeast fermented whey is not placed in a hydrated form, the flavor potentiation is decreased, and in most applications is decreased to the point of significant loss of effectiveness.

While many foods contain more than sufficient water to hydrate the dried yeast fermented whey, other foods will not contain sufficient water to accomplish this hydration. In that case, the dried yeast fermented whey should be applied in a water medium to the food or water should be added to the mixture. In any case, the end result should be that the water naturally contained in the food and the water added by way of medium or to the mixture should be sufficient that the dried yeast fermented whey becomes at least partially hydrated. While the amount of the potentiators used with foods can vary widely on a weight basis, but are preferably between 0.05 and up to 3%.

Example 2: Dried yeast fermented whey (moisture content about 2% by weight) was mixed and blended with cream cheese in the amount of 5% by weight, based on the cream cheese. Samples of the resulting cream cheese were given to a panel of 28 tasters, along with samples of plain cream cheese. The panel did not know which samples contained the potentiators or the number of samples, if any, which contained the potentiators. The panel was asked to select the samples which had better flavor and to note comments, if any, for their selection. 81.5% of the panel picked the samples with the potentiators and the reasons generally noted for that selection was smoother flavor. The 18.5% of the panel who selected the plain cream cheese made their selection on the basis of the sharper flavor. This example illustrates the utility of the potentiators in dairy products.

Bland Whey Product

Whey derived from the fermentation of milk, cream, etc. in the production of cheeses, after separation of the curds from the fermented product, has a characteristic disagreeable odor and flavor. While the odor and flavor vary somewhat with the particular cheese being produced, nevertheless the fermentation process always results in a whey of a disagreeable odor and flavor.

A.J. Luksas; U.S. Patent 3,728,128; April 17, 1973; assigned to Beatrice Foods Co. discovered that whey may be converted into a bland whey product by inoculation with a yeast, supplying air and/or oxygen to the inoculated whey in amounts sufficient that the yeast grows in the whey and deflavors the whey by essentially an assimilation of the objectionable odor and flavor compounds. The process, in its broadest form, is based on the fact that the action of the yeast on the fluid whey is by assimilation of the unwanted compounds therein and not by fermentation.

At this point it is important to clearly distinguish between assimilation and fermentation by the yeast. First of all fermentation proceeds to its final products through an entirely different set of mechanisms as compared with assimilation. Secondly, fermentation, generally, proceeds by acting on sugars or sugar producing components in the whey, while assimilation is not so limited. Assimilation may use a wide variety of components in the whey, with or without using the sugar or sugar producing components. Thirdly fermentation results in the major proportions of CO_2 and ethyl alcohol, among other things such as aldehydes, ketones, fusel oils and the like, while assimilation results in major proportions of CO_2 and water, instead of alcohol and related compounds. Fourthly, assimilation requires that a substantially higher concentration of dissolved oxygen be available, than does a fermentation. Lastly, since the whey starting material, yeast inoculant and temperatures may be essentially the same in an assimilation process as in a fermentation process, it is important that very special and very critical conditions are imposed to insure that assimilation will take place, and that substantially no fermentation takes place. In this latter regard, the oxygen level in the inoculated whey is most critical.

When whey is inoculated with a yeast and maintained under conventional growth conditions, there is an initial time period where the total number of yeast cells in the whey, assuming a conventional level of yeast introduced into the whey by the inoculation, is relatively small and no appreciable change in the whey takes place. This is referred to as the "lag" period and it may vary from about 1 to 5 hours, depending on the inoculation level of yeast cells. Since the yeast growth and hence number of yeast cells with time follows a generally exponential behavior, after this lag period the growth rate and number of yeast cells increase quite rapidly.

Hence, during the lag period, of relatively little growth, the oxygen content of the whey, with or without aeration, may be quite close to the maximum amount of oxygen that can be dissolved by the fluid whey, i.e., approximately 160 parts per million by volume. However, after this lag period and when substantial growth of the yeast cells begins, the oxygen content in ppm drops most rapidly, as the growing yeast cells use the available dissolved oxygen, and within a relatively short time the oxygen content will drop to about 1 ppm under conventional conditions.

In order to move the action of the yeast from essential fermentation, after the lag period, toward significant amounts of assimilation, it is most critical that the whey medium contain at least about 5 ppm of dissolved oxygen. At about 5 ppm of dissolved oxygen a significant amount of assimilation takes place, although fermentation still takes place at this oxygen level. Between 5 and 10 ppm of dissolved oxygen, an unstable region exists where the action is a combination of fermentation and assimilation with increasing proportions of the latter as the dissolved oxygen level goes from about 5 toward the 10 ppm level. At about 10 ppm the action of the yeast is essentially that of assimilation.

Specialized vessels with special defoaming apparatus in which dissolved oxygen contents can be increased without frothing and overfoam are used. With these specialized vessels the assimilation process may be carried out with air as the oxygen source and, by careful control, dissolved oxygen contents of 5 to 10 ppm can be achieved in the fluid yeast/whey medium. For easier operation, it is preferred that at least part of the oxygen source is molecular oxygen, which allows oxygen contents in the yeast/whey medium of 10 ppm or higher. Generally, at least 20% by volume of the air is replaced by molecular oxygen. Conventional fermenters may be used if the air which is normally bubbled through the medium is replaced in a major proportion with molecular oxygen, e.g., about 50% or more by volume.

Conventional yeast may be used in the process, e.g., *S. cerevisiae*, *S. unisporum*, *S. fragilis* and *S. lactis*. Preferably, the whey is first pasteurized to eliminate pathogens or contaminating organisms. The heating can be at 145°F. for 30 minutes, 180°F. for 20 to 30 minutes, 310°F. for 1 to 20 seconds, 300°F. for 15 seconds, or 275°F. for 15 seconds.

The air and/or oxygen is bubbled through the medium throughout the growth period. Growth is continued at 78.8°F. The product is then pasteurized to destroy the yeast and is then dried by spray or roller drying. The product dries easily. The total time required to carry out the assimilation will vary somewhat depending on the lag time and the lag time depends on the level of inoculation, as well as temperature. However, with conventional levels of inoculation, the lag period will generally be between 1 to 5 hours. After the lag period, the rapid growth times will vary between 12 to 24 hours.

This growth period will be essentially complete when there is a distinct increase in oxygen content of the system, which indicates that the yeast has assimilated the unwanted compounds and no longer requires large amounts of oxygen. At this point the oxygen demand of the yeast is greatly reduced and the same oxygen feed rates will therefore simply increase the dissolved oxygen content, which shows that the assimilation is essentially complete. After the growth is essentially complete the oxygen supply should be terminated, since keeping the whey product in a high oxygen content fluid for extended times after growth is complete will cause unwanted off-tastes.

Example: Into a vessel having a gas bubbling mechanism, a high speed impeller, a mechanical defoamer and a Beckman oxygen analyzer attached, is added Cheddar cheese whey. The whey is heated to about 79°F. and inoculated with 3% by volume of 10^3/cc *Saccharomyces fragilis*. Air is slowly introduced. The dissolved oxygen content of the whey is continuously analyzed. After about 15 minutes the oxygen content is about 160 ppm and then begins to slowly drop. The amount of air is increased as the oxygen content continues to drop. After about 2 hours, the oxygen content begins dropping rapidly and the air rate is increased rapidly to maintain a dissolved oxygen level of about 15 ppm.

After about 13 hours the system drops to an oxygen content of about 12 ppm and stabilizes at that point. The rate of air introduced is about 5 volumes of air per minute per volume of whey. At these high air rates the mechanical defoamer on the vessel is used to prevent frothing and overfoaming. After about 15 hours after the lag period, the oxygen content begins to distinctly increase and the air is discontinued. The medium is then pasteurized at 145°F. for 30 minutes. The product is then homogenized at 2,000 psi and spray dried. The dried product is a bland tasting powder and when used in place of milk in a baked bread gives no appreciable taste to it. The product can also be incorporated into vanilla ice cream, candy and reconstituted milk without imparting any appreciable taste to them.

COMPANY INDEX

The company names listed below are given exactly as they appear in the patents, despite name changes, mergers and acquisitions which have, at times, resulted in the revision of a company name.

INVENTOR INDEX

U.S. PATENT NUMBER INDEX

NOTICE

SUGAR SUBSTITUTES AND ENHANCERS 1973

by Roger Daniels

Food Technology Review No. 5

Because of recent restrictions on the use of synthetic sweeteners, there has developed considerable interest and research activity concerning the use of naturally occurring materials to induce or expand the sweetness of natural sugars, so that lower sugar levels can be used for lower caloric content of foods and beverages.

All of these approaches are covered in this volume, as are new synthetic sweeteners with very low toxicity figures, such as chemical modifications of natural amino acids.

The appendix contains important directives by the U.S. Food and Drug Administration on the status, use, and labeling requirements of effective combinations of such sugar substitutes and enhancers.

Except for the appendix, the book is based on 105 recent U.S. patents. Numbers in ( ) indicate the number of processes per topic. Chapter headings are given here, followed by examples of important subtitles.

ISBN 0-8155-0492-6 **275 pages**

PROTEINS FROM HYDROCARBONS 1973

by S. Gutcho

Food Technology Review No. 4

Production of proteins from petroleum hydrocarbons is already a reality in some parts of the world. Interest in this source of proteinaceous food is on the increase while the population explosion continues as one of the world's problems.

Hydrocarbons for this purpose, such as paraffinic alkanes, are available in large supply from the petroleum industry. They are readily transportable to areas where the end product, proteins, is most needed. Also, they are obtainable throughout the year without the risks associated with the growing and harvesting of crops or the rearing of livestock. Furthermore, the rate of growth of microorganisms greatly exceeds the rate for animals and agricultural crops.

There are problems of palatability and residual toxicity, but large scale commercial production should be a reality by the late 1970's.

Describes 64 patent-based processes. A partial and condensed table of contents follows. Numbers in ( ) indicate the number of processes per topic. Chapter headings are given, followed by examples of important subtitles.

ISBN 0-8155-0480-2 **221 pages**

EDIBLE COATINGS AND SOLUBLE PACKAGING 1973

by R. Daniels

Food Technology Review No. 3

Edible and soluble packaging can be used to
— retard moisture transmission
— serve as a barrier to oxygen or other gases
— prevent loss of volatile components and flavor
— resist transfer of oils and grease
— serve as a barrier to bacterial invasion
— provide protection from mechanical injury — or at least minimize crumbling and breakage
— isolate reactive ingredients from each other.

The processes in this book are grouped according to the food products to which the edible or soluble coatings are applied. The subjects of edible and soluble capsules for medicines and of casings for sausages are so vast that they are treated in separate volumes.

118 patent-based processes. A partial and condensed table of contents follows. Numbers in ( ) indicate the number of patents per topic. Chapter headings are given, followed by examples of important subtitles.

ISBN 0-8155-0475-6

360 pages

PREPARED SNACK FOODS 1973

by M. Gutcho

Food Technology Review No. 2

The April 24, 1972 issue of the "Chemical Marketing Reporter" places the value of the U.S. snack industry at $2 billion a year. It is therefore not surprising that many companies are working on improving existing processes or on developing new snack foods altogether.

The dictionary defines a snack as "food eaten between meals," which implies that snacks purchased in a supermarket or other food store should already be prepared, ready to eat. The present book therefore deals to a large extent with what is called "crisps" in the food processing industry, i.e., potato chips, corn chips, popcorn, expanded products, nuts and nut products, and french fried potatoes. The preparation of additional crisps, such as onion rings and apple crisps, is described under miscellaneous snack products.

151 patent-based processes. A partial and condensed table of contents follows. Numbers in ( ) indicate the number of patents per topic. Chapter headings are given, followed by examples of important subtitles.

ISBN 0-8155-0474-8 289 pages

TEXTURED FOODS AND ALLIED PRODUCTS 1973

by M. Gutcho

Food Technology Review No. 1

This book deals primarily with the preparation of meat-like foods from readily available protein, such as soy protein, peanut protein, and wheat gluten. Techniques employed by the textile and plastics industries have been borrowed and applied to plant protein.

Processes have been developed for spinning protein into odorless tasteless filaments, which provide the necessary texture for simulated food products, or as meat extenders. Fibrous textures can also be obtained by nonspinning techniques, such as thermoplastic extrusion.

This new category of food has been designated by the Food and Drug Administration as "textured protein products." These sources of protein, when used as meat, in order to be acceptable to the consumer, must simulate the texture, chewiness, flavor, and appearance of a natural meat product.

Textured plant protein has gained acceptance as a meat extender, especially in school lunch and geriatric programs, which permit a 30% addition of plant protein to meat items. The book reviews 78 U.S. patents, most of them issued quite recently. Numbers in ( ) indicate a plurality of processes per topic. Chapter headings are given, followed by examples of important subtitles.

ISBN 0-8155-0463-2 315 pages

WASTE DISPOSAL CONTROL IN THE FRUIT AND VEGETABLE INDUSTRY 1973

by H. R. Jones

Pollution Technology Review No. 1

The fruit and vegetable processing industry is certainly a major industry. As pointed out in a recent report by the National Canners Association ca. 170,000 persons are employed in ca. 1,800 plants of the canned and frozen fruits industry in the United States.

It is estimated that in the U.S. alone (incl. Hawaii) this industry annually:

Processes 26 million tons of raw product.
Discharges 83 billion gallons of wastewater.
Generates 800 million pounds of biochemical oxygen demand (BOD),
And 392 million pounds of suspended solids,
Together with 8 million tons of solid residue.

While these huge wastes are biodegradable, they add enormously to the garbage waste and disposal problem.

This book, largely based on authoritative government reports and surveys, attempts to clarify the ways and means open to the food processor who must keep his polluting wastes down to a minimum.

In the following partial list of contents chapter headings are given, together with the more important subtitles.

ISBN 0-8155-0466-7 **261 pages**